BIOCHEMISTRY RESEARCH TRENDS

PROKARYOTES

PHYSIOLOGY, BIOCHEMISTRY AND CELL BEHAVIOR

BIOCHEMISTRY RESEARCH TRENDS

Additional books in this series can be found on Nova's website
under the Series tab.

Additional e-books in this series can be found on Nova's website
under the e-book tab.

BIOCHEMISTRY RESEARCH TRENDS

PROKARYOTES

PHYSIOLOGY, BIOCHEMISTRY AND CELL BEHAVIOR

MARINA NISNEVITCH
EDITOR

New York

Library of Congress Cataloging-in-Publication Data

ISBN: 978-1-63321-592-4

Library of Congress Control Number: 2014953194

Published by Nova Science Publishers, Inc. † New York

Contents

Preface		**vii**
Chapter 1	Staphylococcal Biofilm Formation and Development: Related Diseases, Host Immune Responses and Therapy *Livia Visai, Davide Ferrari, E. Magda Barbu, Xiaowen Liang, Vannakambadi K. Ganesh, Magnus Höök and Marcello Imbriani*	**1**
Chapter 2	Bacterial Biofilms and Role of the Second Messenger Cyclic-Di-GMP *Michelle A. Stan and Raymond J. Turner*	**35**
Chapter 3	Multidrug Transporter Proteins of the Major Facilitator Superfamily: Not Only for Drug Efflux *Kamela O. Alegre, Stephanie Paul and Christopher J. Law*	**55**
Chapter 4	Adaptation of Bacterial Cells to Toxic Aromatic Hydrocarbons *Rivka Cahan, Esti Michael and Nissim Swissa*	**77**
Chapter 5	Hfq-Like RNA Chaperones and Small Non-Coding Regulatory RNAs: A New Paradigm of Bacterial Gene Regulation *Jorge H. Leitão, Christian G. Ramos, Joana R. Feliciano, Sílvia A. Sousa, André M. Grilo and Paulo J. P. da Costa*	**91**
Chapter 6	Effects of the Cytolethal Distending Toxin from Gram-Negative Bacteria on Mammalian Cells *Elisabeth Bezine, Julien Vignard and Gladys Mirey*	**109**
Chapter 7	Cytotoxic Properties and Medical Usages of Cyt Toxins from *Bacillus thuringiensis* Subsp. *Israelensis* *Shmuel Cohen*	**149**
Chapter 8	The Novel Plasmids of Botulinum Neurotoxigenic Clostridia *Giovanna Franciosa, Angelo Iacobino and Concetta Scalfaro*	**163**
Chapter 9	Sensitivity of Bacteria to Photodynamic Chemotherapy *Marina Nisnevitch and Faina Nakonechny*	**197**

Chapter 10 Monod's Bacterial Growth Model. Kinetic Arguments for "Simple, but Not Too Simple" **221**
Vlad Tudor Popa

Editor's Contact Information **247**

Index **249**

Preface

For thousands of years, mankind has both used bacteria and suffered from them without actually being aware of them, until 340 years ago when Anton van Leeuwenhoek observed bacteria for the first time. People then realized that they are surrounded by a wonderful and, at the same time frightening, micro-world. Only during the second half of the twentieth century did researchers begin to unravel secrets of bacterial cells, and today the amount of knowledge in the field is in an exponential growth stage. However, we are still far from an overall understanding of bacterial cell processes. This book is another step towards revealing principles and mechanisms of bacterial functioning and contains contemporary data and results in hot topics such as bacterial biofilm formation, multidrug transporter proteins, bacterial cells' adaptation to toxic compounds, bacterial gene regulation, cytotoxins and neurotoxins from Gram-negative and Gram-positive aerobic and anaerobic bacteria, bacterial sensitivity to photodynamic chemotherapy and kinetic analysis of bacterial growth.

Chapter 1 – Various biomolecular and regulatory factors are involved in adhesion, biofilm formation and host immune evasion of staphylococcal species. Adherance of staphylococcal species to either biotic or abiotic surfaces is the critical first event in the establishment of an infection with these serious pathogens. A variety of proteinaceous and non proteinaccous adhesins are konwn to mediate attachment to a multitude of host factors, such as extracellular matrix, and plasma proteins, and human host cells or inter-bacterial cell adhesins, which is essential for biofilm accumulation. Formation of biofilm is a two step process: (1) adherence of the pathogen to a surface and (2) further progression by accumulation of cells to form multilayered cell clusters. Biofilm is one of the important virulence factor of staphylococci that plays a role in many device-related infections such as native valve endocarditis, otitis media, urinary tract infections, cystic fibrosis, and acute septic arthritis, The increasing prevalence of antibiotic resistant strains and the vast ability of bacteria to evade host immune defenses are becoming more and more important for medical infections. The chapter focuses on the role of specific cell surface adhesins, the ica operon, accumulation-associated proteins and quorum-sensing systems in the adhesion and infection processes of the Gram-positive pathogens, *Staphylococcus epidermidis* and *Staphylococcus aureus*. In addition, staphylococcal biofilm associated diseases and host immune responses, along with the current and future therapeutic strategies for staphylococcal biofilm infections including chemotherapeutic, vaccine and the nanothecnology applications, are discussed.

Chapter 2 – Biofilm formation occurs in response to external cues and provides bacteria to live in communities that enhance overall survival to antimicrobials and other stressors.

The transition from motile to biofilm lifestyle appears to be controlled by the bacterial messenger, cyclic dimeric guanosine monophosphate (c-di-GMP). This is synthesized and degraded by c-di- GMP metabolizing proteins. In this chapter some of these proteins that have been implicated in the regulation of biofilm-associated phenotypes and characteristics are examined and described.

Chapter 3 – Multidrug resistance in prokaryotes is due primarily to efflux of offending antimicrobials from the cell by representatives of several different families of integral membrane transporter proteins. Clearly, in evolutionary terms, these proteins did not arise specifically to pump human-made antimicrobials out of the cell and thereby confer resistance. Despite this, often only their role in antibiotic resistance is characterised and highlighted. In recent years, however, a transition from the traditional anthropocentric perception of antibiotic resistance mechanisms in microorganisms has occurred, with naturally produced antimicrobials now generally regarded as physiologically important signalling molecules or sources of nutrition for bacteria rather than antimicrobial agents, and bacterial multidrug efflux proteins not merely as a defensive response to antimicrobials but as important players in fundamental physiological processes such as cellular homeostasis. This emerging perspective supports the notion that a better understanding of the complexities of infection and multidrug resistance in bacteria can be achieved via a more detailed understanding of those physiological processes. In this chapter the 'true' physiological roles of multidrug efflux proteins of the largest non-ATP-hydrolysing family of membrane transporters, the major facilitator superfamily, and explore the evidence for their function in processes such as pH and metal homeostasis, import and export of metabolites and biofilm formation are reviewed.

Chapter 4 – Aromatic hydrocarbons affect the environment, and in particular the fauna and flora. Organic solvents are among the most toxic chemicals. Biodegradation of these molecules has been found under both aerobic and anaerobic conditions. Bacteria utilize these compounds as an energy and carbon source. Microorganisms have evolved various biochemical and physiological adaptation mechanisms which protect them from the lethal effects of organic solvents. The adaptation mechanisms include: alteration of the properties of the bacterial cell membrane, morphological and bacterial proteome changes.

Chapter 5 – The Hfq protein was discovered almost forty years ago as an *Escherichia coli* host factor required for the replication of the RNA genome of the bacteriophage Qβ. However, the exact role of the protein as a global regulator of bacterial physiology and metabolism emerged much later, through the characterization of the broad phenotypes exhibited by hfq mutants in *E. coli* and in the zoonotic pathogen *Brucella abortus*. Recent studies have shown that the protein also plays a critical role on the pathogenicity of several Gram-negative and Gram-positive pathogens. The understanding of the pleiotropic phenotypes exhibited by hfq mutants emerged recently with evidence that Hfq plays a major role in the post-transcriptional regulation of gene expression, by facilitating the interactions between small non-coding regulatory RNAs and their target mRNAs. This knowledge added a new layer of complexity to our understanding of prokaryotic gene regulation. This chapter is focused on the protein Hfq and on its roles on pleiotropic and virulence phenotypes in several bacteria, including recent results from our research group on the roles of Hfq-like proteins on stress resistance and virulence of bacteria of the *Burkholderia cepacia* complex. The role of Hfq as a mediator of interactions of sRNAs with their target mRNAs is also presented.

Chapter 6 – The cytolethal distending toxin (CDT) is produced by many pathogenic Gram-minus bacteria (*Escherichia coli*, *Helicobacter hepaticus*, *Haemophilus ducreyi* and others) colonizing the oral cavity, the intestinal tract, the liver, etc. Depending on the affected organs, these bacteria are responsible for different pathologies. However, as a virulence factor, CDT has been shown to allow bacteria to persistently colonize the body, evade the immune system, induce inflammation and trigger genetic instability. Moreover, in vivo, the production of CDT by *Helicobacter hepaticus* induces the development of dysplastic liver nodules, thus defining CDT as a potential carcinogen. Given the important role of the toxin in the physiopathology of these bacteria, it is essential to accurately identify its mode of action .

In human cells, CDT exposure leads to a singular cytotoxicity associated to a characteristic cell distension, and induces a cell cycle arrest at the G2/M transition that is dependent on the DNA damage response. Thus, CDT has been classified as a cyclomodulin and a genotoxin, as this toxin induces DNA double-strand breaks. It is therefore important to characterize the outcome of these CDT-induced DNA damage and the consequences for the intoxicated cells and organisms. Indeed, the presence of unrepaired damage can lead to cell death, whereas effective repair will allow cells to resume cell cycle. However, improper repair of DNA damage can induce genetic instability and lead to cancer .

After years of study, the CDT mode of action only begins to be unrevealed and many aspects remain to be studied. In this chapter, the bacteria producing CDT as well as their cellular niche are presented. The CDT trafficking in the eukaryotic cells, in order to understand the cell entry and the nucleus targeting of the toxin, are discussed. The host cell behavior in response to CDT intoxication, by focusing on morphological changes, cell cycle modulation and cell death response, is considered. The biochemistry of CDT, specifically regarding its catalytic activities and relate these aspects to the overall physiopathology of the toxin, is examined.

Chapter 7 – *Bacillus thuringiensis* subsp. *israelensis* (*Bti*) is a Gram-positive spore-forming bacterium and a highly potent mosquitocidal strain. During the sporulation phase it synthetizes several pore-forming pre-δ-endotoxins: four of the Cry (Crystal) family (Cry 4Aa, Cry4Ba, Cry10Aa and Cry11Aa) and three of the Cyt (Cytotoxic) family (Cyt1Aa and a tiny amount of Cyt2Ba and Cyt1Ca). These proteins are assembled together into a crystalline-like body which is dissolved in the alkaline environment of the larval midgut, digested and activated by endoproteases. Cry pre-toxins are composed of three domains. One domain is associated with a specific receptor on the midgut epithelial cells and the other domains undergo conformational change and are inserted into the lipid bilayer. Once a number of these toxins are associated together within the membrane, they create pores that interrupt the osmotic balance of the cells and lead to cell destruction. The Cyt toxins have no specific receptor on the cell surface and non-specifically destroy various insect and mammalian cells by a pore-forming or a detergent-like mechanism of membrane lysis. Several studies have shown that Cyt toxins serve as receptors for the Crys (e.g., Cry11Aa). They act synergistically to increase the toxicity of Cry toxins and also prevent the development of larval resistance to Crys. The active forms of Cyt1Aa and CytBa toxins have been recently crystallized and their 3D structure was deciphered and demonstrated to be similar in spatial structure and homology to other pore-forming toxins such as Cyt2Aa, volvatoxin A2 (VVA2) and the Erwinia virulence factor (Evf). This chapter summarizes up-to-date findings on the Cyts family and their homology toxins and discusses the potential use of these toxins in the therapeutic field, especially by specific targeting of Cyts to cancer cells.

Chapter 8 – Twenty years after the first discovery of a plasmid harbouring the gene coding for the botulinum neurotoxin type G in a very rare botulinum neurotoxigenic clostridium species, i.e. *Clostridium argentinense*, large novel plasmids carrying one or even two different botulinum neurotoxin genes are currently being identified in an increasing number of *C. botulinum* strains. Antibiotic resistance megaplasmids have concomitantly been revealed in rare botulinum neurotoxigenic *C. butyricum* strains associated with human botulism . The chapter is dedicated to recent findings on the novel plasmids being identified in botulinum neurotoxigenic clostridia and discusses the potential benefits of those plasmids to their clostridial hosts.

Chapter 9 – Increasing widespread resistance of microbial cells to antibiotics raises a need to develop novel antibacterial technologies. One of the approaches that may lead to improved antimicrobial treatment and replace or least significantly reduce the use of antibiotics is photodynamic chemotherapy. This technique includes visible light excitation of nontoxic components – photosensitizers, which transfer energy to molecular oxygen, resulting in the production of reactive oxygen species that in turn cause irreversible damage to cellular components of bacteria. Photodynamic chemotherapy has been extensively studied towards both Gram-positive and Gram-negative bacteria, including antibiotic-resistant species. Gram-positive and Gram-negative bacteria react differently to photodynamic chemotherapy, at that the former exhibit high sensitivity to the treatment. Gram-negative bacteria are less susceptible to photodynamic chemotherapy due to the presence of phospholipids, complex lipoproteins and polysaccharides in the additional outer cell envelope. From the numerous studies on the effect of photodynamic chemotherapy on different microorganisms no development of bacterial resistance to photosensitizers has been reported so far. Excitation of photosensitizers in the process of bacterial eradication can be performed by several methods, including illumination with external light sources, chemiluminescent illumination and sonodynamic activation. The two latter techniques enable to deliver photodynamic treatment without external illumination in the dark, which opens prospects for combating internal infections that are difficult or not available for treatment by traditional photodynamic chemotherapy. This property turns these techniques into promising novel antimicrobial therapeutic strategies.

Chapter 10 – The Monod-type model was analyzed from the chemical kinetics side (Hinshelwood) in "Langmuir adsorption" terms, or from the microbiology side (Button) in "Michaelis-Menten" enzyme kinetics terms, but it seems that a general kinetic derivation of the Monod growth equation is missing. Also missing is a more or less general answer to the question: Why the model still holds, within surprisingly broad limits concerning bacterial strains, nutrient and environment conditions? The present study proves that the actual form of the Monod equation may be derived in strictly kinetic terms within assumptions/ approximations/methods borrowed from heterogeneous catalysis (Horiuti, Temkin, Happel and Boudart): (quasi)steady state approximation, most abundant reaction intermediate, kinetically significant step, reaction route or path. The main approximation involves considering bacterial individuals (in various stages of their growth) as some quasi-chemical species, the reactivity of which is subject to mass action law. The growth is amenable to a two-step sequence where the term "step" stands for some complex, possibly single-route, and kinetically significant process. Consequently, the growth may be approached in quasi-chemical terms and the derivation of the Monod equation is straightforward for two limiting cases: 1) "starving conditions", with substrate binding as a limiting step; 2) "normal" laboratory conditions with abundant nutrients (excepting possibly the growth controlling or limiting one) and bacterial fission as limiting step. Using the hypothesis of a most abundant growth intermediate, magi, as a kinetic key, several other mechanistic alternatives are presented, involving multiple-substrate binding before and/or after the formation of this key intermediate. Several experimental observations are discussed in connection with the mechanistic expressions obtained for the Monod growth parameters. Application to microcalorimetric growth data in batch conditions involves the decomposition of the complex thermogram via Peakfit processing of the thermogram. The thermal version of the Monod equation describes component, non-steady growth processes surprisingly well. A possible explanation would be the conservation of the *magi* (and thus of the rate equation) over the whole extension of the growth process.

In: Prokaryotes
Editor: Marina Nisnevitch

ISBN: 978-1-63321-592-4

Chapter 1

Staphylococcal Biofilm Formation and Development: Related Diseases, Host Immune Responses and Therapy

Livia Visai[1,6], Davide Ferrari[2,6], E. Magda Barbu[3], Xiaowen Liang[4], Vannakambadi K. Ganesh[4], Magnus Höök[4] and Marcello Imbriani[5,6]*

[1]Department of Molecular Medicine, Center for Tissue Engineering (C.I.T.), University of Pavia, Pavia, Italy
[2]Department of Biochemistry and Molecular Biology, University of Parma, Parma, Italy
[3]Department of Infectious Diseases, The University of Texas MD Anderson Cancer Center, Houston, TX, US
[4]Center for Infectious and Inflammatory Diseases, Institute of Biosciences and Technology, Texas A&M Health Science Center, Houston, TX, US
[5]Department of Public Health, Experimental Medicine and Forensics, University of Pavia, Pavia, Italy
[6]Department of Occupational Medicine, Ergonomy and Disability, Lab of Nanotechnology, Salvatore Maugeri Foundation, IRCCS, Pavia, Italy

Abstract

Various biomolecular and regulatory factors are involved in adhesion, biofilm formation and host immune evasion of staphylococcal species. Adherence of staphylococcal species to either biotic or abiotic surfaces are a critical first event in the establishment of infection with these serious pathogens. A variety of proteinaceous and non-proteinaceous adhesins are known to mediate attachment to a multitude of host factors, such as extracellular matrix, and plasma proteins, and human host cells or inter-

* Corresponding author: Email: livia.visai@unipv.it.

bacterial cell adhesins, which is essential for biofilm accumulation. Formation of the biofilm is a two-step process: (1) adherence of the pathogen to a surface and (2) further progression by accumulation of cells to form multilayered cell clusters. Biofilm is one of the important virulence factor of staphylococci that plays a role in many device-related infections such as native valve endocarditis, otitis media, urinary tract infections, cystic fibrosis, and acute septic arthritis, The increasing prevalence of antibiotic resistant strains and the vast ability of bacteria to evade host immune defenses are becoming more and more important for medical infections. The present review focuses on the role of specific cell surface adhesins, the ica operon, accumulation-associated proteins and quorum-sensing systems in the adhesion and infection processes of the Gram-positive pathogens, *Staphylococcus epidermidis* and *Staphylococcus aureus*. In addition, we will discuss staphylococcal biofilm-associated diseases and host immune responses, along with the current and future therapeutic strategies for staphylococcal biofilm infections including chemotherapeutic, vaccine and the nanotechnology applications.

Keywords: Staphylococcal biofilms, biofilm related disease, vaccine, nanotecnology

Introduction

To date, staphylococcal species remain one of the major causes of both health care associated (HA) as well as community-associated (CA) infections. *Staphylococcus aureus* (*S. aureus*) is commonly associated with tissue infections, such as endocarditis (Fitzsimmons, 2010) and osteomyelitis (Zuluaga, 2006) whereas *Staphylococcus epidermidis* (*S. epidermidis*) is a frequent etiological agent of biofilm infections on medical devices, including indwelling catheters and prostheses (Fitzpatrick, 2005a; Otto, 2008; Fey, 2010a). Staphylococcal adherence to an either biotic or abiotic surfaces is the critical first event in the establishment of an infection with these oppurtunistic deadly pathogens. In particular, *S. aureus* harbours a variety of proteinaceous and non-proteinaceous adhesins that mediate attachment to a multitude of host factors, such as extracellular matrix and plasma proteins, human host cells, and promote pathogen-pathogen intercellular adhesion, which is essential for biofilm accumulation (Heilmann, 2011).

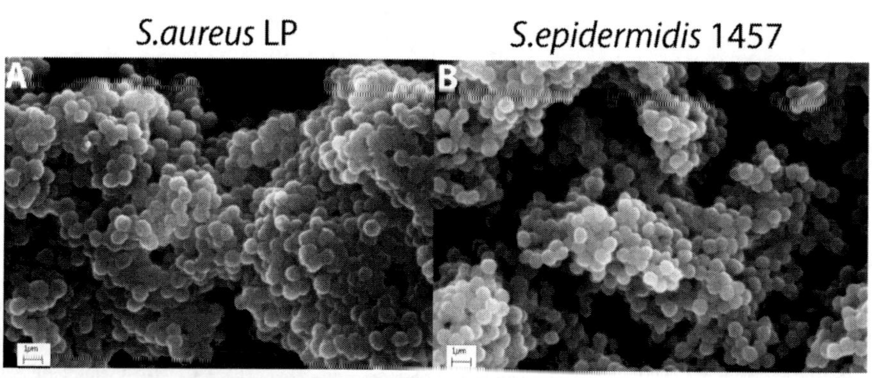

Figure 1. SEM images of staphylococcal biofilms. *S. aureus* LP (panel A) and *S. epidermidis* 1457 (Panel B) biofilms were grown and prepared for SEM observations as previously reported (Sharma, 2008). Magnification, x 5,000.

In contrast to *S. aureus*, virulence properties associated with *S. epidermidis* are fewer and biofilm formation is the defining virulence factor associated with colonization and infection, as demonstrated by animal models of biomaterial-related infections (Fey, 2010b). Increasing attention has been focused on understanding bacterial biofilms (Figure 1) and this growth modality's relation to human disease (Jones, 2001;Fey, 2010a).

In this review we will focus on the mechanisms governing biofilm formation with a special focus on the involvement of bacterial adhesins and the molecular mechanism of staphylococcal biofilm dispersal. In addition, we will discuss staphylococcal biofilm associated diseases and host immune responses, along with the current and future therapeutic strategies for staphylococcal biofilm infections including chemotherapeutic, vaccine and the nanotechnology applications.

Staphylococcal Biofilm Formation

Attachment: Mechanism and Factors Involved

It is widely accepted that biofilm formation is a two-stage process implicating a primary attaching (or nucleating) step succeeded by a maturation phase (O'Toole, 2000) (Figure 2).

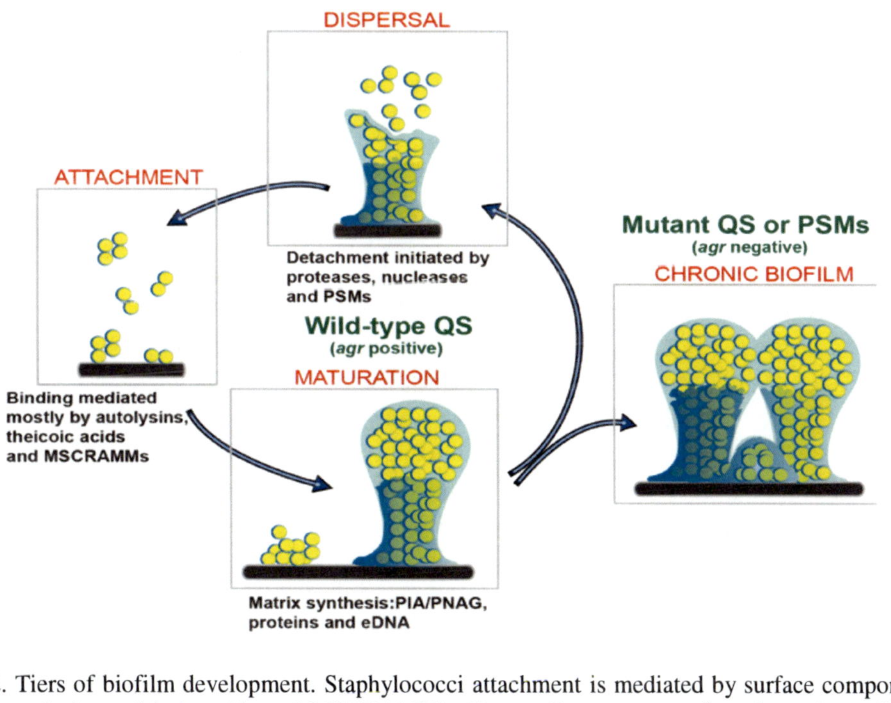

Figure 2. Tiers of biofilm development. Staphylococci attachment is mediated by surface components, such as autolysins, teichoic acids and MSCRAMMs. Upon adherence to surface, bacteria proliferate and engulf themselves in a matrix composed of exopolysaccharides, proteins and eDNA. A subsequent maturation step ensures the formation of characteristic mushroom-like structures. Once the biofilm reaches the end-stage of growth (high cell density), the possibility of dispersal becomes available. Quorum sensing (QS; encoded by the *agr* locus) activates the expression of factors involved in dispersal, such as proteases, endonucleases and surfactant Phenol Soluble Modulins (PSMs). Alternatively, QS and PSMs mutants form thick biofilms that lack the ability to detach.

A subsequent dispersal stage involving the detachment of individual cells by several mechanical or physico-chemical mechanisms which allows bacteria to disseminate and colonize distant sites (O'Toole, 2000). Attachment is driven, at least in part, by gravitational and hydrodynamic surrounding forces (Donlan, 2002; Beloin, 2008). Once staphylococci encounter a substrate, overall surface hydrophobicity determines the degree and velocity of adherence (Dunne, 2002). Concomitantly, specific extracellular molecular determinants act to withstand shear forces and ensure a steady bond with the surface (Whiteley, 2001; Beloin, 2004). In both *S. aureus* and *S. epidermidis*, teichoic acids and the non-covalently attached surface autolysins seem to confer overall surface hydrophobicity and mediate primary adherence to abiotic surfaces (Heilmann, 1998; Gross, 2001; Rupp, 2001). However, more recent studies have linked the major autolysin AtlE activity to extracellular DNA (eDNA) release, a biofilm matrix component initially associated with biofilm maturation, suggesting significant imbrications among the factors involved in bacterial adherence and proliferation of the biofilm (Qin, 2007; Houston, 2011).

Although staphylococci are notorious for their innate ability to colonize inert surfaces, their attachment to tissues or indwelling medical devices is facilitated by the interaction of bacteria with mammalian matrix proteins, which abundantly cover fixtures soon after insertion (Fleer, 1990; Patti, 1993; Foster, 1998; Clarke, 2006). In both *S. aureus* and *S. epidermidis*, irreversible attachment is mainly attained by a class of surface-anchored proteins referred to as microbial surface components recognizing adhesive matrix molecules (MSCRAMMs), which have also been shown to play a role in infection (Foster, 1998). These molecules share a common domain organization that includes an LPXTG cell-wall anchoring motif, a transmembrane segment and a positively charged cytoplasmic tail at the C-terminus (Foster, 1998; Ponnuraj, 2003). In a subset of MSCRAMMs, a B-region containing repeated β-sandwich modules of unknown function accompanies the R repeat subdomain composed of multiple serine-aspartate dipeptide repeats (Foster, 1998). A ligand-binding A-region containing two (e.g. collagen-binding MSCRFAMMS Ace, Cne, and Cnm have only N1-N2 domain) or three subdomains (N1, N2, and N3 of CNA), adopts an IgG-like fold, and is exposed at the amino terminus (Ponnuraj, 2003; Ganesh, 2008; Ganesh, 2011). *S. aureus* strains display a wider variety of LPXTG cell-wall-anchored MSCRAMMs (~22) relative to *S. epidermidis* (~12) (McCrea, 2000; Gill, 2005). In *S. aureus* fibrinogen-binding proteins (clumping factor A (ClfA), and B (ClfB)) (McDevitt, 1994; Ni Eidhin, 1998), fibronectin-binding proteins (FnBPA, FnBPB) (Schwarz-Linek, 2003; Keane, 2007), and collagen-binding protein (Cna) (Patti, 1993), initiate adherence, and as a result, appear to favor biofilm formation. Similarly, the *S. epidermidis* SdrG mediates bacterial interaction with the β-chain of fibrinogen, but its contribution to biofilm formation has not yet been clearly established. In addition, the GehD lipase appears to play a role in bacterial adhesion to collagen (Bowden, 2002).

Maturation

Upon attachment, staphylococci proliferate and synthesize an encasing scaffolding matrix (glycocalix or slime) composed of exopolysaccharides (Mack,1996; Cramton,1999), proteins(Cucarella, 2001; Corrigan, 2007; O'Neill, 2008; Schroeder, 2009; Merino, 2009;

Abraham, 2012) and eDNA (Izano, 2008) that enables biofilm growth into multilayered mushroom-like structures. Early *in vitro* studies have shown that the most prominent component of *S. aureus* and *S. epidermidis* biofilms is the intercellular polysaccharide adhesin (PIA) composed of b-1,6-linked N-acetylglucosamine (PNAG) (Mack, 1996; Cramton, 1999). The synthesis of PIA polymer is mediated by the products of the *ica* gene locus composed of the two *icaA* and *icaD* N-acetylglucosamine transferases, the *icaB* PIA deacetylase, the *icaC* putative PIA exporter, and the regulatory gene *icaR* (Heilmann, 1996; Gerke, 1998; Vuong, 2004a). Of functional importance, the deacetylation process appears to confer PNAG a positively charged character, which presumably enables its interaction with the negatively charged surface teichoic acids to promote intercellular interactions. Interestingly, exopolysaccharides similar to PIA have been detected in other biofilm-forming pathogens, suggesting that these polymers may be functionally conserved in bacterial biofilm formation (Darby, 2002; Kaplan, 2004b; Wang, 2004). Surprisingly, *ica* operon presence has also been associated with *S. epidermidis* invasiveness (Li, 2005).

Role of Biomacromolecules in Biofilm

Biomacromolecules, polysaccharides, proteins and nucleic acid molecules contribute to the formation and progression of biofilms. The role of each molecules and the contribution of these types of molecules are dependent on various factors.

Although recognized as fundamental virulence factor in *S. epidermidis* (Rupp, 1999a; Rupp, 1999b; Rupp, 2001; Vuong, 2004a; Fluckiger, 2005), the role of PIA in biofilm formation and biofilm-associated infection is not exclusive, as biofilm formation in the absence of PIA has been demonstrated (Rohde, 2005). Moreover, a subset of strains isolated from biofilm-associated infections does not contain the *ica* genes which are mediators of PIA synthesis (Arciola, 2006). Further investigations revealed that i*ca* operon was absent in 27% of the biofilm-forming *S. epidermidis* strains. The intercellular interactions were in part mediated by the accumulation associated protein (Aap) (Rohde, 2007), which promotes intercellular interactions by forming fibrils on the cell surface that seemingly act as a network holding bacteria together (Banner, 2007). In addition, Aap homologs SSP-1 and SSP-2 proteins have been implicated in biofilm development but their specific role is biofilm is not well understood (Veenstra, 1996). Similarly, the biofilm-forming capacity of the *S. aureus* clinical isolate UAMS-1 (University of Arkansas Medical System–1) was not affected by the absence of *ica* operon (Beenken, 2004). In addition, the biofilm formation by the MRSA strain BH1CC was found to be almost exclusively dependent of the FnBP MSCRAMMs (Fitzpatrick, 2005). Notably, epidemiological studies have indicated that PIA–dependent biofilm accumulation correlates with methicillin susceptibility, whereas methicillin-resistant isolates are capable of biofilm formation when *icaADBC* genes are deleted (O'Neill, 2007). Taken together, these data suggest that the contribution of polysaccharides and surface proteins to staphylococcal biofilm maturation is strain-dependent.

In the case of *S. aureus*, the MSCRAMMs that are implicated in biofilm growth include biofilm-associated protein (Bap) (Cucarella, 2001), ClfB (Abraham, 2012), FnBPA, FnBPB (O'Neill, 2008), *S. aureus* surface protein C (SasC) (Schroeder, 2009), SasG (Corrigan, 2009), and surface protein A (SpA) (Merino, 2009). The involvement of Bap has been found only in bovine mastitis isolates (Cucarella, 2001). ClfB, FnBPA, FnBPB and SpA are found in numerous strains, whereas SasC and SasG are present only in a subset of isolates (Schroeder, 2009; Corrigan, 2009). The molecular mechanism by which these surface anchored proteins contribute to intercellular interactions is largely unknown, but recent studies established a metal-ion dependent homophilic interactions between MSCRAMMs present on neighboring cells (Geoghegan, 2010; Conrady, 2008; Abraham, 2012; Conrady, 2013) that could contribute to biofilm.

More recently, it has been shown that eDNA released from decaying cells is critical for biofilm stabilization (Whitcurch, 2002; Yang, 2007; Mann EE, 2009). These conclusions are also supported by evidence that DNase treatment eliminates biofilm *in vitro* and the synergistic use of DNase-antibiotic therapy was clinically successfull in the treatment of cystic fibrosis patients (Gibson, 2003). Accumulating evidence suggest that a holin/antiholin-like system encoded by the *cidA* and *lrgA* genes control cell death during biofilm development (Rice, 2007). To add another level of complexity, the cell lysis and subsequent genomic DNA release is also attributed to autolysins activity during proliferation (Heilmann, 1997; Heilmann, 1998).

Mechanisms Governing Biofilm Formation

The transition from planktonic growth to biofilm appears to be the result of cues, such as unfavorable environmental conditions (low-nutrient levels, osmolarity, and oxygen availability) or permissive surface availability (hydrophobic substrate, receptor presence), and depends on various conserved and/or species-specific factors (Monds, 2009). Upon attachment, numerous regulatory networks translate into co-ordinate gene expression adjustments that endow bacteria with properties required for sessile growth (Whiteley, 2001; Donlan, 2002; Beloin, 2008). Structural and functional factors involved in biofilm formation are subject to complex control mechanisms, in which multiple *cis*- and trans-acting factors function coordinately or independently to affect transcription. The two major staphylococcal virulence gene expression regulators, accessory regulator (*sarA*) and accessory gene regulator (*agr*), have also been implicated in biofilm formation (Novick, 2008). The intricate interplay between these global regulators as well as their direct or indirect influence on the downstream two-component regulatory systems and structural effectors involved in biofilm formation is still under investigation. In staphylococci one gene can often be controlled by multiple regulators or regulatory systems, which influence each other to ensure precise temporal and spatial expression. In this review, we will focus here on the systems that have been thoroughly described in the literature.

Regulation of PIA Synthesis

Since PIA has been long equated with the matrix, the regulation of PNAG synthesis is probably the best studied event involved in biofilm maturation (reviewed in O'Gara, 2007). Initially identified in *S. epidermidis* using transposon mutagenesis to isolate defective in biofilm formation (Mack, 1994; Heilmann, 1996; Heilmann, 1998), the *ica* locus was later shown to also play a role in *S. aureus* biofilm formation (Cramton, 1999). In *S. epidermidis*, the expression of the *ica* operon appears to be upregulated *in vivo* (McKenney, 1999; Fluckiger, 2005). In contrast, PNAG synthesis and biofilm production are tightly regulated *in vitro* by oxygen and iron limitation, and increase osmolarity (Ammendolia, 1999; McKenney, 1999; Cramton, 2001; Stepanovic, 2003; Fitzpatrick, 2005 and 2006). Environmental changes are in part intercepted and translated by the TetR family transcriptional repressor *icaR,* which is located upstream of the icaADBC operon in both *S. epidermidis* and *S. aureus* (Conlon, 2002a). Osmolarity fluctuations or addition of ethanol to the culture medium activate regulatory pathways in an *icaR*-dependent manner (Conlon, 2002a and b; Conlon, 2004). In addition, TcaR, a MarR family of transcriptional regulators, was identified as a weak repressor of *ica* operon transcription. Moreover, deletion of *icaR* and *icaR* had a synergistic negative effect on *S. aureus* PNAG and biofilm production (Jefferson, 2004). In addition to IcaR and TcaR, the reversible insertion sequence element IS256 into the *ica* locus negatively regulates biofilm development in *S. epidermidis* (Ziebuhr, 1999). Subsequently it was demonstrated that IS256 insertions lacks specificity for the *ica* locus and has a pleiotropic effect, perturbing multiple virulence pathways in staphylococci (Conlon, 2004; Handke, 2004).

Early reports highlighted the importance of *sarA* locus for *ica* operon transcription, PIA/PNAG production and biofilm formation in *S. aureus* (Beenken, 2003; Valle, 2003). *sarA* transcription is upregulated in *in vitro* biofilms relative to planktonic cultures (Valle, 2003). Consequently, stains bearing *sarA* mutations lose their ability to form biofilm (Beenken, 2003). Interestingly, biofilm formation by *sarA* mutants can be rescued by a coexisting deletion of the *nuc* thermostable nuclease or exogenous addition of protease inhibitors (Tsang, 2008). It appears that by inhibiting nuclease and protease activities, *sarA* also prevents degradation of secondary structural matrix components such as eDNA and proteins. This ultimately allows bacteria to form robust and enduring multicellular communities. Although the SarA proteins of *S. epidermidis* and *S. aureus* share significant amino acid identity (~84%) and that the three *S. aureus sarA* promoters P1, P2 and P3 are also present in *S. epidermidis*, their organization differs (Fluckiger, 1998), suggesting that *sarA* regulation and activity may not be identical. For example, a recent study identified *sarA* as a negative regulator of biofilm formation by modulating the expression of extracellular matrix binding protein Empb (Christner, 2012). Given that SarA also regulates *agr* activity (Cheung, 1994), further investigations aiming to dissect these interactions indicated that *sarA* affects biofilm formation in an *agr*-independent manner (Beenken, 2003; Valle, 2003).

Additional levels of biofilm formation control are achieved through the global stress response regulators *sigB* (σB) and RsbU (a positive regulator of σB), which mainly affecting *ica* operon transcription by repressing transcription of *icaR* in *S. epidermidis*, but not *S. aureus* (Knobloch, 2001; Valle, 2003). Since σB controls gene expression in response to environmental stress, studies have focused on understanding how environmental stressors are processed in its absence. Unexpectedly, biofilm formation by *S. epidermidis rsbU* deletion mutants is impaired in elevated osmolarity, whereas ethanol-mediated biofilm induction remains unaffected (Knobloch, 2001; Conlon, 2004). At the same time, ethanol-induced biofilm development by a σB mutant could be rapidly dispersed relative to the wild-type strain. Together, these observations emphasize the importance σB for biofilm stability and dissemination of bacteria from these communities (Jager, 2005). Moreover, the species-specific σB activity grants more substance to the realization that *S. epidermidis* and *S. aureus* are distinct at least at the level of matrix composition.

Although PIA expression is not directly regulated by the major quorum-sensing (QS) *agr* system in *S. epidermidis*, the LuxS secondary QS mechanism appears to be required for the synthesis of AI-2 autoinducer molecule and *ica* expression (Xu, 2006). Significantly, a *luxS* mutant exhibited enhanced virulence in a rat model of biofilm-associated infection (Xu, 2006). Unlike *S. epidermidis*, *luxS* deletion in *S. aureus* was not associated with an altered biofilm-forming capacity.

Regulation of Surface Proteins Important for Staphylococcal Biofilm Development

It has been long considered that the rapid and precise control of gene expression achieved by QS in a wide variety of bacteria is evolutionarily important; its distinct role in staphylococcal biofilm formation and dispersion has been at the center of numerous studies. Overall, these reports establish that the *agr* operon encodes for a secreted post-translationally modified peptide that acts as an effector for a typical autoactivation of two-component regulatory system placed within the same locus (reviewed in Novick and Geisinger, 2008). Agr activation is triggered by the increase in cell density and ultimately results in a concerted gene expression shift correlated with the switch from exponential to early stationary phase. The agr locus (Figure 3) is expressed from two divergent promoters. The P2 transcript contains the AgrA and C two-component signaling system and the AgrB and D products encoding for the 7–9 amino acid autoinducer peptide (AIP) that contains a five amino acid thiolactone ring. Similar to most QS systems described, the AIP binds to and activates AgrC, which in turn activates AgrA and the activated AgrA upregulates its own promoter (Novick, 1995). In addition to P2, AgrA also activates the P3 promoter driving the expression of a regulatory RNA, RNAIII, which controls several hundred genes involved in a variety of physiological processes and virulence (Dunman, 2001). The Agr system downregulates adhesive factors such as MSCRAMMs and upregulates the expression of toxins, proteases and phenol-soluble modulins (PSMs) (Janzon, 1989). As a result, by inhibiting cell adherence, Agr acts as a limiting factor of biofilm formation.

Equally important, concomitant activation of proteases and surfactants marks the maturation of biofilm and initiates detachment of cells and possibly dispersal to remote sites. Conversely, repression of *agr* activity augments the expression of MSCRAMMs and inhibits the secretion of surfactant PSMs leading to excessive biofilm growth that lacks the capacity to disperse (Jarraud, 2000). Therefore, the staphylococcal global regulators of gene expression act in an opposing manner, where *sarA* induces attachment and early biofilm formation presumably by activating PIA expression and concomitantly repressing nucleolytic and proteolytic extracellular enzymes.

Once biofilms have reached maturation (or certain cell density), *agr* activation leads to expression of numerous factors involved in biofilm dispersal and cell dissemination. In addition to Agr, adhesive surface protein expression is positively regulated by σB (Ji, 1997; Ji, 2005a and 2005b). In contrast, σB abates the expression molecules that promote single cell growth or biofilm dispersal, such as β-hemolysin, serine protease (SplA), cysteine protease (SplB), the metalloprotease aureolysin, staphopain and leukotoxin D (Kahl, 2000). In agreement, studies revealed that a σB-deficient *S. aureus* strain is impaired in biofilm formation (Kalkum, 2003).

Despite intense efforts, the importance of these findings for biofilm formation *in vivo* is still mostly unknown. In a recent review, Joo and Otto (2012) coined a model that correlates these results with different types of virulence.

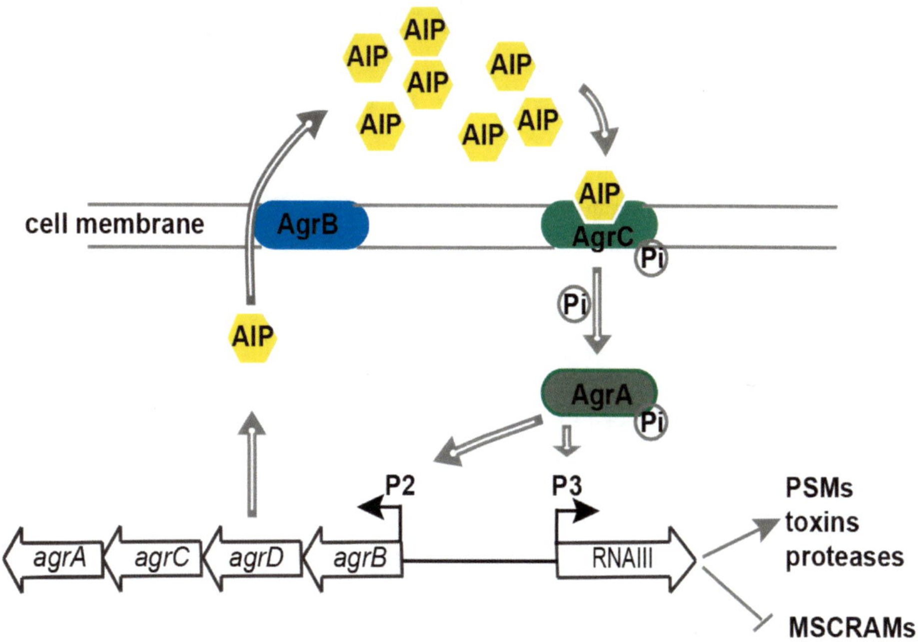

Figure 3. Quorum sensing regulation in staphylococci. AgrD encodes for a pheromone-like secreted translationally-modified peptide (AIP) that accumulates in the extracellular milieu directly proportional with cell density. AIP activates a classic two-component system in an auto feedback loop, which subsequently activates its own expression and that of a regulatory RNA (RNAIII) responsible for up-regulation of factors responsible for matrix degradation. In contrast, proteins (MSCRAMMs) involved in intercellular interactions are down-regulated.

Specifically, homeostatic QS expression is important for acute infection and formation of easily dispersible biofilms that allow bacteria to disseminate to distant sites. In contrast, downregulation and/or QS mutation resembles chronic biofilm-associated infections characterized by development of thick, long-lasting biofilms (Joo and Otto, 2012). Collectively, these observations highlight the complexity of biofilm formation and regulation in staphylococci.

The Molecular Mechanism of Staphylococcal Biofilm Dispersal

As biofilm matures, clusters of cells begin to detach from these communities by either passive or active mechanisms. These mechanisms likely involve degradation of extracellular matrix and concerted regulatory changes that equips bacteria with properties facilitating their return to planktonic lifestyle. In addition to mechanical dispersal triggered by shear forces, the decision to return to single cell life style is influenced by environmental cues, such as nutrient availability, oxygen gradient and toxic products accumulation (Karatan, 2009). Importantly, several lines of evidence suggest that biofilm dispersal plays a key role in biofilm-associated infections with serious consequences for patient's health. For example, dissemination of bacteria colonizing indwelling devices can result in sepsis, or in the case of detachment from endotracheal tubes, result in severe hospital-acquired pneumonia (Adair, 1999; Wang, 2011).

Controlled detachment implicates cessation of slime production and activation of matrix-degrading factors and surfactants, processes that will allow bacteria to dislodge and disseminate. As described earlier, the Agr system controls the production of matrix degrading enzymes in a cell density-dependent manner (Novick and Geisinger, 2008). Upon reaching a threshold, AIP (nanomolar range), activates the so-called Agr proteome composed of numerous proteases, toxins and PSMs. Hence, *S. aureus* and *S. epidermidis agr* deficiencies inhibit biofilm maturation (Boles, 2011; Beenken, 2010). In the agreement, *sarA* and σB inactivation triggers an increase in extracellular protease production that springs biofilm disassembly (Lauderdale, 2009; Beenken, 2010; Tsang, 2008). Recent work supporting the role of proteases in biofilm dispersion showed that the *S. epidermidis* extracellular protease Esp inhibits *S. aureus* biofilm formation and nasal colonization (Iwase, 2010).

In *S. aureus*, thermonuclease *nuc* is another factor that has been implicated in cell dissemination from biofilm. Consistently, exogenous addition of DNases or restriction enzymes to mature biofilms results in dose-dependent biofilm inhibition. Moreover, *S. aureus nuc* mutants display an increased biofilm-forming phenotype, highlighting the importance of eDNA release for scaffold formation (Tsang, 2008).

Lastly, QS-controlled PSMs, a class of surfactant peptides that have first been described as virulence factors of *S. epidermidis* and *S. aureus*, facilitate cells release from the degrading matrix (reviewed in Peschel and Otto, 2013). The latter species contains three PSMs operons located at different sites in the genome (Wang, 2007). The *psmα* encodes four peptides (PSMα1–PSMα4). PSMβ1 and PSMβ2 are part of the *psmβ* operon; and δ-toxin is encoded within the sequence for RNAIII, the RNA effector molecule of the Agr QS system (Wang, 2007). *S. epidermidis* produces six such peptides, PSMα, PSMβ1, PSMβ2, PSMδ, PSMε and δ-toxin, encoded in four operons (Mehlin, 1999; Vuong, 2004b; Yao, 2005). When grouped according to their length: the α-type peptides (*S. aureus* PSMα1–PSMα4 and δ-toxin, and *S. epidermidis* PSMα, PSMδ, PSMε and δ-toxin) are ~20–25 amino acids long, whereas the β-type peptides (*S. aureus* and *S. epidermidis* PSMβ1 and PSMβ2) are ~45amino acids in length. Conforming to their physicochemical attributes, all PSMs contain an amphipathic α-helix responsible for their chromatographical behavior and biochemical properties. While the α-helix spans the entire length of short α-type PSMs, in β-type PSMs the α-helical region is located in the carboxyl-terminal region. Of importance, the pronounced amphipathy of β-type PSMs enables these peptides to act as detergent-like molecules in structuring and detachment of biofilms (Wang, 2011; Periasamy, 2012). Consistently, staphylococcal strains harboring a *psmβ* operon deletion form thick and more compact biofilms that do not easily disperse compared to the isogenic wild-type strains. Further investigations into the physiological importance of these peptides in the staphylococcal pathogenesis prompted numerous researchers to speculate that PSMs may play key roles in the colonization of the skin or nutrients scavenging from the environment. More recent work suggests that PSMs may have a dual function in biofilm formation (Schwartz, 2012). On one hand, it appears that ordered aggregation of soluble PSM peptides into amyloid-like fibers during proliferation stage may aid biofilm formation and abrogate single cell detachment. On the other hand, as the biofilm matures, monomeric synthesis of PSMs enables biofilm disassembly. However, the regulatory mechanisms controlling this emerging paradigm are still elusive. Nonetheless, a seminal study by Wang et al. (2011) demonstrated that PSMs promotes biofilm dispersal *in vivo* and enable the dissemination of *S. epidemidis* from infected catheters.

Staphylococcal Biofilm-Related Diseases

S. aureus and CoNS (coagulase-negative staphylococci, i.e. *S. epidermidis* and most other staphylococci other than *S. aureus*) are known as the most frequently isolated nosocomial pathogens from intensive care unit patients (National Nosocomial Infections Surveillance System, 2004). Furthermore, extremely high percentage of these isolates are resistant to antibiotics. Besides the specific antibiotic resistance (due the acquisition of chromosomally, or more often plasmid-encoded genetic resistance factors), staphylococci have non-specific mechanisms of resistance based on biofilm formation. Staphylococci are recognized as the most frequent causes of biofilm-associated infections. This exceptional status among biofilm-associated pathogens is due to the fact that staphylococci are frequent commensal bacteria on the human skin and mucous surfaces (and those of many other mammals). To this end, we will briefly report on *S.aureus* and *S.epidermidis* biofilm-related diseases.

S.aureus has re-emerged as a clinically relevant pathogen due to its resistance to antibiotics and the increased use of indwelling medical devices (Wenzel, 2001; Rubin, 1999). *S. aureus* biofilms, once established, are recalcitrant to antimicrobial treatment and the host response, and therefore are the etiological agent of many recurrent infections (Jones, 2001). Osteomyelitis (Lew, 2004; Ziran, 2007; Zuluaga, 2006), indwelling medical device infection (Baldoni, 2009; Costerton, 2005; Herrmann, 1988), periodontitis and peri-implantitis (Heitz-Mayfield, 2010; Lang, 2011), Chronic wound infection (Gjødsbøl, 2006; Hansson, 1995; Davis, 2008; Bowling, 2009; Schierle, 2009), chronic rhinosinusitis (Foreman, 2010; Ferguson, 2005; Psaltis, 2008; Stephenson, 2010), endocarditis (Røder, 1999; Fowler, 2005), ocular infection (Murugan, 2010; Leid, 2002) and polymicrobial biofilm infections are an incomplete list of *S. aureus* diseases with a demonstrated biofilm component (Peters, 2010; Sudbery, 2004; Archer, 2011). *S. aureus* stages of colonization, infection and maturation of the biofilm is mimicked in all the previous indicated diseases. In summary, bacteria can be introduced through a hematogenous route or direct inoculation during surgery, trauma or an overlying infection. While in the blood, *S. aureus* upregulates the production of adhesins for many host adhesive matrix molecules (e.g., fibrinogen, fibrin, osteopontin, fibronectin, collagen, elastin, etc.) (Resch, 2006). Once in the tissue, *S. aureus* attaches to localized areas of trauma forming an early biofilm and constitutively expressing a low level of the auto-inducing peptide (AIP) of the quorum sensing system. Then, the agr quorum sensing system is activated, resulting in the production of toxins, immunoavoidance factors and other agr-controlled virulence factors with a concomitant downregulation in microbial adhesins (Dunman, 2001). Simultaneously, the innate inflammatory response effectors (e.g., PMNs) are recruited to respond to a number of microbial inflammatory mediators (e.g., formylmethionyl peptides, low level hemolysins, lipotechoic acids, peptidoglycan, PSMs and staphylococcal DNA). However, PMNs attack the early biofilm in a non specific way resulting in the host cell lysis and frustrated phagocytosis (Anvar, 2009). At the end the formation of the *S.aureus* mature biofilm shows detaching cells and tissue damage due to PMN-induced collateral host tissue damage providing the devitalized surface for further biofilm development. At this point the infection can be resolved only by surgical removal of biofilm.

S. epidermidis is known as an opportunistic pathogen because it predominantly causes infection in immuno-compromised individuals such as intravenous drug abusers, AIDS patients, patients receiving immuno-suppressive therapy and premature newborns (Vadyvaloo, 2005). In healthy patients, *S. epidermidis* causes infection only after penetration of the skin or mucous membranes, which can occur by trauma, inoculation, or implantation of medical devices. Among CoNS, *S. epidermidis* clearly causes the greatest number of infections (Uckay, 2009; Rogers, 2009; Fey, 2010a).

Host Response to Staphylococcal Biofilm Infection

Biofilm-associated infections result in chronic disease that are resistant to the host immune response. Although a myriad of information exists on biofilm development, little is known how the immune system reacts to infection. Recent research studies are focuses on elucidating the host responses to staphylococcal biofilm infections.

The innate immune response to numerous microbial infections, including staphylococcal species, is mediated, in part, by preexisting soluble factors that recognize and destroy pathogens or target them for killing by macrophages, polymorphonuclear neutrophils (PMN) and dendritic cells. For example, both the antimicrobial peptides (AMPs) and the complement pathway (Medzhitov, 2000; Medzhitov, 2007) represent a first line of defense against invading microbes. AMPs are universal innate defense molecules in humans and other higher organisms. AMPs are peptides showing narrow or broad-spectrum activity against bacteria, fungi, viruses, and/or parasites. Thanks to their net-positive charge, amphipathicity, and small size, AMPs determine disruption of microbial membranes and also inhibit cell wall, nucleic acid, and protein biosynthesis (Yeaman and Yount, 2003; Brogden, 2005). Recent studies have shown the importance of a number of AMPs against *S. aureus* biofilms by either inhibiting biofilm formation or down regulating the expression of genes involved in biofilm development (Lopez-Lebanetal, 2010; Dean, 2011; Menousek, 2012).

The complement system consists of blood-derived proteins that are activated in an enzymatic cascade to eliminate pathogens from the host. S. *epidermidis* planktonic cells and biofilms behave in a different way: PMN killing of planktonic cells is complement dependent (Clark, 1986; Kristian, 2008; Fredheim, 2011) whereas biofilms avoiding complement deposition cause an increased resistance to opsonization and phagocyte-mediated killing (Kristian, 2008; Fredheim, 2011).

PMNs are known as the "first line of defense" against bacterial infection. Numerous bactericidal activities have been identified including phagocytosis, production of reactive oxygen species, or nitric oxide and degranulation with release of bactericidal entities such as defensins, lysozyme, lactoferrin or cathelicidins (Stuart et al. 2005; Kobayashi et al., 2005;

Fuchs et al., 2007; Segal, 2005; Nathan, 2006; Nauseef, 2007). PMNs produce large amounts of reactive oxygen intermediates (ROI) catalyzed by NADPH oxidase. PMNs also secrete several proinflammatory cytokines, including TNF-α and IL-1β as well as chemokines, such as CXCL2 (macrophage inflammatory protein-2; MIP-2) and CXCL1 (KC, keratinocyte-derived chemokine; both are functional mouse homologs of IL-8) and CCL3 (MIP-1β) (Cassatella, 1995; Witko-Sarsat et al., 2000; Nathan, 2006). A variety of receptors have been identified to which PMN can interact with the bacteria. Particularly well-studied host receptors are pattern-recognition receptors (PRRs) of the Toll-like (TLRs) family. PPRs interact with widespread pathogen associated molecular patterns (PAMPs) (Kawai and Akira, 2011). Thirteen TLRs have been described in the human and 10 in the mouse (Kopp and Medzhitov, 2003; Kawai and Akira, 2011). Staphylococcal lipoteichoic acid (LTA), peptidoglycan (PGN) (Morath et al., 2002; Dziarski, 2003; Weber et al., 2003), and lipoproteins (Lpp) have been identified as potent TLR2 ligands (Hashimoto et al., 2006 a,b; Kurokawa et al., 2009). Likewise, PIA and PSM expression in *S. epidermidis* promotes biofilm formation and can be recognized by TLR2 (Hajjar et al., 2001; Stevens et al., 2009). TLR9 is an intracellular receptor that recognizes unmethylated CpG motifs characteristic of bacterial DNA (Hemmi et al., 2000; Bauer et al., 2001). It is known that extracellular DNA (eDNA) can also trigger TLR9-dependent activation, which is relevant to biofilms due to the extensive amount of eDNA within the matrix (Whitchurch et al., 2002; Allesen-Holm et al., 2006; Rice et al., 2007).

Innate immune cells, including macrophages, neutrophils (PMNs), and dendritic cells, express TLR2 and TLR9 and are competent to respond to both Lpp/PGN/LTA and eDNA, respectively, which culminates in the induction of a wide array of classical pro-inflammatory mediators and bactericidal activity (Takeuchi et al., 1999; Bauer et al., 2001; Hertz et al., 2001; Jones et al., 2001; Kirschning and Schumann, 2002; Hayashi et al., 2003). Recent reports have determined that one mechanism utilized by biofilms to evade host immunity is by circumventing TLR2 and TLR9 recognition (Bernthal et al., 2011; Thurlow et al., 2011). The mechanism(s) responsible for TLR2/TLR9 evasion by *S. aureus* biofilms are not known but could also be explained by the fact that complex polysaccharide polymers, known as components of the biofilm matrix (Flemming, 2010; Thurlow, 2011), may interfere with optimal engagement of potential ligands with TLRs.

Immune recognition of staphylococcal biofilms may be also performed by other molecules. A recent report has revealed a role for IL-1β in controlling early bacterial burdens in a post-arthroplasty *S. aureus* biofilm infection model (Bernthal, 2011). Other studies have recently investigated the role of myeloid differentiation factor 88 (MyD88), the common downstream adaptor utilized by both the IL-1 receptor (IL-1R) and TLRs in regulating *S. aureus* biofilm growth (Hanke, 2012). Collectively, these studies reveal a role for IL-1β, potentially mediated through MyD88 signaling, in *S. aureus* biofilm containment. Although it is evident that these pathways influence biofilm development, they are not sufficient to eradicate staphylococcal biofilms, since these infections persist in an immunocompetent host.

Although PMNs exert *S. aureus* bactericidal activity under planktonic conditions (Fournicr, 2005; Graham, 2007), the direct role of these cells in modulating *S. aureus* biofilm growth has not yet been examined. Several studies investigating innate immunity to staphylococcal species have focused on PMNs (Kobayashi, 2003; Anwar, 2009; Graves, 2010), whereas macrophage responses have received relatively less attention. Several differences need to be kept in mind between PMNs and macrophages. PMNs show a short life span, are constantly recruited into infection sites and produce fewer proinflammatory mediators that are critical for immune cell recruitment and activation into infection sites (Yamashiro, 2001; Borregaard, 2010; Mantovani, 2011). Macrophages are exactly the opposite (Serbina, 2008; Gonzalez-Mejia, 2009; Silva, 2010 and 2011; Furze, 2008).

Recently various evidences seem to indicate that macrophages skew the immune response away from bacterial killing to favor fibrosis (Thurlow, 2011). Macrophage's response to bacterial infections involves the induction of genes related to proinflammatory M1 polarization that is usually associated with protection during acute infections. Other studies showed that bacterial persistence is often linked to macrophage reprogramming toward an anti-inflammatory M2 state (Benoit, 2008). In comparison to PMNs (Günther, 2009a and 2009b; Graves, 2010; Thurlow, 2011), macrophages show limited ability to ingest S. *aureus* when organized within a biofilm compared to planktonic bacteria (Thurlow, 2011). This could be due to the failure to physically engulf the complex biofilm structure, or (2) inability to opsonize intact biofilms. Further studies need to be performed in this direction in order to better understand the role of macrophage in persistent bacterial infections.

Therapy and Prophylaxis of Staphylococcal Biofilm Infections

Biofilm formation is an important mode of growth for *S. aureus* and *S. epidermidis* in many environments and is a significant contributor to the persistence of chronic infections. However, no concomitant advances in therapy or prevention have been forthcoming. Resolution of *S. aureus* biofilm infections continues to require surgical removal of the nidus of infection due to innate resistance to antimicrobial and host response factors (O'May, 2010).

In principle, several experimental approaches can be used to combat biofilms: a) strategies interfering with biofilm development and b) strategies focused on dispersal and disruption of mature biofilm (Speziale, 2008; Archer, 2011; Arciola, 2012).

Regarding the first type of strategies, prophylactic approach to inhibit biofilm development includes the use of drugs that interfere with the selective biosynthesis of precursors of biofilm components or treatment with compounds that block the initial attachment, accumulation and aggregation of bacteria in multicellular communities (Speziale, 2008; Archer, 2011; Arciola, 2012). Regarding the second type of strategies, preformed and mature biofilms can be challenged by chemical or enzymatic treatments that detach and disperse surface bound microorganisms. When bacteria are dispersed from a biofilm they become rapidly susceptible to chemotherapy (Speziale, 2008; Archer, 2011; Arciola, 2012).

Furthermore, the role of biomaterial science in combating biofilm-related infections is becoming more and more important. Different strategies may be employed to construct biofilm-resistant materials. Achievement of infection-resistant materials can be based on different strategies: (i) modification of the biomaterial surface to give anti-adhesive properties, (ii) doping the material with antimicrobial substances, and (iii) combining anti-adhesive and antimicrobial effects in the same coating. Furthermore, in orthopaedics another requirement to be achieved is (iv) the realization of material able to oppose biofilm formation and, at the same time, to support bone repair (Arciola, 2012) (Figure 4). The substantial progress made over the last few years in understanding the functional and structural factors involved in Staphylococcus biofilm formation and the regulatory circuits that control their expression, are undoubtedly opening the way to new strategies to combat implant infections.

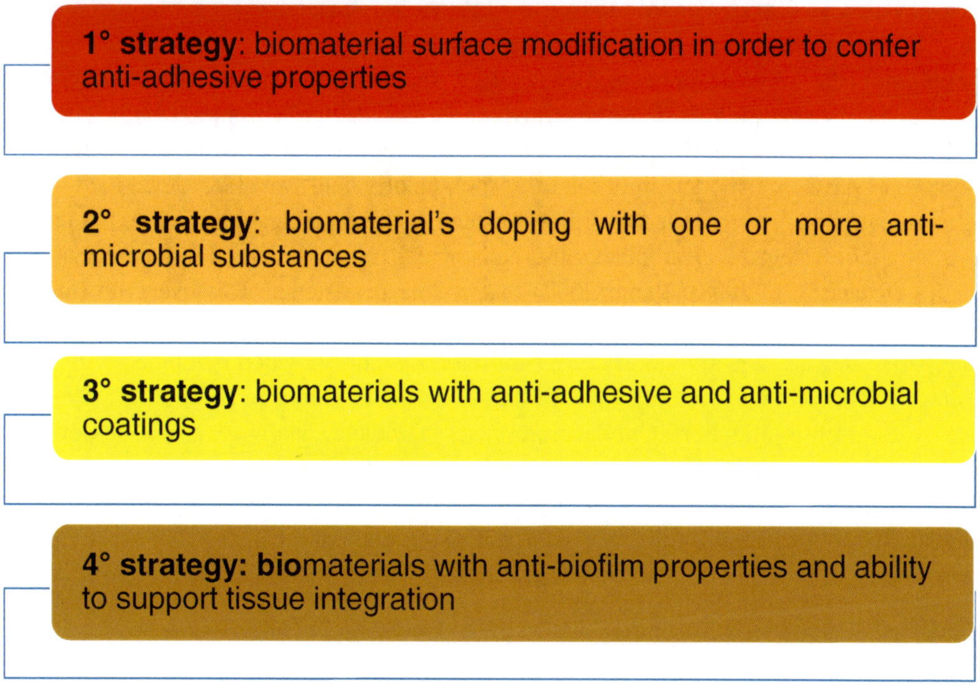

Adapted from Arciola (Arciola et al., 2012).

Figure 4. Strategies to be adopted in combating biofilm-related infections involving biomaterials.

Vaccine

Although literature and scientific data regarding staphylococcal vaccines for the prevention of *S. aureus* infections is comprehensive, the development of an efficacious and efficient vaccine is still problematic due to several unique challenges (Kurlenda, 2012; Proctor, 2012b). These can be associated with the fact that *S. aureus* is part of the normal flora as many people are carriers (transient or constant) on the mucus of the nasal or oral cavity. In addition, in contrast with other pathogens for which an efficient vaccine has already been found, *S. aureus* carries roughly 70 different virulence factors (Harro, 2010) including surface proteins that allow adhesion to host components such as fibrinogen (Clumping Factors, ClfA and B), fibronectin (fibronectin binding proteins, FnBP A and B), (Clarke, 2006, Heilmann, 2011), cytokeratin (ClfB) and proteins which scavenge nutrients that are normally sequestered in vivo (such as iron-responsive surface determinants, IsdA and B) (Cassa, 2012). Each virulence factor has different infection-mode and expression patterns depending on which disease is considered, which complicates the development of a vaccine because we must consider not only the pathogen but also the type of disease to be prevented. For these reasons all *S. aureus* vaccines, targeting single antigens that underwent clinical trials so far have failed (Bagnoli, 2012). The scientific communities now advocate a multivalent vaccine approach and researchers are now seeking the right combinations of antigens (Daum, 2012; Andersonn, 2012).

This type of approach was recently applied to *S. aureus* infections with a biofilm mode of growth where gene expression and protein production differ greatly from the planktonic mode of growth (Brady, 2006). For this reason a vaccine developed for preventing infection in planktonic models may not be effective against biofilm infection.

Following this idea Brady et al. developed a vaccine targeting four different antigens known to be present in *S. aureus* biofilms grown *in-vitro*: glucosaminidase, SA0486 (a hypothetical lipoprotein), SA0037 (a conserved hypothetical protein of unknown function), SA0688 (an ABC transporter lipoprotein). The quadrivalent vaccine, developed against biofilm upregulated protein, failed to promote effective bacterial clearance due to the inability of the immune system to clear planktonic cells which possess several immuno-avoidance strategies (Richardson, 2008; Wang, 2007) that enable persistence. However, the treatment was effective in combination with antimicrobial agents. Compared to their biofilm-embedded counterparts, they are sensitive to effective antimicrobial agents when combined with a dose of 40 mg vancomycin/kg twice daily for 10 days, and the therapy can abrogate all signs of *S. aureus* osteomyelitis infection. Only in rare cases complete clearance is not achieved but bacterial levels and severity of the disease were markedly decreased.

In conclusion, the combined use of a prophylactic multivalent vaccine against *S. aureus* biofilms infection and antibiotic treatment against the planktonic cells holds significant promise and hope for those patients who at present rely only on surgical intervention.

Nanotechnology Application

The use of nanotechnology is a promising new strategy against biofilm-mediated infection and, more in general, to overcome microbial drug resistance (Pelgrift, 2013; Taylor, 2011). Microorganisms are very creative and well adapted in developing antimicrobial drugs resistance. For instance they: (1) express pumps that are able to increase the efflux of drugs (Jayaraman, 2009), (2) express an altered version of the drug target to make the drug ineffective (Knetsch, 2011), (3) inactivate the drug by covalent modification (Taylor, 2011; Poole, 2002), (4) increasing the production of a competitive inhibitor of the drug (Iliades, 2004), (5) switch to a metabolically inactive mode enabling the survival of persistent (Taylor, 2011) and, last but not least, the production of biofilm (Knetsch, 2011) which is the topic of this review.

The extracellular polymeric substances (EPS) secreted by the biofilm surround the bacteria and protect them from very high concentration of antibiotics. Bacteria embedded in biofilms are up to 1000 fold more resistant to antibiotics than planktonic cells, often leading to chronic infection (Huang, 2011). In addition, it is likely that EPS promotes the development of drug-resistant variants as the concentration of drugs inside the biofilm will be well below the required inhibitor concentration, a condition that might promote the adaptation of the bacteria to the antibiotic (Knetsch, 2011; Hindi, 2009) and develop into deadly drug-resistant variants.

Nanoparticles (NPS) may be a promising agent to combat microbial resistance in general and biofilm based infection in particular. Several types of nanoparticles have simultaneous mechanism of toxicity toward bacteria, but not toward mammalian cells, thus making the development of resistance very unlikely. Indeed, bacteria would have to mutate several genes jointly in order to acquire resistance (Knetsch, 2011; Hindi, 2009; Blecher, 2011).

Among the different nanoparticles, the most promising ones have shown to be: nitric oxide-releasing NPs, chitosan containing NPs, and metal (mainly Ag, Zn and Bi) containing NPs (Figure 5).

Nitric oxide-releasing NPs exert their antimicrobial activity through reactive nitrogen oxide intermediates (RNOS) which are able to kill bacteria by damaging their DNA and protein molecules (Schairer, 2012). One particular application called hydrogen/glass composite NO, when applied topical, decreases significantly the size of methicillin-resistant *Staphylococcus aureus* (MRSA) infected abscesses if compared with vancomycin treatment only (Martinez, 2012; Han, 2009; Schairer, 2012). NO NPs have also been shown effectively not only against biofilm formation but were also able to kill *S. aureus* present in already formed biofilms (Hetrick, 2009).

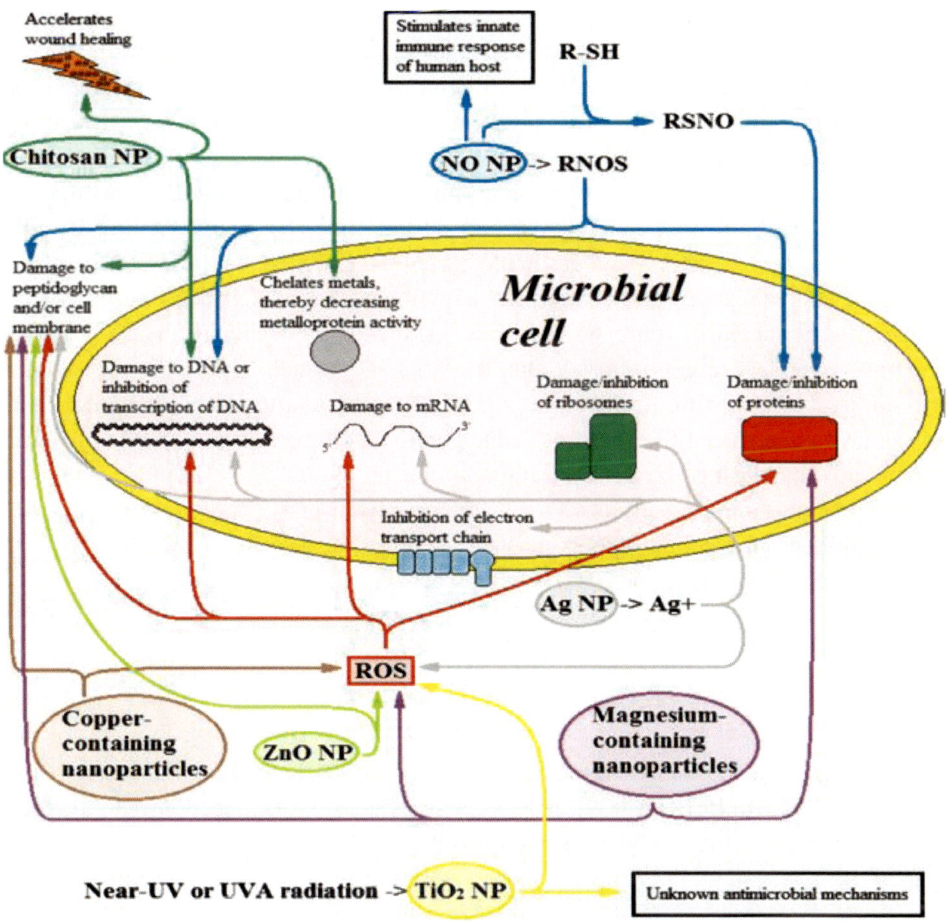

Figure 5. Multiple mechanisms of antimicrobial action of nitric oxide-releasing nanoparticles (NO NPs), chitosan-containing nanoparticles (chitosan NPs), silver-containing nanoparticles RNOS refers to reactive nitrogen oxide intermediates. ROS refers to reactive oxygen species. R-SH refers to a compound with a thiol (SH) group. RSNO refers to S-nitrosothiol. With the permission from (Pelgrift, 2013).

Chitosan-containing nanoparticles (chitosan NP) also use multiple mechanisms to combat microbes (Friedman,2013) mainly exploiting its positive charge, along the chitosan polymer molecule, which may bound to cell walls causing permeability, osmotic passage and flow of cytoplasmic contents out of the cell (Blecher, 2011; Friedman, 2013).

Chitosan encapsulated into nanoparticles has shown greater efficacy against *S. aureus* than chitosan alone and different antibiotics making it a promising tool for nanotechnological application against MRSA. Metal containing NPs (MNPs) are varies in properties depending on which metal they are carrying. Silver containing NPs exploit the well known antibacterial activity of silver as numerous mechanisms rely on the interaction of Ag^+ with different chemical groups located on the membrane proteins (Knetsch, 2011; Lara, 2010) and, once inside the cells, Ag^+ exerts several additional antimicrobial effects like electron transfer inhibition, DNA damages and ribosome denaturation (Lara,2010). The effect of these antimicrobial mechanisms is well described in a recent work where silver NPs are shown to increase the antibacterial effect of various antibiotics (Blecher, 2011). Another promising metal MNP is zinc oxide containing NPs. Like silver NPs they act by using several antimicrobial mechanisms (Hajipour, 2012; Blecher, 2011). ZnO NPs have been shown to exert antimicrobial activity against MRSA (Hajipour, 2012) and such effect is inversely proportional to the size of the metal particles (Raghupathi, 2011). Another interesting type of MNPs is bismuth-containing NPs which can be conjugated with X-ray treatment for a synergistic and potent effect against drug-resistant bacteria. Unlike topical treatment, X-ray can reach bacterial infection deep inside the dermis. Upon irradiation with X-rays Bi emits electrons and free radicals that can heavily damage bacterial DNA. Conjugating Bi NPs to antibodies specific for the target pathogen will shorten the distance that the free radicals must travel from the Bi to bacteria significantly increasing the bactericidal effect (Luo, 2013).

Ideal antibacterial NPs would be delivered efficiently and specifically to the site of infection and selectively target bacterial. One way to achieve such specificity is to conjugate NPs with an antibody specific to the target site of infection. On the same line the NPs may also be conjugated with antibiotic drugs, which will therefore be delivered to the site of infection greatly increasing the concentration of the antibiotic. For instance, vancomycin encapsulated in chitosan NPs is effective against vancomycin-resistant *S. aureus* (Huh, 2011). Ideally, the antimicrobial action of NPs must be activated at the site of infection by certain stimuli, like reactive oxygen species (ROS) or low pH.

Acknowledgment

LV would like to acknowledge financial support from MIUR, PRIN 2010 project entitled “Nanomed” (prot. 2010FPTBSH_009). We are also grateful to D. Picenoni (Politecnico di Milano, Milano, Italy) for technical assistance in SEM studies.

References

Abraham, N.M., Jefferson, K.K., (2012) *Staphylococcus aureus* clumping factor B mediates biofilm formation in the absence of calcium. *Microbiology* 158(Pt 6): 1504-1512.

Adair, C.G., Gorman, S.P., Feron, B.M., Byers, L.M., Jones, D.S., Goldsmith, C.E., Moore, J.E., Kerr, J.R., Curran, M.D., Hogg, G., Webb, C.H., McCarthy, G.J., Milligan, K.R., (1999). Implications of endotracheal tube biofilm for ventilator-associated pneumonia. *Intensive Care Med* 25(10): 1072-1076.

Allesen-Holm, M., Barken, K.B., Yang, L., Klausen, M., Webb, J.S., Kjelleberg, S., Molin, S., Givskov, M., Tolker-Nielsen, T. (2006). A characterization of DNA release in *Pseudomonas aeruginosa* cultures and biofilms. *Mol.Microbiol.* 59: 1114–1128.

Ammendolia, M.G., Di Rosa, R., Montanaro, L., Arciola, C. R., Baldassarri, L. (1999). Slime production and expression of the slime-associated antigen by staphylococcal clinical isolates. *J Clin Microbiol* 37(10): 3235-3238.

Anderson, A.S., Miller, A.A., Donald, R.G.K., Scully, I.L., Nanra, J.S., Cooper, D., Jansen, K.U., (2012) Development of a multicomponent *Staphylococcus aureus* vaccine designed to counter multiple bacterial virulence factors. *Human Vaccines & Immunotherapeutics* 8(11): 1585–1594.

Anwar, S, Prince, L.R., Foster, S.J., Whyte, M.K., Sabroe, I. The rise and rise of *Staphylococcus aureus*: laughing in the face of granulocytes. *Clin Exp Immunol* 2009; 157:216-24.

Archer, N.K., Mazaitis, M.J., Costerton, J.W., Leid, J.G., Powers, M.E., Shirtliff, M.E. (2011) *Staphylococcus aureus* biofilms: properties, regulation, and roles in human disease. *Virulence* Sep-Oct;2(5):445-59.

Arciola, C.R., Campoccia, D., Baldassarri, L., Donati, M.E., Pirini, V., Gamberini, S., Montanaro, L., (2006). Detection of biofilm formation in *Staphylococcus epidermidis* from implant infections. Comparison of a PCR-method that recognizes the presence of ica genes with two classic phenotypic methods. *J Biomed Mater Res A* 76(2): 425-430.

Arciola, C.R., Campoccia, D., Speziale, P., Montanaro, L., Costerton, J.W, (2012). Biofilm formation in Staphylococcus implant infections. A review of molecular mechanisms and implications for biofilm-resistant materials. *Biomaterials* Sep;33(26):5967-82.

Bagnoli, F., Bertholet, S., Grandi, G., (2012) Inferring Reasons for the Failure of *Staphylococcus aureus* Vaccines in Clinical Trials. *Front. Cell. Infect. Microbiol.* 2(16): 1-4.

Baldoni, D., Haschke, M., Rajacic, Z., Zimmerli, W., Trampuz, A., (2009) Linezolid alone or combined with rifampin against methicillin-resistant Staphylococcus aureus in experimental foreign-body infection. *Antimicrob. Agents. Chemother.* 53: 1142-8.

Banner, M.A., Cunniffe, J.G., Macintosh, R.L., Foster, T.J., Rohde, H., Mack, D., Hoyes, E., Derrick, J., Upton, M., Handley, P.S., (2007). Localized tufts of fibrils on *Staphylococcus epidermidis* NCTC 11047 are comprised of the accumulation-associated protein. *J. Bacteriol.* 189(7): 2793-2804.

Bauer, S., Kirschning, C.J., Häcker, H., Redecke, V., Hausmann, S., Akira, S., Wagner, H., Lipford, G.B. (2001). Human TLR 9 confers responsiveness to bacterial DNA via species-specific CpG motif recognition. *Proc.Natl.Acad.Sci.U.S.A.* 98: 9237–9242.

Beenken, K.E., Dunman, P.M., McAleese, F., Macapagal, D., Murphy, E., Projan, S.J., Blevins, J.S., Smeltzer, M.S., (2004). Global gene expression in *Staphylococcus aureus* biofilms. *J. Bacteriol.* 186(14): 4665-4684.

Beenken, K.E., Mrak, L.N., Griffin, L.M., Zielinska, A.K., Shaw, L.N., Rice, K.C., Horswill, A.R., Bayles, K.W., Smeltzer, M.S., (2010). Epistatic relationships between *sarA* and *agr* in *Staphylococcus aureus* biofilm formation. *PLoS One* 5(5): e10790.

Beloin, C., Roux, A., Ghigo, J.M., (2008). *Escherichia coli* biofilms. *Curr. Top. Microbiol. Immunol.* 322: 249-289.

Beloin, C., Valle, J., Latour-Lambert, P., Faure, P., Kzreminski, M., Balestrino, D., Haagensen, J.A., Molin, S., Prensier, G., Arbeille, B., Ghigo, J.M., (2004). Global impact of mature biofilm lifestyle on *Escherichia coli* K-12 gene expression. *Mol. Microbiol.* 51(3): 659-674.

Benoit, M., Desnues, B., Mege, J.L. (2008). Macrophage polarization in bacterial infections. *J.Immunol.* 181: 3733–3739.

Bernthal, N.M., Pribaz, J.R., Stavrakis, A.I., Billi, F., Cho, J.S., Ramos, R. I., Francis, K.P., Iwakura, Y., Miller, L.S., (2011). Protective role of IL-1β against post-arthroplasty *Staphylococcus aureus* infection. *J. Orthop.Res.* 29: 1621–1626.

Blecher, K., Nasir, A., Friedman, A., (2011) The growing role of nanotechnology in combating infectious disease. *Virulence* 2:(5): 395–401.

Boles, B.R., Horswill, A.R., (2011). *Staphylococcal* biofilm disassembly. *Trends. Microbiol.* 19(9): 449-455.

Borregaard, N., (2010). Neutrophils, from marrow to microbes. *Immunity* 33: 657–670.

Bowden, M.G., Visai, L., Longshaw, C.M., Holland, K.T., Speziale, P., Hook, M., (2002). Is the GehD lipase from *Staphylococcus* epidermidis a collagen binding adhesin? *J. Biol. Chem.* 277(45): 43017-43023.

Bowling, F.L., Jude, E.B., Boulton, A.J., (2009) MRSA and diabetic foot wounds: contaminating or infecting organisms? *Curr. Diab. Rep.* 9: 440-4.

Brady, R.A., Leid, J.G., Camper, A.K., Costerton, J.W., Shirtliff, M.E., (2006) Identification of *Staphylococcus aureus* proteins recognized by the antibody-metiated immune response to a biofilm infection. *Infect. Immun.*74: 3415-3426.

Brady, R.A., O'May, G.A., Leid, J.G., Prior, M.L., Costerton, J.W., Shirtliff, M.E., (2011) Resolution of *Staphylococcus aureus* biofilm infection using vaccination and antibiotic treatment. *Infect. Immun.*79: 1797-1803.

Brogden, K.A. (2005). Antimicrobial peptides: pore formers or metabolic inhibitors in bacteria? *Nat.Rev. Microbiol.* 3, 238–250.

Cassat, J.E., Skaar, E.P., (2012) Metal ion acquisition in *Staphylococcus aureus*: overcoming nutritional immunity. *Semin. Immunopathol.* 34: 215-235.

Cassatella, M.A.(1995).The production of cytokines by polymorphonuclear neutrophils. *Immunol. Today* 16: 21–26.

Clark, L.A., Easmon, C.S., (1986). Opsonic requirements of *Staphylococcus epidermidis. J. Med. Microbiol.* 22: 1–7.

Clarke, S.R., Foster, S.J., (2006) Surface Adhesins of *Staphylococcus aureus*. Robert KP, ed. *Advances in Microbial Physiology*: Academic Press. 187-224.

Cheung, A.L., Projan, S.J., (1994). Cloning and sequencing of *sarA* of *Staphylococcus aureus*, a gene required for the expression of *agr. J. Bacteriol.* 176(13): 4168-4172.

Christner, M., Heinze, C., Busch, M., Franke, G., Hentschke, M., Bayard Duhring, S., Buttner, H., Kotasinska, M., Wischnewski, V., Kroll, G., Buck, F., Molin, S., Otto, M., Rohde, H., (2012). *sarA* negatively regulates *Staphylococcus epidermidis* biofilm formation by modulating expression of 1 MDa extracellular matrix binding protein and autolysis-dependent release of eDNA. *Mol. Microbiol.* 86(2): 394-410.

Conlon, K.M., Humphreys, H., O'Gara, J.P., (2002a). icaR encodes a transcriptional repressor involved in environmental regulation of ica operon expression and biofilm formation in *Staphylococcus epidermidis*. *J. Bacteriol.* 184(16): 4400-4408.

Conlon, K.M., Humphreys, H., O'Gara, J.P., (2002b). Regulation of icaR gene expression in *Staphylococcus epidermidis*. *FEMS Microbiol. Lett.* 216(2): 171-177.

Conlon, K.M., Humphreys, H., O'Gara, J.P., (2004). Inactivations of rsbU and *sarA* by IS256 represent novel mechanisms of biofilm phenotypic variation in *Staphylococcus epidermidis*. *J. Bacteriol.* 186(18): 6208-6219.

Conrady, D.G., Brescia, C.C., Horii, K., Weiss, A.A., Hassett, D.J., Herr, A.B. (2008). A zinc-dependent adhesion module is responsible for intercellular adhesion in staphylococcal biofilms. *Proc. Natl. Acad. Sci. USA* 105(49): 19456-19461.

Conrady, D.G., Wilson, J.J., Herr, A.B. (2013). Structural basis for Zn^{2+}-dependent intercellular adhesion in staphylococcal biofilms. *Proc. Natl. Acad. Sci. USA* 110(3): E202-211.

Corrigan, R.M., Rigby, D., Handley, P., Foster, T.J., (2007). The role of *Staphylococcus aureus* surface protein SasG in adherence and biofilm formation. *Microbiology* 153(Pt 8): 2435-2446.

Costerton, J.W., Montanaro, L., Arciola, C.R., (2005). Biofilm in implant infections: its production and regulation. *Int. J. Artif. Organs.* 28:1062-8;

Cramton, S.E., Gerke, C., Schnell, N.F., Nichols, W.W., Gotz, F., (1999). The intercellular adhesion (ica) locus is present in *Staphylococcus aureus* and is required for biofilm formation. *Infect. Immun.* 67(10): 5427-5433.

Cramton, S.E., Ulrich, M., Gotz, F., Doring, G., (2001). Anaerobic conditions induce expression of polysaccharide intercellular adhesin in *Staphylococcus aureus* and *Staphylococcus epidermidis*. *Infect. Immun.* 69(6): 4079-4085.

Cucarella, C., Solano, C., Valle, J., Amorena, B., Lasa, I., Penades, J.R., (2001). Bap, a *Staphylococcus aureus* surface protein involved in biofilm formation. *J. Bacteriol.* 183(9): 2888-2896.

Darby, C., Hsu, J.W., Ghori, N., Falkow, S., (2002). *Caenorhabditis elegans*: plague bacteria biofilm blocks food intake. *Nature* 417(6886): 243-244.

Daum, R., Spellberg, B., (2012). Progress Toward a *Staphylococcus aureus* Vaccine. *Vaccines* 54, 560-567.

Davis, S.C., Ricotti, C., Cazzaniga, A., Welsh, E., Eaglstein, W.H., Mertz, P.M., (2008). Microscopic and physiologic evidencefor biofilm-associated wound colonization in vivo. *Wound Repair Regen.* 16:23-9.

Dean, S.N., Bishop, B.M., andvan Hoek, M.L. (2011). Natural and synthetic cathelicidin peptides with anti-microbial and anti-biofilm activity against *Staphylococcus aureus*. *BMC Microbiol.* 11: 114.

Donlan, R.M., (2002). Biofilms: microbial life on surfaces. *Emerg. Infect. Dis.* 8(9): 881-890.

Dunman, P.M., Murphy, E., Haney, S., Palacios, D., Tucker-Kellogg, G., Wu, S., Brown, E.L., Zagursky, R.J., Shlaes, D., Projan, S.J., (2001). Transcription profiling-based identification of *Staphylococcus aureus* genes regulated by the *agr* and/or *sarA* loci. *J. Bacteriol.* 183(24): 7341-7353.

Dunman, P.M., Murphy, E., Haney, S., Palacios, D., Tucker- Kellogg, G., Wu, S., Brown, E.L., Zagursky, R.J., Shlaes, D., Projan, S.J., (2001). Transcription profiling-based identification of *Staphylococcus aureus* genes regulated by the agr and/or sarA loci. *J. Bacteriol.* 183:7341-53.

Dunne, W.M., Jr. (2002). Bacterial adhesion: seen any good biofilms lately? *Clin. Microbiol. Rev.* 15(2): 155-166.

Dziarski, R. (2003). Recognition of bacterial peptidoglycan by the innate immune system. *Cell. Mol. Life Sci.* 60, 1793–1804.

Ferguson, B.J., Stolz, D.B., (2005). Demonstration of biofilm in human bacterial chronic rhinosinusitis. *Am. J. Rhinol.* 19:452-7;.

Fey, P.D., (2010a). Modality of bacterial growth presents unique targets: how do we treat biofilm-mediated infections? *Curr. Opin. Microbiol.* 13:610–615.

Fey, P.D., Olson, M.E., (2010b). Current concepts in biofilm formation of *Staphylococcus Epidermidi.s Future Microbiol.* June ; 5(6): 917–933.

Fitzsimmons, K., Bamber, A., Smalley, H.B. (2010). Infective endocarditis: changing aetiology of disease. *Br. J. Biomed.Sci.* 67: 35–41.

Fitzpatrick, F., Humphreys, H., O'Gara, J.P., (2005a). The genetics of staphylococcal biofilm formation– will a greater understanding of pathogenesis lead to better management of device-related infection? *Clin. Microbiol. Infect.* Dec;11(12): 967-73.

Fitzpatrick, F., Humphreys, H., O'Gara, J.P., (2005b). Evidence for icaADBC-independent biofilm development mechanism in methicillin-resistant *Staphylococcus aureus* clinical isolates. *J. Clin. Microbiol.* 43(4): 1973-1976.

Fitzpatrick, F., Humphreys, H., O'Gara, J.P., (2006). Environmental regulation of biofilm development in methicillin-resistant and methicillin-susceptible *Staphylococcus aureus* clinical isolates. *J. Hosp. Infect.* 62(1): 120-122.

Fleer, A., Timmerman, C.P., Besnier, J.M., Pascual, A., Verhoef, J., (1990). Surface proteins of coagulase-negative staphylococci: their role in adherence to biomaterials and in opsonization. *J. Biomater. Appl.* 5(2): 154-165.

Flemming, H.C., Wingender, J., (2010). The biofilm matrix. *Nat. Rev.Microbiol.* 8: 623–633.

Fluckiger, U., Ulrich, M., Steinhuber, A., Doring, G., Mack, D., Landmann, R., Goerke, C., Wolz, C., (2005). Biofilm formation, icaADBC transcription, and polysaccharide intercellular adhesin synthesis by staphylococci in a device-related infection model. *Infect. Immun.* 73(3): 1811-1819.

Fluckiger, U., Wolz, C., Cheung, A.L., (1998). Characterization of a sar homolog of *Staphylococcus epidermidis*. *Infect. Immun.* 66(6): 2871-2878.

Foster, T.J., Hook, M., (1998). Surface protein adhesins of *Staphylococcus aureus*. *Trends Microbiol.* 6(12): 484-488.

Fournier, B., Philpott, D.J., (2005). Recognition of *Staphylococcus aureus* by the innate immune system. *Clin.Microbiol.Rev.* 18: 521–540.

Foreman, A., Wormald, P.J., (2010). Different biofilms, different disease? A clinical outcomes study. *Laryngoscope* 120: 1701-6;

Fowler, V.G. Jr., Miro, J.M., Hoen, B., Cabell, C.H., Abrutyn, E., Rubinstein, E., Corey, G.R., Spelman, D,. Bradley, S.F., Barsic, B., Pappas, P.A., Anstrom, K.J., Wray, D., Fortes, C.Q., Anguera, I., Athan, E., Jones, P., van der Meer, J.T., Elliott, T.S., Levine, D.P., Bayer, A.S., (2005) *Staphylococcus aureus* endocarditis: a consequence of medical progress. *JAMA* 293: 3012–3021.

Fredheim, E.G., Granslo, H.N., Flægstad, T., Figenschau, Y., Rohde, H., Sadovskaya, I., Mollnes, T.E., Klingenberg, C., (2011). *Staphylococcus epidermidis* polysaccharide intercellular adhesin acti- vates complement. *FEMS Immunol. Med.Microbiol.* 63: 269–280.

Friedman, A.J., Phan, J., Schairer, D.O., Champer, J., Qin, M., Pirouz, A., Blecher-Paz, K., Oren, A., Liu, P.T., Modlin, R.L., Kim, J., (2013) Antimicrobial and anti-inflammatory activity of chitosan–alginate nanoparticles: a targeted therapy for cutaneous pathogens. *J. Investig. Dermatol.* 133: 1231–1239.

Fuchs, T.A., Abed, U., Goosmann, C., Hurwitz, R., Schulze, I., Wahn, V., Weinrauch, Y., Brinkmann, V., Zychlinsky, A., (2007) Novel cell death program leads to neutrophil extracellular traps. *J. Cell. Biol.* 176(2): 231-241.

Furze, R.C., Rankin, S.M.(2008). The role of the bone marrow in neutrophil clearance under homeostatic conditions in the mouse. *FASEB J.* 22: 3111–3119.

Ganesh, V. K., Barbu, E.M., Deivanayagam, C.C., Le, B., Anderson, A.S., Matsuka, Y.V., Lin, S.L. Foster, T.J., Narayana, S.V., Hook, M., (2011). Structural and biochemical characterization of *Staphylococcus aureus* clumping factor B/ligand interactions. *J. Biol. Chem.* 286(29): 25963-25972.

Ganesh, V. K., Rivera, J.J., Smeds, E., Ko, Y.P., Bowden, M.G., Wann, E.R., Gurusiddappa, S., Fitzgerald, J.R., Hook, M., (2008). A structural model of the *Staphylococcus aureus* ClfA-fibrinogen interaction opens new avenues for the design of anti-staphylococcal therapeutics. *PLoS Pathog* 4(11): e1000226.

Geoghegan, J.A., Corrigan, R.M., Gruszka, D.T., Speziale, P., O'Gara, J.P., Potts, J.R., Foster, T.J., (2010). Role of surface protein SasG in biofilm formation by *Staphylococcus aureus. J. Bacteriol.* 192(21): 5663-5673.

Gerke, C., Kraft, A., Sussmuth, R., Schweitzer, O., Gotz, F., (1998). Characterization of the N-acetylglucosaminyltransferase activity involved in the biosynthesis of the *Staphylococcus epidermidis* polysaccharide intercellular adhesin. *J. Biol. Chem.* 273(29): 18586-18593.

Gibson, R.L., Burns, J.L., Ramsey, B.W., (2003). Pathophysiology and management of pulmonary infections in cystic fibrosis. *Am. J. Respir. Crit. Care Med.* 168(8): 918-951.

Gill, S.R., Fouts, D.E., Archer, G.L., Mongodin, E.F., Deboy, R.T., Ravel, J., Paulsen, I.T., Kolonay, J.F., Brinkac, L., Beanan, M., Dodson, R.J., Daugherty, S.C., Madupu, R., Angiuoli, S.V., Durkin, A.S., Haft, D.H., Vamathevan, J., Khouri, H., Utterback, T., Lee, C., Dimitrov, G., Jiang, L., Qin, H., Weldman, J., Tran, K., Kang, K., Hance, I.R., Nelson, K.E., Fraser, C,M., (2005). Insights on evolution of virulence and resistance from the complete genome analysis of an early methicillin-resistant *Staphylococcus aureus* strain and a biofilm-producing methicillin-resistant *Staphylococcus epidermidis* strain. *J. Bacteriol.* 187(7): 2426-2438.

Gjødsbøl, K., Christensen, J.J., Karlsmark, T., Jorgensen, B., Klein, B.M., Krogfelt, K.A., (2006) Multiple bacterial species reside in chronic wounds: a longitudinal study. *Int. Wound J.* 3: 225-31.

Gonzalez-Mejia, M.E., Doseff, A.I., (2009).Regulation of monocytes and macrophages cell fate. *Front. Biosci.* 14: 2413–2431.

Graham, D.B., Robertson, C.M., Bautista, J., Mascarenhas, F., Diacovo, M.J., Montgrain, V., Lam, S.K., Cremasco, V., Dunne, W.M., Faccio, R., Coopersmith, C.M., Swat, W., (2007). Neutrophil- mediated oxidative burst and host defense are controlled by a Vav-PLCgamma 2 signaling axis in mice. *J. Clin.Invest.* 117: 3445–3452.

Graves,S.F.,Kobayashi,S.D.,and DeLeo,F.R.(2010). Community-associated methicillin-resistant *Staphylococcus aureus* immune evasion and virulence. *J.Mol.Med. (Berl.)* 88: 109–114.

Gross, M., Cramton, S.E., Gotz, F., Peschel, A., (2001). Key role of teichoic acid net charge in *Staphylococcus aureus* colonization of artificial surfaces. *Infect. Immun.* 69(5): 3423-3426.

Guenther, F., Stroh, P., Wagner, C., Obst, U., Hänsch, G.M., (2009a). Phagocytosis of staphylococci biofilms by polymorphonuclear neutrophils: *S. aureus* and *S. epidermidis* differ with regard to their susceptibility towards the host defense. *Int.J.Artif.Organs* 32: 565–573.

Günther,F., Wabnitz, G.H., Stroh, P., Prior, B., Obst, U., Samstag, Y., Wagner, C., Hänsch, G.M., (2009b). Host defence against *Staphylococcus aureus* biofilms infection: phagocytosis of biofilms by polymorphonuclear neutrophils (PMN). *Mol.Immunol.* 46: 1805–1813.

Hajipour, M.J., Fromm, K.M., AkbarAshkarran, A., Jimenez de Aberasturi, D., Larramendi, I.R.D., Rojo, T., Serpooshan, V., Parak, W.J., Mahmoudi, M. (2012) Antibacterial properties of nanoparticles. *Trends* 30(10): 499–511.

Hajjar, A.M., O'Mahony, D.S., Ozinsky, A., Underhill, D.M., Aderem, A., Klebanoff, S.J., Wilson, C.B. (2001). Cutting edge: functional interactions between toll-like receptor(TLR)2 and TLR1 or TLR6 in response to phenol-soluble modulin. *J.Immunol.* 166: 15–19.

Han, G., Martinez, L.R., Mihu, M.R., Friedman, A.J., Friedman, J.M., Nosanchuk, J.D., (2009) Nitric oxide releasing nanoparticles are therapeutic for *Staphylococcus aureus* abscesses in a murine model of infection. *PLoS One* 4(11): e7804.

Handke, L.D., Conlon, K.M., Slater, S.R., Elbaruni, S., Fitzpatrick, F., Humphreys, H., Giles, W.P., Rupp, M.E., Fey, P.D., O'Gara, J.P., (2004). Genetic and phenotypic analysis of biofilm phenotypic variation in multiple *Staphylococcus epidermidis* isolates. *J. Med. Microbiol.* 53(Pt 5): 367-374.

Hanke, M.L., Kielian, T., (2012). Deciphering mechanisms of *staphylococcal* biofilm evasion of host immunity. *Front. Cell. Infect. Microbiol.* May 8(2):62.

Hansson, C., Hoborn, J., Moller, A., Swanbeck, G., (1995). Themicrobial flora in venous leg ulcers without clinical signs of infection. Repeated culture using a validated standardised microbiological technique. *Acta Derm. Venereol.* 75: 24-30.

Harro, J.M., Peters, B.M., O'May, G.A., Archer, N., Kerns, P., Prabhakara, R., Shirtliff, M.E., (2010). Vaccine development in *Staphylococcus aureus*: taking the biofilm phenotype into consideration. *FEMS Immunol.Med. Microbiol.* 59(3): 306-323.

Hashimoto, M., Tawaratsumida, K., Kariya, H., Aoyama, K., Tamura, T., Suda, Y., (2006a). Lipoprotein is a predominant Toll-like receptor 2 ligand in *Staphylococcus aureus* cell wall components. *Int.Immunol.* 18: 355–362.

Hashimoto, M., Tawaratsumida, K., Kariya, H., Kiyohara, A., Suda, Y., Krikae, F., Kirikae, T., Götz, F., (2006b). Not lipoteichoic acid but lipoproteins appear to be the dominant immunobiologically active compounds in *Staphylococcus aureus*. *J.Immunol.* 177: 3162–3169.

Hayashi, F., Means, T.K., Luster, A.D., (2003). Toll-like receptors stimulate human neutrophil function. *Blood* 102:2660–2669.

Heilmann, C., (2011a). Mechanisms of *Staphylococci* Bacterial Adhesion. Linke D, Goldman A, eds.: Springer Netherlands. 105-123.

Heilmann C., (2011b). Adhesion mechanisms of *Staphylococci. Adv. Exp. Med. Biol.* 715:105-23.

Heilmann, C., Gotz, F., (1998). Further characterization of *Staphylococcus epidermidis* transposon mutants deficient in primary attachment or intercellular adhesion. *Zentralbl Bakteriol* 287 (1-2): 69-83.

Heilmann, C., Hussain, M., Peters, G., Gotz, F., (1997). Evidence for autolysin-mediated primary attachment of *Staphylococcus epidermidis* to a polystyrene surface. *Mol. Microbiol.* 24(5): 1013-1024.

Heilmann, C., Schweitzer, O., Gerke, C., Vanittanakom, N., Mack, D., Gotz, F., (1996). Molecular basis of intercellular adhesion in the biofilm-forming *Staphylococcus epidermidis*. *Mol. Microbiol.* 20(5): 1083-1091.

Hemmi, H., Takeuchi, O., Kawai, T., Kaisho, T., Sato, S., Sanjo, H., Matsumoto, M., Hoshino, K., Wagner, H., Takeda, K., Akira, S., (2000). A Toll-like receptor recognizes bacterial DNA. *Nature* 408: 740–745.

Heitz-Mayfield, L.J., Lang, N.P., (2010). Comparative biology of chronic and aggressive periodontitis vs. peri-implantitis. *Periodontol. 2000* 53:167-81.

Hertz, C.J., Kiertscher, S.M., Godowski, P.J., Bouis, D.A., Norgard, M.V., Roth, M.D., Modlin, R.L. (2001). Microbial lipopeptides stimulate dendritic cell maturation viaToll-like receptor 2. *J. Immunol.* 166: 2444–2450.

Herrmann, M., Vaudaux, P.E., Pittet, D., Auckenthaler, R., Lew, P.D., Schumacher-Perdreau, F., Peters, G., Waldvogel, F.A., (1988). Fibronectin, fibrinogen and laminin act as mediators of adherence of clinical staphylococcal isolates to foreign material. *J. Infect. Dis.* 158: 693-701.

Hetrick, E.M., Shin, J.H., Paul, H.S., Schoenfisch, M.H., (2009). Anti-biofilm efficacy of nitric oxide-releasing silica nanoparticles. *Biomaterials* 30(14): 2782–2789.

Hindi, K.M., Ditto, A.J., Panzner, M.J., Medvetz, D.A., Han, D.S., Hovis, C.E., Hilliard, J.K., Taylor, J.B., Yang, H.Y., Cannon, C.L., Youngs, W.J., (2009). The antimicrobial efficacy of sustained release silver–carbene complex-loaded L-tyrosine polyphosphate nanoparticles: characterization, in vitro and in vivo studies. *Biomaterials* 30(22): 3771–3779.

Houston, P., Rowe, S.E., Pozzi, C., Waters, E.M., O'Gara, J.P., (2011). Essential role for the major autolysin in the fibronectin-binding protein-mediated *Staphylococcus aureus* biofilm phenotype. *Infect. Immun.* 79(3): 1153-1165.

Huh, A.J., Kwon, Y.J., (2011). "Nanoantibiotics": a new paradigm for treating infectious diseases using nanomaterials in the antibiotics resistant era. *J. Control. Release* 156(2): 128–145.

Huang, L., Dai, T., Xuan, Y., Tegos, G.P., Hamblin, M.R., (2011). Synergistic combination of chitosan acetate with nanoparticle silver as a topical antimicrobial: efficacy against bacterial burn infections. *Antimicrob. Agents Chemother.* 55(7): 3432–3438.

Iliades, P., Meshnick, S.R., Macreadie, I.G., (2004). Dihydropteroate synthase mutations in Pneumocystis jiroveci can affect sulfamethoxazole resistance in a *Saccharomyces cerevisiae* model. *Antimicrob. Agents Chemother.* 48(7): 2617–2623.

Iwase, T., Uehara, Y., Shinji, H., Tajima, A., Seo, H., Takada, K., Agata, T., Mizunoe, Y., (2010). *Staphylococcus epidermidis* Esp inhibits *Staphylococcus aureus* biofilm formation and nasal colonization. *Nature* 465(7296): 346-349.

Izano, E.A., Amarante, M.A., Kher, W.B., Kaplan, J.B., (2008). Differential roles of poly-N-acetylglucosamine surface polysaccharide and extracellular DNA in *Staphylococcus aureus* and Staphylococcus epidermidis biofilms. *Appl. Environ. Microbiol.* 74(2): 470-476.

Jager, S., Mack, D., Rohde, H., Horstkotte, M.A., Knobloch, J.K., (2005). Disintegration of *Staphylococcus epidermidis* biofilms under glucose-limiting conditions depends on the activity of the alternative sigma factor sigmaB. *Appl. Environ. Microbiol.* 71(9): 5577-5581.

Janzon, L., Lofdahl, S., Arvidson, S., (1989). Identification and nucleotide sequence of the delta-lysin gene, hld, adjacent to the accessory gene regulator *(agr)* of *Staphylococcus aureus*. Mol. Gen. Genet. 219(3): 480-485.

Jarraud, S., Lyon, G.J., Figueiredo, A.M., Lina, G., Vandenesch, F., Etienne, J., Muir, T.W., Novick, R.P., (2000). Exfoliatin-producing strains define a fourth *agr* specificity group in *Staphylococcus aureus*. J. Bacteriol. 182(22): 6517-6522.

Jayaraman, R., (2009). Antibiotic resistance: an overview of mechanisms and a paradigm shift, *Curr. Sci.* 96(11): 1475–1484

Jefferson, K.K., Pier, D.B., Goldmann, D.A., Pier, G.B., (2004). The teicoplanin-associated locus regulator (TcaR) and the intercellular adhesin locus regulator (IcaR) are transcriptional inhibitors of the ica locus in *Staphylococcus aureus*. *J. Bacteriol.* 186(8): 2449-2456.

Ji, G., Beavis, R., Novick, R.P., (1997). Bacterial interference caused by autoinducing peptide variants. *Science* 276(5321): 2027-2030.

Ji, G., Pei, W., Zhang, L., Qiu, R., Lin, J., Benito, Y., Lina, G., Novick, R.P., (2005). *Staphylococcus intermedius* produces a functional *agr* autoinducing peptide containing a cyclic lactone. *J. Bacteriol.* 187(9): 3139-3150.

Jones, B.W., Means, T.K., Heldwein, K.A., Keen, M.A., Hill, P.J., Belisle, J.T., Fenton, M.J., (2001). Different Toll-like receptor agonists induced istinct macrophage responses. *J. Leukoc.Biol.* 69, 1036–1044.

Jones, S.M., Morgan, M., Humphrey, T.J., Lappin-Scott, H., (2001). Effect of vancomycin and rifampicin on meticillinresistant *Staphylococcus aureus* biofilms. *Lancet* 357:40-1.

Joo, H.S., Otto, M., (2012). Molecular basis of in vivo biofilm formation by bacterial pathogens. *Chem. Biol.* 19(12): 1503-1513.

Kahl, B.C., Goulian, M., van Wamel, W., Herrmann, M., Simon, S.M., Kaplan, G., Peters, G., Cheung, A.L., (2000). *Staphylococcus aureus* RN6390 replicates and induces apoptosis in a pulmonary epithelial cell line. *Infect. Immun.* 68(9): 5385-5392.

Kalkum, M., Lyon, G.J., Chait, B.T., (2003). Detection of secreted peptides by using hypothesis-driven multistage mass spectrometry. *Proc. Natl. Acad. Sci. USA* 100(5): 2795-2800.

Kaplan, J.B., Ragunath, C., Velliyagounder, K., Fine, D.H., Ramasubbu, N., (2004). Enzymatic detachment of *Staphylococcus epidermidis* biofilms. Antimicrob. *Agents Chemother.* 48(7): 2633-2636.

Karatan, E., Watnick, P., (2009). Signals, regulatory networks, and materials that build and break bacterial biofilms. *Microbiol. Mol. Biol. Rev.* 73(2): 310-347.

Keane, F.M., Loughman, A., Valtulina, V., Brennan, M., Speziale, P., Foster, T.J., (2007). Fibrinogen and elastin bind to the same region within the A domain of fibronectin binding protein A, an MSCRAMM of *Staphylococcus aureus*. *Mol. Microbiol.* 63(3): 711-723.

Kirschning, C.J., Schumann, R.R., (2002). TLR2, cellular sensor for microbial and endogenous molecular patterns. *Curr.Top.Microbiol. Immunol.* 270: 121–144.

Knetsch, M.L., Koole, L.H., (2011). New strategies in the development of antimicrobial coatings: the example of increasing usage of silver and silver nanoparticles. *Polymers* 3(1): 340–366.

Knobloch, J.K., Bartscht, K., Sabottke, A., Rohde, H., Feucht, H.H., Mack, D., (2001). Biofilm formation by *Staphylococcus epidermidis* depends on functional RsbU, an activator of the sigB operon: differential activation mechanisms due to ethanol and salt stress. *J. Bacteriol.* 183(8): 2624-2633.

Kobayashi, S.D., Voyich, J.M., Burlak, C., DeLeo, F.R., (2005). Neutrophils in the innate immune response. *Arch. Immunol. Ther. Exp. (Warsz)*. 53(6): 505-517.

Kopp, E., Medzhitov, R., (2003). Recognition of microbial infection by Toll-like receptors. *Curr. Opin. Immunol.* 15: 396–401.

Kristian, S.A., Birkenstock, T.A., Sauder, U., Mack, D., Götz, F., Landmann, R., (2008). Biofilm formation induces C3a release and protects *Staphylococcus epidermidis* from IgG and complement deposition and from neutrophil-dependent killing. *J. Infect.Dis.* 197: 1028–1035.

Kurlenda, J., Grinholc, M., (2012). Alternative therapies in *Staphylococcus aureus* diseases. *Acta Bioch. Polonica* 59(2), 171-184.

Kurokawa, K., Lee, H., Roh, K.B., Asanuma, M., Kim, Y.S., Nakayama, H., Shiratsuchi, A., Choi, Y., Takeuchi, O., Kang, H.J., Dohmae, N., Nakanishi, Y., Akira, S., Sekimizu, K., Lee, B.L., (2009). The triacylated ATP binding cluster transporter substrate-binding lipoprotein of Staphylococcus aureus functions as a native ligand for toll-like receptor 2. *J. Biol.Chem.* 284, 8406–8411.

Lang, N.P., Berglundh, T., (2011). Periimplant diseases: where are we now? Consensus of the Seventh European Workshop on Periodontology. Working Group 4 of Seventh European Workshop on Periodontology. *J. Clin. Periodontol.* Mar;38 Suppl 11:178-81.

Lara, H.H., Ayala-Núnez, N.V., Ixtepan Turrent, L.D.C., Rodríguez Padilla, C., (2010). Bactericidal effect of silver nanoparticles against multidrug-resistant bacteria.*World J.Microbiol. Biotechnol.* 26(4): 615–621.

Lauderdale, K.J., Boles, B.R., Cheung, A.L., Horswill, A.R., (2009). Interconnections between Sigma B, agr, and proteolytic activity in *Staphylococcus aureus* biofilm maturation. *Infect. Immun.* 77(4): 1623-1635.

Leid, J.G., Costerton, J.W., Shirtliff, M.E., Gilmore, M.S., Engelbert, M., (2002). Immunology of *Staphylococcal* biofilm infections in the eye: new tools to study biofilm endophthalmitis. *DNA Cell Biol.* 21:405-13.

Lew, D.P., Waldvogel, F.A., (2004) Osteomyelitis. *Lancet* 364:369-79.

Li, H., Xu, L., Wang, J., Wen, Y., Vuong, C., Otto, M., Gao, Q., (2005). Conversion of *Staphylococcus epidermidis* strains from commensal to invasive by expression of the ica locus encoding production of biofilm exopolysaccharide. *Infect. Immun.* 73(5): 3188-3191.

Lopez-Leban, F., Kiran, M.D., Wolcott, R., Balaban, N., (2010). Molecular mechanisms of RIP, an effective inhibitor of chronic infections. *Int. J. Artif. Organs* 33: 582–589.

Luo, Y., Hossain, M., Wang, C., Qiao, Y., An, J., Ma, L., Su, M., (2013). Targeted nanoparticles for enhanced X-ray radiation killing of multidrug-resistant bacteria. *Nanoscale* 5: 687–694.

Mack, D., Fischer, W., Krokotsch, A., Leopold, K., Hartmann, R., Egge, H., Laufs, R., (1996). The intercellular adhesin involved in biofilm accumulation of *Staphylococcus epidermidis* is a linear beta-1,6-linked glucosaminoglycan: purification and structural analysis. *J. Bacteriol.* 178(1): 175-183.

Mann, E.E., Rice, K.C., Boles, B.R., Endres, J.L., Ranjit, D., Chandramohan, L., Tsang, L.H., Smeltzer, M.S., Horswill, A.R., Bayles, K.W., (2009). Modulation of eDNA release and degradation affects *Staphylococcus aureus* biofilm maturation. *PLoS One* 4(6): e5822.

Mantovani, A.,Cassatella, M.A., Costantini, C., Jaillon, S., (2011). Neutrophils in the acti vation and regulation of innate and adaptive immunity. *Nat.Rev. Immunol.* 11: 519–531.

Martinez, L.R., (2012). Nitric oxide releasing nanoparticles: challenges and future prospects of therapeutic delivery. *Ther. Deliv.* 3: 1139–1142.

McCrea, K.W., Hartford, O., Davis, S., Eidhin, D.N., Lina, G., Speziale, P., Foster, T.J., Hook, M., (2000). The serine-aspartate repeat (Sdr) protein family in *Staphylococcus epidermidis*. *Microbiology* 146 (Pt 7): 1535-1546.

McDevitt, D., Francois, P., Vaudaux, P., Foster, T.J., (1994). Molecular characterization of the clumping factor (fibrinogen receptor) of *Staphylococcus aureus*. *Mol. Microbiol.* 11(2): 237-248.

McKenney, D., Pouliot, K.L., Wang, Y., Murthy, V., Ulrich, M., Doring, G., Lee, J.C., Goldmann, D.A., Pier, G.B., (1999). Broadly protective vaccine for *Staphylococcus aureus* based on an in vivo-expressed antigen. *Science* 284(5419): 1523-1527.

Menousek, J., Mishra, B., Hanke, M.L., Heim, C.E., Kielian, T., Wang, G. (2012). Data base screening and *in vivo* efficacy of antimicrobial peptides against meticillin-resistant *Staphylococcus aureus* USA300. *Int. J. Antimicrob. Agents* 39: 402–406.

Merino, N., Toledo-Arana, A., Vergara-Irigaray, M., Valle, J., Solano, C., Calvo, E., Lopez, J.A., Foster, T.J., Penades, J.R., Lasa, I., (2009). Protein A-mediated multicellular behavior in *Staphylococcus aureus*. *J. Bacteriol* 191(3): 832-843.

Monds, R.D., O'Toole, G.A., (2009). The developmental model of microbial biofilms: ten years of a paradigm up for review. Trends Microbiol 17(2): 73-87.

Morath, S., Stadelmaier, A., Geyer, A., Schmidt, R.R., Hartung., T., (2002). Synthetic lipoteichoic acid from *Staphylococcus aureus* is a potent stimulus of cytokine release. *J. Exp Med.* 195: 1635–1640.

Nathan, C., (2006). Neutrophilsand immunity:challenges and opportunities. *Nat. Rev. Immunol.* 6, 173–182.

Murugan, K., Usha, M., Malathi, P., Al-Sohaibani, A.S., Chandrasekaran, M., (2010). Biofilm forming multi drug resistant *Staphylococcus spp.* among patients with conjunctivitis. *Pol. J. Microbiol.* 59:233-9.

National Nosocomial Infections Surveillance System, National Nosocomial Infections Surveillance (NNIS) System Report, data summary from January 1992 through June 2004, (2004). *Am J Infect Control*. Dec;32(8):470-85

Nauseef, W.M., (2007). How human neutrophils kill and degrade microbes: an integrated view. *Immunol. Rev.* 219, 88–102.

Ni Eidhin, D., Perkins, S., Francois, P., Vaudaux, P., Hook, M., Foster, T.J., (1998). Clumping factor B (ClfB), a new surface-located fibrinogen-binding adhesin of *Staphylococcus aureus*. *Mol. Microbiol.* 30(2): 245-257.

Novick, R.P., Geisinger, E., (2008). Quorum sensing in *Staphylococci*. *Annu. Rev. Genet.* 42: 541-564.

Novick, R.P., Projan, S.J., Kornblum, J., Ross, H.F., Ji, G., Kreiswirth, B., Vandenesch, F., Moghazeh, S., (1995). The *agr* P2 operon: an autocatalytic sensory transduction system in *Staphylococcus aureus*. *Mol. Gen. Genet.* 248(4): 446-458.

O'Gara, J.P., (2007). ica and beyond: biofilm mechanisms and regulation in *Staphylococcus* epidermidis and *Staphylococcus aureus*. *FEMS Microbiol. Lett* 270(2): 179-188.

O'May, G.A., Brady, R.A., Prabhakara, R., Leid, J.G., Calhoun, J.H., Shirtliff, M.E., (2010) Eds. Osteomyelitis. In Biofilm Infections: New York: Springer 2010.

O'Neill, E., Pozzi, C., Houston, P., Humphreys, H., Robinson, D.A., Loughman, A., Foster, T.J., O'Gara, J.P., (2008). A novel *Staphylococcus aureus* biofilm phenotype mediated by the fibronectin-binding proteins, FnBPA and FnBPB. *J. Bacteriol.* 190(11): 3835-3850.

O'Neill, E., Pozzi, C., Houston, P., Smyth, D., Humphreys, H., Robinson, D.A., O'Gara, J.P., (2007). Association between methicillin susceptibility and biofilm regulation in *Staphylococcus aureus* isolates from device-related infections. *J. Clin. Microbiol.* 45(5): 1379-1388.

O'Toole, G., Kaplan, H.B., Kolter, R., (2000). Biofilm formation as microbial development. *Annu. Rev. Microbiol.* 54: 49-79.

Otto, M., (2008). *Staphylococcal* biofilms. *Curr.Top.Microbiol. Immunol.* 322: 207–228.

Patti, J.M., Allen, B.L., McGavin, M.J., Hook, M., (1994). MSCRAMM-mediated adherence of microorganisms to host tissues. *Annu. Rev. Microbiol.* 48: 585-617.

Patti, J.M., Boles, J.O., Hook, M., (1993). Identification and biochemical characterization of the ligand binding domain of the collagen adhesin from *Staphylococcus aureus*. *Biochemistry* 32(42): 11428-11435.

Pelgrift, R.Y., Friedman, A.J., (2013). Nanotechnology as a therapeutic tool to combat microbial resistance. *Advanced Drug Delivery Reviews* ADR-12486, in press.

Periasamy, S., Joo, H.S., Duong, A.C., Bach, T.H., Tan, V.Y., Chatterjee, S.S., Cheung, G.Y., Otto, M., (2012). How *Staphylococcus aureus* biofilms develop their characteristic structure. *Proc. Natl. Acad. Sci. USA* 109(4): 1281-1286.

Peschel, A., Otto, M., (2013). Phenol-soluble modulins and staphylococcal infection. *Nat. Rev. Microbiol.* 11(10): 667-673.

Peters, B.M., Jabra-Rizk, M.A., Scheper, M.A., Leid, J.G., Costerton, J.W., Shirtliff, M.E., (2010). Microbial interactions and differential protein expression in *Staphylococcus aureus-Candida albicans* dual-species biofilms. *FEMS Immunol. Med. Microbiol.* 59: 493-503.

Poole, K., (2002). Mechanisms of bacterial biocide and antibiotic resistance. *J. Appl.Microbiol.* 92, 55S–64S.

Ponnuraj, K., Bowden, M.G., Davis, S., Gurusiddappa, S., Moore, D., Choe, D., Xu, Y., Hook, M., Narayana, S.V., (2003). A "dock, lock, and latch" structural model for a staphylococcal adhesin binding to fibrinogen. *Cell* 115(2): 217-228.

Proctor, RA., (2012). Is there a future for a *Staphylococcus aureus* vaccine? *Vaccine* 19;30(19): 2921-2927.

Psaltis, A.J., Weitzel, E.K., Ha, K.R., Wormald, P.J., (2008). The effect of bacterial biofilms on post-sinus surgical outcomes. *Am. J. Rhinol.* 22: 1-6.

Qin, Z., Ou, Y., Yang, L., Zhu, Y., Tolker-Nielsen, T., Molin, S., Qu, D., (2007). Role of autolysin-mediated DNA release in biofilm formation of *Staphylococcus epidermidis*. *Microbiology* 153(Pt 7): 2083-2092.

Raghupathi, K.R., Koodali, R.T., Manna, A.C., (2011). Size-dependent bacterial growth inhibition and mechanism of antibacterial activity of zinc oxide nanoparticles. *Langmuir.* 27(7): 4020–4028.

Resch, A., Leicht, S., Saric, M., Pasztor, L., Jakob, A., Gotz, F., Nordheim, A., (2006). Comparative proteome analysis of Staphylococcus aureus biofilm and planktonic cells and correlation with transcriptome profiling. *Proteomics* 6:1867-77.

Rohde, H., Burandt, E.C., Siemssen, N., Frommelt, L., Burdelski, C., Wurster, S., Scherpe, S., Davies, A.P., Harris, L.G., Horstkotte, M.A., Knobloch, J.K., Ragunath, C., Kaplan, J.B., Mack, D., (2007). Polysaccharide intercellular adhesin or protein factors in biofilm accumulation of *Staphylococcus epidermidis* and *Staphylococcus aureus* isolated from prosthetic hip and knee joint infections. *Biomaterials* 28(9): 1711-1720.

Rohde, H., Burdelski, C., Bartscht, K., Hussain, M., Buck, F., Horstkotte, M.A., Knobloch, J.K., Heilmann, C., Herrmann, M., Mack, D., (2005). Induction of *Staphylococcus epidermidis* biofilm formation via proteolytic processing of the accumulation-associated protein by staphylococcal and host proteases. *Mol. Microbiol.* 55(6): 1883-1895.

Rice, K.C., Mann, E.E., Endres, J.L., Weiss, E.C., Cassat, J.E., Smeltzer, M.S., Bayles, K.W. (2007). The cidA murein hydrolase regulator contributes to DNA release and biofilm development in *Staphylococcus aureus*. *Proc. Natl. Acad. Sci.U.S.A.* 104, 8113–8118.

Richardson, A. R., Libby, S. J., Fang, F. C., (2008). A nitric oxide-inducible lactate dehydrogenase enables *Staphylococcus aureus* to resist innate immunity. *Science* 319, 1672-*1676.*

Røder, B.L., Wandall, D.A., Frimodt-Moller, N., Espersen, F., Skinhoj, P., Rosdahl, V.T., (1999). Clinical features of *Staphylococcus aureus* endocarditis: a 10-year experience in Denmark. *Arch. Intern. Med.* 159:462-9;

Rogers, K.L., Fey, P.D., Rupp, M.E., (2009). Coagulase-negative staphylococcal infections. *Infect. Dis. Clin. North. Am.* 23:73–98.

Rubin, R.J., Harrington, C.A., Poon, A., Dietrich, K., Greene, J.A., Moiduddin, A., (1999). The economic impact of *Staphylococcus aureus* infection in New York City hospitals. *Emerg. Infect. Dis.* 5:9-17.

Rupp, M.E., Fey, P.D., Heilmann, C., Gotz, F., (2001). Characterization of the importance of *Staphylococcus epidermidis* autolysin and polysaccharide intercellular adhesin in the pathogenesis of intravascular catheter-associated infection in a rat model. *J. Infect. Dis.* 183(7): 1038-1042.

Rupp, M.E., Ulphani, J.S., Fey, P.D., Bartscht, K., Mack, D., (1999). Characterization of the importance of polysaccharide intercellular adhesin/hemagglutinin of *Staphylococcus epidermidis* in the pathogenesis of biomaterial-based infection in a mouse foreign body infection model. *Infect. Immun.* 67(5): 2627-2632.

Rupp, M.E., Ulphani, J.S., Fey, P.D., Mack, D., (1999). Characterization of *Staphylococcus epidermidis* polysaccharide intercellular adhesin/hemagglutinin in the pathogenesis of intravascular catheter-associated infection in a rat model. *Infect. Immun.* 67(5): 2656-2659.

Schairer, D.O., Chouake, J.S., Nosanchuk, J.D., Friedman, A.J., (2012). The potential of nitric oxide releasing therapies as antimicrobial agents. *Virulence* 3:(3), 0–1

Schairer, D.O., Martinez, L.R., Blecher, K., Chouake, J.S., Nacharaju, P., Gialanella, P., Friedman, J.M., Nosanchuk, J.D., Friedman, A.J., (2012). Nitric oxide nanoparticles: pre-clinical utility as a therapeutic for intramuscular abscesses. *Virulence* 3:(1), 62–67.

Schierle, C.F., De la Garza, M., Mustoe, T.A., Galiano, R.D., (2009). Staphylococcal biofilms impair wound healing by delaying reepithelialization in a murine cutaneous wound model. *Wound Repair Regen.* 17:354-9

Schroeder, K., Jularic, M., Horsburgh, S.M., Hirschhausen, N., Neumann, C., Bertling, A., Schulte, A., Foster, S., Kehrel, B.E., Peters, G., Heilmann, C., (2009). Molecular characterization of a novel *Staphylococcus aureus* surface protein (SasC) involved in cell aggregation and biofilm accumulation. *PLoS One* 4(10): e7567.

Schwartz, K., Syed, A.K., Stephenson, R.E., Rickard, A.H., Boles, B.R., (2012). Functional amyloids composed of phenol soluble modulins stabilize *Staphylococcus aureus* biofilms. *PLoS Pathog.* 8(6): e1002744.

Schwarz-Linek, U., Werner, J.M., Pickford, A.R., Gurusiddappa, S., Kim, J.H., Pilka, E.S., Briggs, J.A., Gough, T.S., Hook, M., Campbell, I.D., Potts, J.R., (2003). Pathogenic bacteria attach to human fibronectin through a tandem beta-zipper. *Nature* 423(6936): 177-181.

Segal, A.W., (2005). How neutrophils kill microbes. *Annu. Rev. Immunol.* 23:197-223.

Serbina, N.V., Jia, T., Hohl, T.M., Pamer, E.G., (2008). Monocyte-mediated defense against microbial pathogens. *Annu. Rev. Immunol.* 26: 421–452.

Sharma, M., Visai, L., Bragheri, F., Cristiani, I., Gupta, P.K., Speziale P., (2008). Toluidine blue-mediated photodynamic effects on staphylococcal biofilms. *Antimicrob Agents Chemother*. 52(1):299-305.

Silva M.T., (2010). When two is better than one: macrophages and neutrophils work in concert in innate immunity as complementary and cooperative partners of a myeloid phagocyte system. *J. Leukoc.Biol.* 87, 93–106.

Silva, M.T., (2011). Macrophage phagocytosis of neutrophils at inflamma tory/infectious foci: a cooperative mechanism in the control of infection and infectious inflammation. *J. Leukoc.Biol.* 89: 675–683.

Speziale, P., Visai, L., Rindi, S., Pietrocola, G., Provenza, G., Provenzano, M., (2008). Prevention and treatment of *Staphylococcus* biofilms. *Curr. Med. Chem.* 15(30): 3185-95.

Stephenson, M.F., Mfuna, L., Dowd, S.E., Wolcott, R.D., Barbeau, J., Poisson, M., James, G., Desrosiers, M., (2010). Molecular characterization of the polymicrobial flora in chronic rhinosinusitis. *J. Otolaryngol. Head. Neck. Surg.* 39:182-7.

Stepanovic, S., Djukic, N., Djordjevic, V., Djukic, S., (2003). Influence of the incubation atmosphere on the production of biofilm by staphylococci. *Clin. Microbiol. Infect.* 9(9): 955-958.

Stevens, N.T., Sadovskaya, I., Jabbouri, S., Sattar, T., O'Gara, J.P., Humphreys, H., Greene, C.M., (2009). *Staphylococcus* epidermidis polysaccharide intercellular adhesin induces IL-8 expression in human astrocytes via a mechanism involving TLR2. *Cell. Microbiol.* 11: 421–432.

Stuart, L.M., Ezekowitz, R.A., (2005). Phagocytosis: elegant complexity. *Immunity* 22(5):539-550.

Sudbery, P., Gow, N., Berman, J., (2004). The distinct morphogenic states of *Candida albicans*. *Trends Microbiol.* 12: 317-24.

Takeuchi, O., Hoshino, K., Kawai, T., Sanjo, H., Takada, H., Ogawa, T., Takeda, K., Akira, S., (1999). Differential roles of TLR2 and TLR4 in recognition of gram negative and gram-positive bacterial cell wall components. *Immunity* 11: 443–451.

Taylor, E., Webster1, T.J., (2011). Reducing infections through nanotechnology and nanoparticles. *Intern. J. of Nanomed.* 6: 1463-1473.

Thurlow, L.R., Hanke, M.L., Fritz, T., Angle, A., Williams, S.H., Engebretsen, I.L., Bayles, K.W., Horswill, A.R., Kielian,T., (2011). *Staphylococcus aureus* biofilms prevent macrophage phagocytosis and attenuate inflammation *in vivo*. *J.Immunol.* 186: 6585–6596.

Tsang, L.H., Cassat, J.E., Shaw, L.N., Beenken, K.E., Smeltzer, M.S., (2008). Factors contributing to the biofilm-deficient phenotype of *Staphylococcus aureus sarA* mutants. *PLoS One* 3(10): e3361.

Uckay, I., Pittet, D., Vaudaux, P., Sax, H., Lew, D., Waldvoge, F., (2009). Foreign body infections due to *Staphylococcus* epidermidis. *Ann. Med.* 41:109–119.

Vadyvaloo, V., Otto, M., (2005). Molecular genetics of *Staphylococcus* epidermidis biofilms on indwelling medical devices. *Int. J. Artif. Organs* 28:1069–1078.

Valle, J., Toledo-Arana, A., Berasain, C., Ghigo, J.M., Amorena, B., Penades, J.R., Lasa. I., (2003). SarA and not sigmaB is essential for biofilm development by *Staphylococcus aureus*. *Mol. Microbiol.* 48(4): 1075-1087.

Veenstra, G.J., Cremers, F.F., van Dijk, H., Fleer, A., (1996). Ultrastructural organization and regulation of a biomaterial adhesin of *Staphylococcus epidermidis*. *J. Bacteriol.* 178(2): 537-541.

Vuong, C., Durr, M., Carmody, A.B., Peschel, A., Klebanoff, S.J., Otto, M., (2004). Regulated expression of pathogen-associated molecular pattern molecules in *Staphylococcus epidermidis*: quorum-sensing determines pro-inflammatory capacity and production of phenol-soluble modulins. *Cell Microbiol.* 6(8): 753-759.

Vuong, C., Kocianova, S., Voyich, J.M., Yao, Y., Fischer, E.R., DeLeo, F.R., Otto, M., (2004). A crucial role for exopolysaccharide modification in bacterial biofilm formation, immune evasion, and virulence. *J. Biol. Chem.* 279(52): 54881-54886.

Wang, R., Braughton, K.R., Kretschmer, D., Bach, T.H.L., Queck, S.Y., Li, M., Kennedy, A.D., Dorward, D.W., Klebanoff, S.J., Peschel, A.P., DeLeo, F.R., Otto, M., (2007). Identification of novel cytolytic peptides as key virulence determinants for community-associated MRSA. *Nat. Med.* 13: 1510-1514.

Wang, R., Braughton, K.R., Kretschmer, D., Bach, T.H., Queck, S.Y., Li, M., Kennedy, A.D. Dorward, D.W., Klebanoff, S.J., Peschel, A., DeLeo, F.R., Otto, M., (2007). Identification of novel cytolytic peptides as key virulence determinants for community-associated MRSA. *Nat. Med.* 13(12): 1510-1514.

Wang, R., Khan, B.A., Cheung, G.Y., Bach, T.H., Jameson-Lee, M., Kong, K.F., Queck, S.Y., Otto, M., (2011). *Staphylococcus epidermidis* surfactant peptides promote biofilm maturation and dissemination of biofilm-associated infection in mice. *J. Clin. Invest.* 121(1): 238-248.

Wang, X., Preston, J.F. 3rd, Romeo. T., (2004). The pgaABCD locus of *Escherichia coli* promotes the synthesis of a polysaccharide adhesin required for biofilm formation. *J. Bacteriol.* 186(9): 2724-2734.

Weber, J.R., Moreillon, P., Tuomanen, E.I., (2003). Innate sensors for Gram positive bacteria. *Curr. Opin. Immunol.* 15: 408–415.

Wenzel, R.P., Edmond, M.B., (2001). The impact of hospital acquired bloodstream infections. *Emerg. Infect. Dis.* 7: 174-7.

Whitchurch, C.B., Tolker-Nielsen, T., Ragas, P.C., Mattick, J.S., (2002). Extracellular DNA required for bacterial biofilm formation. *Science* 295:1487.

Whiteley, M., Bangera, M.G., Bumgarner, R.E., Parsek, M.R., Teitzel, G.M., Lory, S., Greenberg, E.P., (2001). Gene expression in *Pseudomonas aeruginosa* biofilms. *Nature* 413(6858): 860-864.

Witko-Sarsat, V., Rieu, P., Descamps-Latscha, B., Lesavre,P., Halbwachs-Mecarelli,L., (2000). Neutrophils: molecules, functions and pathophysiological aspects. *Lab.Invest.* 80: 617–653.

Xu, L., Li, H., Vuong, C., Vadyvaloo, V., Wang, J., Yao, Y., Otto, M., Gao, Q., (2006). Role of the luxS quorum-sensing system in biofilm formation and virulence of *Staphylococcus epidermidis*. *Infect. Immun.* 74(1): 488-496.

Yamashiro, S., Kamohara, H., Wang, J.M., Yang, D., Gong, W.H., Yoshimura,T., (2001). Phenotypic and functional change of cytokine-activated neutrophils: inflammatory neutrophils are heterogeneous and enhance adaptive immune responses. *J. Leukoc.Biol.* 69: 698–704.

Yang, L., Barken, K.B., Skindersoe, M.E., Christensen, A.B., Givskov, M., Tolker-Nielsen, T., (2007). Effects of iron on DNA release and biofilm development by *Pseudomonas aeruginosa*. *Microbiology* 153(Pt 5): 1318-1328.

Yao, Y., Sturdevant, D.E., Otto, M., (2005). Genomewide analysis of gene expression in *Staphylococcus epidermidis* biofilms: insights into the pathophysiology of *S. epidermidis* biofilms and the role of phenol-soluble modulins in formation of biofilms. *J. Infect. Dis.* 191(2): 289-298.

Yeaman, M.R., Yount, N.Y., (2003). Mechanisms of antimicrobial peptide action and resistance. *Pharmacol. Rev.* 55: 27–55.

Ziebuhr, W., Heilmann, C., Gotz, F., Meyer, P., Wilms, K., Straube, E., Hacker, J., (1997). Detection of the intercellular adhesion gene cluster (ica) and phase variation in *Staphylococcus epidermidis* blood culture strains and mucosal isolates. *Infect. Immun.* 65(3): 890-896.

Ziran, B.H., (2007) Osteomyelitis. *J. Trauma* 62:59-60.

Zuluaga, A.F., Galvis, W., Saldarriaga, J.G., Agudelo, M., Salazar, B.E., Vesga, O., (2006). Etiologic diagnosis of chronic osteomyelitis: a prospective study. *Arch. Intern. Med.* 166: 95–100.

In: Prokaryotes
Editor: Marina Nisnevitch
ISBN: 978-1-63321-592-4

Chapter 2

Bacterial Biofilms and Role of the Second Messenger Cyclic-Di-GMP

Michelle A. Stan and Raymond J. Turner*
Department of Biological Sciences,
Biofilm Research Group,
Faculty of Science, University of Calgary,
Calgary, Alberta Canada

Abstract

Biofilm formation occurs in response to external cues and provides bacteria living in biofilm communities enhanced overall survival to antimicrobials and other stressors. The transition from a motile to biofilm lifestyle appears to be controlled by the bacterial messenger, cyclic dimeric guanosine monophosphate (c-di-GMP). This intracellular molecule is synthesized and degraded by c-di-GMP metabolizing proteins. Here, we examine some of these metabolizing proteins and their regulation of biofilm-associated phenotypes and characteristics.

Keywords: Biofilm, cyclic-di-GMP, exopolysaccharides

Introduction

The discovery that bacteria in nature typically are found in surface-adherent multicellular communities rather than as free-swimming, planktonic cells led to a major paradigm shift in microbiology. It has been estimated that > 99% of bacterial species grow in these microbial communities known as biofilms, and thus, they have become the focus of intense research over the past few decades (Costerton 2004; Costerton et al., 1995; Gilbert et al., 2002).

* Corresponding author: Email: turnerr@ucalgary.ca.

Biofilms are differentiated bacterial cells surrounded by a self-produced protective slime layer that adhere to surfaces or aggregate for growth. They have been implicated by the CDC (Centers for Disease Control and Prevention) as the cause of roughly 65% of infectious disease (Lewis 2007) and have become increasingly problematic for public health and industry (Hall-Stoodley et al., 2004). The biofilm represents a sessile mode of growth in which aggregates of bacterial cells develop into complex heterogeneous communities, distinct from their planktonic counterparts, that allows for enhanced survival in hostile environments (Costerton et al., 1999). The antimicrobial protection provided to microorganisms in biofilms has become ubiquitous with this multicellular life style.

The relationship between biofilm-associated resistance to stress, antimicrobials, and host immunological defenses has become a well-recognized physiological characteristic of biofilm bacteria (Harrison et al., 2007; Harrison et al., 2005). However, the mechanisms underlying this innate resistance are not fully understood. It has been hypothesized that biofilm-mediated resistance and/or tolerance is multifactorial; and is likely the outcome of many different mechanisms unique to this lifestyle.

Proposed mechanisms that can account for this marked resilience include: a protective role for the biofilm matrix; the creation of metabolic and nutritional gradients that can result in quiescent cell populations within the biofilm; biofilm-specific phenotypes that result from distinct physiological changes in surface-attached cells; differential gene expression, including resistance genes exclusive to the biofilm phenotype; the presence of persister cells and other phenotypic variants that can resist killing by antimicrobials and contribute to population diversity within biofilms (Drenkard & Ausubel, 2002; Gilbert et al., 2002; Harrison et al., 2007; Harrison et al., 2005; Lewis 2007; Patel 2005; Spoering & Lewis, 2001; Stewart & Costerton, 2001).

Thus, the resistance afforded to biofilm bacteria is complex and likely involves the additive benefits of different physiological and phenotypic factors characteristic of surface-adherent bacteria.

Development of the Biofilm

Biofilm formation is a developmental process that involves the coordination of both bacterial physiological and social behaviours (Harrison et al., 2005; Stoodley et al., 2002). In brief, biofilm development occurs when motile free-swimming bacteria transition to a sessile, attached state through a step-wise process that involves 1) the sensing and adsorption of cells to a surface, 2) irreversible attachment and extracellular matrix production, 3) growth and division of surface-attached bacteria to form mature microcolonies encased in a slime matrix, and finally 4) dispersal of cells from the biofilm to complete the cycle (Harrison et al., 2005; O'Toole et al., 2000) (Figure 1).

As will be discussed below, the development of biofilm formation requires the complex integration and regulation of bacterial processes such as motility, cell-cell and cell-surface adhesion factors, and extracellular polysaccharide synthesis. Many of the processes that contribute to biofilm development have been suggested to be regulated by the bacterial second messenger bis-(3'-5')-cyclic dimeric guanosine monophosphate (c-di-GMP).

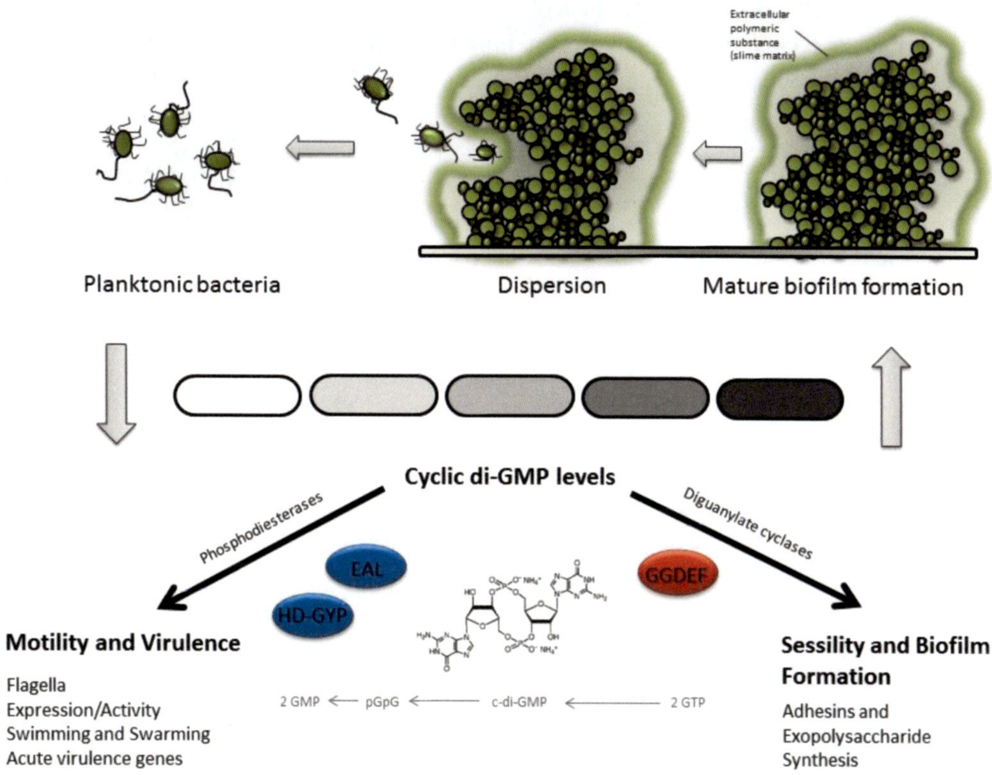

Figure 1. The physiological functions of c-di-GMP. Overview of the physiological functions of c-di-GMP in bacteria. Low levels of c-di-GMP promote motility and virulence through the regulation of flagella and acute virulence gene expression. Increased levels of c-di-GMP promote sessility and biofilm formation by regulating adhesin and exopolysaccharide production in surface-attached cells. Progression of the biofilm lifestyle involves fluxes in messenger, with decreased levels of c-di-GMP promoting dispersal of free-swimming, planktonic cells from the biofilm. The metabolism of c-di-GMP requires the opposing action of diguanylate cyclases with GGDEF domains and phosphodiesterases with EAL or HD-GYP domains, which make and break c-di-GMP respectively.

The Secondary Messenger Cyclic-Di-GMP

The GGDEF/EAL Domain Proteins

The transition from a motile, free-swimming state to a biofilm mode of life appears to be controlled (at least in part) by a cytoplasmic cyclic nucleotide molecule. C-di-GMP is a prokaryotic second messenger, found amongst >85% of all bacteria investigated to date, and is associated with the promotion of aggregative growth and the production of exopolysaccharides necessary for assembly of the 'slimy', self-produced biofilm matrix (Romling & Amikam, 2006). This molecule has also emerged as an integral intracellular messenger in global signalling networks that orchestrate the control of prokaryotic cellular function in response to environmental cues, including physiological behaviour and stress response (Hengge 2009; Romling & Amikam, 2006).

Cyclic-di-GMP is one of several nucleotide messengers found in the bacterial cell. These nucleotides act as second messengers or signalling molecules and include cAMP, cGMP, and ppGpp(p). As a secondary messenger, c-di-GMP concentration is tightly controlled through the antagonistic enzymatic action of the c-di-GMP turnover proteins; the diguanylate cyclases (DGCs) and phosphodiesterases (PDEs) that make and break c-di-GMP, respectively (Figure 1). These putative proteins, referred to as GGDEF and EAL, are named after their conserved domain sequences or motifs (Galperin et al., 2001). DGC activity resides in the GGDEF domain, which synthesizes c-di-GMP from GTP, while PDE activity resides in the EAL domain, which degrades c-di-GMP into its component GMP nucleotides in the cytosol (Christen et al., 2005; Hengge 2009; Paul et al., 2004; Ryan et al., 2006; Ryjenkov et al., 2005; Schirmer & Jenal, 2009; Schmidt et al., 2005; Simm et al., 2004; Tamayo et al., 2005). The active site (A site) of the DGC requires an intact GGDEF motif, any amino acid substitution (except for GG[D/E]EF) abrogates activity (Malone et al., 2007; Ryjenkov et al., 2005). GGDEF domain encoding proteins also typically contain an 'I site', a feedback inhibition site that allows for secondary binding of c-di-GMP and thus allosteric control over c-di-GMP synthesis (Christen et al., 2006). The monomeric EAL domain, as well as the unrelated HD-GYP domain, can linearize (hydrolytically cleave) c-di-GMP through a catalytic process that requires Mg^{2+} or Mn^{2+} (Christen et al., 2005; Galperin et al., 2001; Rao et al., 2008; Schmidt et al., 2005; Tamayo et al., 2005).

The model organism *E. coli* K-12 has 29 genes (identified based on their signature motifs) that encode proteins involved in c-di-GMP metabolism. Of these 29 genes, 12 encode GGDEF domain proteins, 10 encode EAL domain proteins, and 7 proteins contain both domains in a single polypeptide (Povolotsky & Hengge, 2012; Sommerfeldt et al., 2009). Proteins in which both domains are present can act either as a PDE or DGC, but do not typically have both enzymatic functions (Christen et al., 2005; Lacey et al., 2010). Often, one of the domains present is not entirely conserved, and thus it has been proposed that the degenerate domain may serve a regulatory function (amend the enzymatic activity of its partner domain) (Christen et al., 2005).

In addition to their C-terminal GGDEF/EAL domains, these regulatory proteins also contain putative N-terminus sensory domains that may be involved in sensing stimuli from outside the cell. These sensory domains (PAS, PAC, MASE1, MASE2, GAF, BLUF, HAMP, and REC to name a few) are of both known and unknown function (Galperin 2005; Povolotsky & Hengge, 2012). For instance, the REC or two-component receiver domain is associated with proteins found in two-component signal transduction systems, and thus allows for the phosphorylation of proteins in these signalling pathways (Hengge 2009). However, in only a few cases has the sensory signal of these "input" domains been identified. In *E. coli*, the PAS domain of the Dos protein (a GGDEF + EAL protein) can bind molecular oxygen, and in conjunction with YddV, control c-di-GMP activity in response to oxygen levels (Delgado-Nixon et al., 2000; Tuckerman et al., 2009). The BLUF domain of the EAL-only YcgF can act as a blue-light sensory domain, although the C-terminal EAL domain is degenerate in this protein (Tschowri et al., 2009). Of particular interest is the identification of c-di-GMP metabolizing proteins that respond to antibiotics, such as the Arr EAL protein of *P. aeruginosa*, which either directly or indirectly interacts with the aminoglycoside antibiotic tobramycin (Hoffman et al., 2005).

Hence, the modular nature of DGCs and PDEs likely allows for the integration of environmental information into the cellular response; as well as regulatory control of the "output" functions of the c-di-GMP metabolic domains of these proteins. Many of the c-di-GMP turnover proteins (19 of the 29 in *E. coli* K-12) also contain transmembrane segments or domains that allows for the extracellular positioning of their N-terminal signalling (sensory) domain (i.e. periplasmic orientation). Such an arrangement can facilitate the transduction of an environmental stimulus, such as oxygen levels, via c-di-GMP mediated intracellular signalling cascades in the cell.

The c-di-GMP Control Module and the Generation of Signal Specificity of GGDEF/EAL Proteins

The sheer abundance of GGDEF and EAL domains in the genomes of bacteria originally drew the attention of researchers to these proteins (Galperin 2005; Galperin et al., 2001). Based on bioinformatic studies, the genome of *Vibrio cholerae* is believed to encode for a total of 72 GGDEF/EAL and HD-GYP domains (Galperin 2004; Galperin et al., 2001); whereas the *P. aeruginosa* genome encodes for 41 GGDEF/EAL/HD-GYP domains (Galperin et al., 2001), and *Salmonella enterica* serovar Typhimurium encodes for 19 GGDEF/EAL domain proteins (Simm et al., 2004). The multiplicity of these proteins begs the question of how signalling specificity can be achieved within the bacterial cell with such high levels of redundancy (Jonas et al., 2009; Povolotsky & Hengge, 2012). Furthermore, c-di-GMP signalling clearly has many different physiological outputs; such as control of EPS production, flagellar motility, cell cycle progression, and virulence factor expression (Hengge 2009). How is this achieved if all GGDEF/EAL proteins are contributing to a freely-diffusible, common cellular pool of c-di-GMP?

One way in which signalling specificity can be achieved is through the c-di-GMP control module (reviewed in Hengge 2009). Many of the GGDEF and/or EAL proteins of *E. coli* appear to operate within distinct c-di-GMP control modules that allow for the spatial and temporal restriction of their activity. Such control modules can have very different functional outputs, such as the regulation of flagellar or adhesive factor expression, and generate regulatory cascades that may operate in a parallel but non-convergent manner. The c-di-GMP control module, at its most basic level, is comprised of at least one DGC and PDE that can metabolize c-di-GMP in response to an extracellular input signal, an effector that senses and directly binds this c-di-GMP, and target components whose functional output is directly regulated by interaction with the effector (Hengge 2009; Hengge 2010). Effectors can be proteins (such as the PilZ domain containing proteins which directly bind c-di-GMP), or RNA riboswitches, that can translate the signal to specific output targets within the cell (Hengge 2010; Hengge 2009). Targets can be DNA (the promoter region of a gene if the effector is a transcription factor), other enzymes, or complex biosynthetic structures like those involved with EPS and flagellar synthesis and activity (Hengge 2009; Hengge 2010). A good example of non-converging c-di-GMP control modules are the signalling cascades that regulate both curli and cellulose expression in *E. coli* (Figure 2).

There also appears to be many other mechanisms by which c-di-GMP signalling can be controlled within the bacterial cell. Complex regulation can result from transcriptional control of GGDEF/EAL encoding genes by 1) global regulators, including the stress response sigma factor RpoS (σ^S) (Weber et al., 2005) and the RNA-binding protein CsrA (Jonas et al., 2008), 2) two-component regulatory systems and response regulators, such as the biofilm regulator

CsgD (Brombacher et al., 2006; Ogasawara et al., 2011; Romling et al., 2000; Simm et al., 2004) and the two-component CpxAR system that responds to cell envelope and copper stress (Price & Raivio, 2009; Raivio 2005; Yamamoto & Ishihama, 2006; Yamamoto & Ishihama, 2005), 3) quorum sensing (in *Vibrio cholerae*) (Waters et al., 2008), and 4) cAMP signalling (Jonas et al., 2009). Post-translational control of GGDEF/EAL proteins adds further complexity to c-di-GMP signalling and can be achieved through the direct sensing of specific signals by the N-terminal sensory domains (which activate the enzymatic domains of these proteins), as well as through protein stability, localization, and direct protein-protein interactions (Jonas et al., 2009). Thus, c-di-GMP signalling specificity is likely brought about through the multilayered control of complex hierarchical signalling circuits at multiple levels. This is also indicative that the GGDEF/EAL proteins can have unique physiological functions within the cell, despite their multiplicity and common output function (the synthesis or degradation of c-di-GMP).

Regulation of E. Coli Biofilm Formation by c-di-GMP

Biofilms are multicellular communities that provide resistance against many environmental stresses. The regulation of biofilm formation and the functional processes that contribute to biofilm growth (cell adhesion and aggregation, motility, and exopolysaccharide production) are tightly regulated by c-di-GMP in response to environmental and physiological cues. In general, c-di-GMP promotes the biofilm lifestyle by repressing flagellar motility and subsequently increasing the expression of extracellular adhesion factors/proteins and polysaccharides (Figure 1). Although, the majority of the c-di-GMP turnover proteins are still of unknown function, many studies have begun to assign biofilm-associated phenotypes to these proteins based on their deletion and overproduction in the cell (Hengge 2010; Sommerfeldt et al., 2009).

The repression of motility by c-di-GMP occurs through its action on the PilZ domain protein YcgR (Ryjenkov et al., 2006). This effector protein can act as a molecular brake of flagellar-mediated motility through its interactions with FliM and FliG of the flagellar motor switch complex (Paul et al., 2010). Deletion of the EAL encoding *yhjH* gene also impairs motility, presumably through an increase in c-di-GMP levels (Pesavento et al., 2008). Interestingly, the control cascades for flagellar function and motility (FlhDC) and adhesion expression (σ^S/YdaM/YciR/CsgD) appear to be inversely regulated in *E. coli* (for a review see Hengge 2009; Hengge 2010). The RpoS (σ^S) general stress response regulator inhibits the motility control cascade, while promoting the transcription of the *csgDEFG* operon encoding CsgD (curli subunit gene D), a LuxR family transcriptional regulator (Brombacher et al., 2006; Gerstel et al., 2003; Robbe-Saule et al., 2006). This occurs in a c-di-GMP dependent fashion through the positive σ^S-regulation of GDDEF/EAL encoding genes (*ydaM*, *yciR*, *yegE*, *yedQ*) (Pesavento et al., 2008; Sommerfeldt et al., 2009; Weber et al., 2006).

In *E. coli* and *Salmonella*, regulation of adhesive curli fimbriae is also under CsgD and c-di-GMP-dependent control (Sommerfeldt et al., 2009). The c-di-GMP mediated regulation of curli expression is an important step in the transition from motility towards sessility and hence, biofilm growth. Additional c-di-GMP control modules can act downstream of CsgD expression to modulate both curli fimbriae and cellulose biosynthesis (Hengge 2010). Curli fimbriae are proteinaceous fibres that mediate cell-cell and cell-surface adhesion (and thus biofilm development) in *E. coli* and *Salmonella* species (Barnhart & Chapman, 2006; Prigent-Combaret, et al., 2000; Romling et al., 1998). They are a major structural element of the extracellular matrix and in conjunction with cellulose, are important mediators of multicellular behaviour (Prigent-Combaret, et al., 2000; Romling 2005; Simm et al., 2007; Simm et al., 2004). Proteinaceous curli adhesins have also been implicated in bacterial-host interactions and have been shown to contribute to the virulence of pathogenic species (Barnhart & Chapman, 2006; Gophna et al., 2001; Prigent-Combaret, et al., 2000; Wang et al., 2006). In addition to their role in biofilm formation and pathogenesis, curli variants may also contribute to stress resistance (either negatively or positively) dependent on environment (Carter et al., 2011).

Curli fimbriae production is reliant on the activation of the curli structural operon *csgBAC* by the biofilm regulator CsgD (Arnqvist et al., 1994; Hammar et al., 1995). Thus, through the various DGC/PDE control modules that regulate CsgD (i.e. YdaM/YciR), as well as positive σ^S control (Arnqvist et al., 1994; Sommerfeldt et al., 2009), curli-mediated adhesion and biofilm formation is dependent on both the general stress response and c-di-GMP levels in the cell. Although the regulation of these systems (motility and curli fimbriae expression) has not been described here in their entirety, the examples given demonstrate the hierarchical arrangement and cross-talk that can occur in pathways that integrate c-di-GMP signalling to coordinate the switch from a motile to sessile mode of growth. In the following sections, we will focus on the expression of two major exopolysaccharides (EPS) of the *E. coli* biofilm matrix, PNAG and cellulose, and their regulation by c-di-GMP.

The Regulation of Exopolysaccharide Production by c-di-GMP Signalling

C-di-GMP appears to influence EPS production as increases in this messenger concomitantly leads to increased EPS synthesis in many bacterial species. Further, these increases in EPS can be connected to other biofilm-associated behaviours such as adherence, aggregation, and antibiotic resistance (Kirisits et al., 2005; Starkey et al., 2009). The slimy, self-produced matrix of bacterial biofilms is comprised of many components, including nucleic acids, proteins, and exopolysaccharides (Ryder et al., 2007). The matrix serves to provide structural integrity and support to the biofilm, and can also contribute to surface adhesion and protection against environmental stressors, including antimicrobial compounds (Danese et al., 2000; Karatan & Watnick, 2009; Ryder et al., 2007; White et al., 2006). In the past, it has been suggested that the matrix may serve as a restrictive barrier towards the diffusion of antibiotics into the inner layers of the biofilm structure (Gilbert et al., 2002). However, more recent studies have shown that antibiotics do penetrate into the biofilm (Anderl et al., 2000; Coquet et al., 1998; Patel 2005). Nevertheless, the matrix and the molecules that comprise it perform important functions in the development and maturation of biofilms as well as the interaction of bacterial biofilms within the host environment.

The *E. coli* biofilm matrix is typically comprised of proteins, polysaccharides (PNAG, cellulose, and colanic acid), adhesins (curli fimbriae), and eDNA (Karatan & Watnick, 2009; Van Houdt & Michiels, 2005). The regulation, expression, and synthesis of extracellular polysaccharides is quite complex. For example, the biosynthesis of colanic acid requires the expression of 19 genes found within the *wca* operon in *E. coli* (Stevenson et al., 1996). Currently, it is not known whether c-di-GMP signalling regulates the production of colanic acid. However, a recent finding has linked the activity of the degenerate EAL protein YcgF to the production of the biofilm matrix substance colanic acid through the activation of a signalling cascade that involves the RcsC/RcsD/RcsB two-component phosphorelay system (Tschowri et al., 2009). C-di-GMP is also believed to inversely regulate eDNA concentrations in both *E. coli* and *P. aeruginosa* (Sanchez-Torres et al., 2011; Ueda & Wood, 2010); extracellular DNA has been shown to contribute to the antibiotic resistance of *P. aeruginosa* biofilms (Mulcahy et al., 2008).

The Exopolysaccharide PNAG

The exopolysaccharide PNAG (PGA or PIA) is a linear, unbranched polymer of β-1,6-*N*-acetyl-D-glucosamine that is found in the biofilm matrices of both Gram-negative and positive organisms; including *E. coli*, *Yersinia pestis*, *Staphylococcus epidermis*, and *Staphylococcus aureus* (Cramton et al., 1999; Itoh et al., 2005; Wang et al., 2004). PNAG is as a key polysaccharide, and unlike cellulose, seems to be required for the adhesion of cells to surfaces as well as to each other (Wang et al., 2004). Chemical disruption of PNAG can lead to dissolution of the biofilm via the dispersal of single cells (Wang et al., 2004). Therefore, PNAG appears to be an important matrix component for both structural integrity and formation of complex multilayered biofilms (Itoh et al., 2005; Wang et al., 2004).

Production and secretion of PNAG requires multiple proteins encoded by the *pgaABCD* operon in Gram-negative species (*hmsHFRS* in *Yersinia* species, *icaABCD* in the staphylococcal species) (Karatan & Watnick, 2009). The inner membrane protein PgaD encoded by this operon has been implicated in biofilm formation; as deletion of *pgaD* leads to a reduction in biofilm formation as well as surface adhesion (Wang et al., 2004). Similar to cellulose, the regulation of PNAG synthesis can occur at multiple levels. In response to high salt concentrations, transcription of the *pgaABCD* operon is activated by the NhaR protein (Goller et al., 2006). At the post-transcriptional level, expression of the *pgaABCD* locus is strongly inhibited by the carbon storage regulator CsrA, by binding and subsequent degradation of its mRNA (Cerca & Jefferson, 2008; Itoh et al., 2008; Wang et al., 2005; Wang et al., 2004). Further post-translational control of PNAG synthesis is achieved through the integration of c-di-GMP metabolic proteins into the PNAG biosynthetic pathway. The RNA-binding CsrA, in addition to its control of the *pgaABCD* operon, also represses the expression of two GGDEF encoding genes, *ydeH* and *ycdT* (Jonas et al., 2008).

YdeH is a known DGC (*in vitro* confirmation (Boehm et al., 2009; Zahringer et al., 2011)), which can positively control PNAG production through the synthesis of c-di-GMP (Figure 2). YdeH is thought to stabilize the PgaD protein via a yet unknown mechanism and is also required for the induction of PNAG production in response to aminoglycosides and other translation-inhibitory antibiotics (Boehm et al., 2009). YdeH expression is negatively regulated by the σ^S (RpoS) subunit (Sommerfeldt et al., 2009), and thus induction of PNAG production can occur independently of the stringent response (ppGpp signalling molecule) (Boehm et al., 2009). YdeH expression can also be induced by the CpxAR two component system (Price & Raivio, 2009; Yamamoto & Ishihama, 2006; Yamamoto & Ishihama, 2005).

The Cpx system has been shown to sense the presence of misfolded periplasmic proteins as well as cell envelope stress (Vogt & Raivio, 2011).

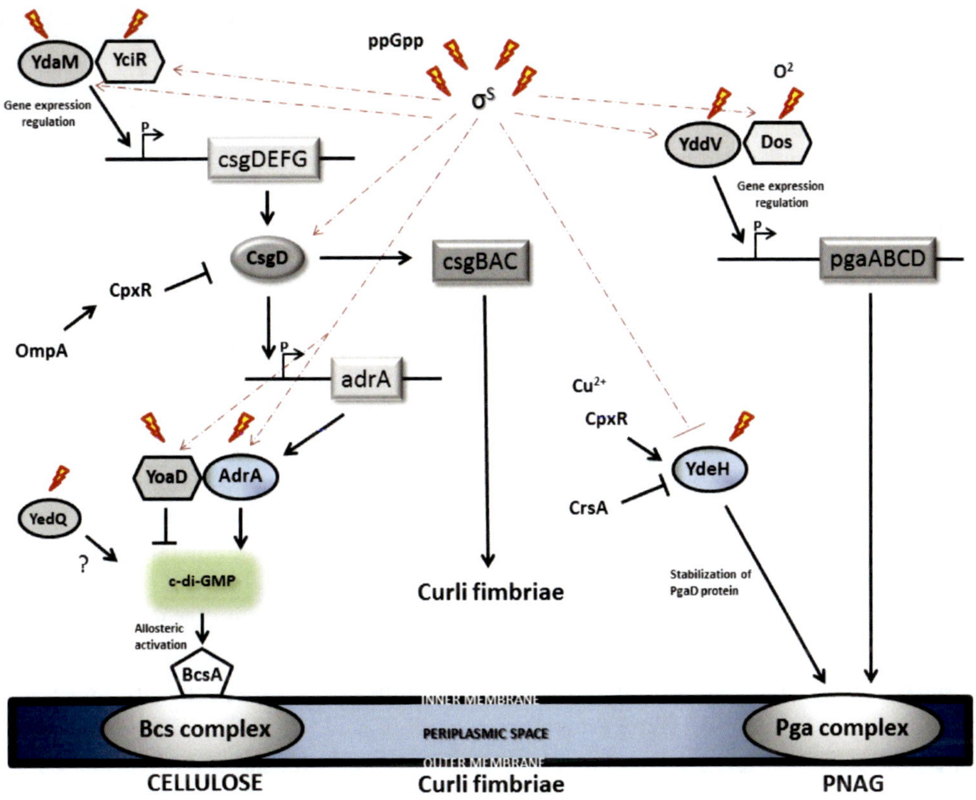

Figure 2. Regulation of cellulose and PNAG production by c-di-GMP. Abbreviated model of the DGC/PDE control cascades directing cellulose and PNAG synthesis in *E. coli* K-12. YdaM and YciR are arranged in a hierarchy, with YdaM influencing expression of the *csgDEFG* operon. The biofilm regulator CsgD in turn activates expression of cellulose and curli fimbriae via *adrA* (*yaiC*) and the *csgBAC* operon. The DGC activity of AdrA is required for the allosteric activation of the Bcs complex by c-di-GMP. YddV influences the regulation of both the *pgaABCD* and *csgBAC* operons, which are required for PNAG and curli synthesis, respectively. The DGC activity of YdeH appears to be required for stabilization of the PgaD protein of the PNAG biosynthetic complex. For greater detail and references refer to text. Ovals represent GGDEF proteins with DGC activity, whereas hexagons represent EAL proteins with PDE activity. GGDEF proteins of particular interest are in blue. Arrowheads and barred lines indicate positive and negative regulation, respectively. Red hashed arrows and lines indicate σ^S regulation. Lightning bolts denote environmental signal input.

In addition to YdeH, another GGDEF only protein, YddV, positively regulates PNAG production by activating the expression of the *pgaABCD* operon (Figure 2) (Tagliabue et al., 2010). This regulation is dependent on the DGC activity of the YddV protein; thus, Tagliabue and colleagues (2010) have suggested that this process may require the interaction of c-di-GMP (generated by YddV) with an yet unknown effector protein, similar to the c-di-GMP mediated control of cellulose biosynthesis.

Cellulose

Cellulose is a major polysaccharide of plant cell walls and is also secreted by microorganisms. Cellulose is a common matrix component of biofilm bacteria; including *Pseudomonas fluorescence*, *E. coli*, and the closely related *Salmonella* species (Romling 2005; Solano et al., 2002; Spiers & Rainey, 2005; Zogaj et al., 2001). Its production was first described in the organism *Gluconoacetobacter xylinum* and led to the recognition of c-di-GMP as a major regulatory signalling network in Gram-negative bacteria (Ross et al., 1987; Weinhouse et al., 1997). Cellulose was originally proposed to act as a cell to -cell and -surface adhesion factor in *E. coli* (Romling et al., 2000). However, further work has demonstrated an unclear function for this exopolysaccharide in biofilm formation. It has been suggested that the EPS cellulose plays a role in resistance to environmental stresses, such as desiccation, rather than as a biofilm adhesion factor (Gualdi et al., 2008; Wang et al., 2006; White et al., 2006). Thus, Landini (2009) has suggested that the c-di-GMP-dependent extracellular structures, curli fimbriae and cellulose, fulfil two possible functions: 1) to mediate the transition from an unattached planktonic state to a biofilm lifestyle via curli fimbriae expression, and 2) to contribute to EPS-mediated stress resistance.

The hierarchical control of cellulose biosynthesis in *E. coli* and *Salmonella* involves the coordinated activities of several c-di-GMP (DGC-PDE) control modules, including YdaM/YciR and YaiC/YoaD (AdrA/AdrB in *Salmonella*, and subsequently renamed in *E. coli* due to its homologous functions) (Figure 2) (Brombacher et al., 2006; Kader et al., 2006; Pesavento et al., 2008; Simm et al., 2007; Weber et al., 2006; Zogaj et al., 2001). In the *Enterobacteriaceae*, the biosynthesis of cellulose requires proteins encoded by the highly conserved bacterial cellulose synthesis operons, *bcsABZC* and *bcsEFG* (Solano et al., 2002; Zogaj et al., 2001). These proteins comprise the biosynthetic machinery necessary for the synthesis and export of cellulose outside of the cell (Zogaj et al., 2001). BcsA and BcsB together form the cellulose synthase complex, whereas BcsZ acts as a cellulase (Romling 2002).

The DGC, YaiC (AdrA) is responsible for the allosteric activation of the Bcs complex and subsequently, biosynthesis of the exopolysaccharide cellulose in *Enterobacteriaceae* species (Antoniani et al., 2010; Ma & Wood, 2009; Romling et al., 2000; Zogaj et al., 2001). The putative PDE YoaD appears to have the inverse function on cellulose production (Figure 2) (Brombacher et al., 2006). The BcsA protein of the cellulose biosynthetic complex binds c-di-GMP produced by YaiC to initiate its function as a cellulose synthase, and is thus an example of a PilZ-like c-di-GMP effector (Amikam & Galperin, 2006; Zogaj et al., 2001). The transcription of *yaiC* (*adrA*) is controlled by the regulator CsgD, which is also involved in the control of curli fimbriae expression (Figure 2) (Brombacher et al., 2006; Romling et al., 2000; Simm et al., 2004). As mentioned previously, expression of *csgD* is in itself regulated by the inverse function of the YdaM/YciR control module. Each of these proteins can also have their own unique signal input, theoretically allowing for the integration of environmental signals at various points along the hierarchical GGDEF/EAL control cascade.

Regulation by CsgD links the production of cellulose with that of curli fimbriae; as a result these extracellular components are frequently found to be co-expressed. Biosynthesis of the extracellular components cellulose and curli is associated with the multicellular rdar (red dry and rough) morphotype of bacteria grown on solid surfaces (Zogaj et al., 2001). This morphotype has been linked to enhanced biofilm formation, persistence, and stress resistance

(Scher et al., 2005; White et al., 2008; White et al., 2006). The expression of *yaiC* and *yoaD* is also up-regulated during growth on a solid surface; occurs primarily at temperatures below 30°C, and is under the control of the σ^S stationary phase and general stress response regulator (Sommerfeldt et al., 2009).

The production of cellulose and PNAG, two major polysaccharides of the biofilm matrix, is tightly regulated by c-di-GMP signalling in *E. coli*. In addition to their converse regulation by the σ^S (RpoS) subunit, cellulose synthesis is also inhibited by the *E. coli* outer membrane protein OmpA and the CpxR system, via its repression of *csgD* expression (Ma & Wood, 2009). Further, both exopolysaccharides exhibit divergent temperature-dependent regulation as cellulose expression is more predominant at lower growth temperatures (Gualdi et al., 2008), whereas PNAG production is up-regulated at higher temperatures, 37°C (Cerca & Jefferson, 2008). This opposing regulation of cellulose and PNAG production may indicate that the synthesis of different matrix components occur dependent on the needs of the bacterium in a particular environment, and may also result from specific responses to an extracellular signal or cue.

A Caveat on GGDEF and EAL Protein Overexpression

Many of the physiological functions assigned to c-di-GMP turnover proteins have been based on the overexpression of GGDEF-encoding genes to artificially increase c-di-GMP levels (Antoniani et al., 2010; Mendez-Ortiz et al., 2006). A problem with the overexpression of GGDEF proteins may be that influxes of messenger lead to unspecific responses in the cell. That is to say, that the manufactured changes in c-di-GMP may result in non-physiological cross-talk between non-converging c-di-GMP-dependent pathways. This may be evidenced by the interchangeability of heterologous DGCs and PDEs as well as the generation of unrelated phenotypes with deletion or overexpression of the same gene encoding a GGDEF protein (Boehm et al., 2010; Garcia et al., 2004; Jenal & Malone, 2006; Jonas et al., 2008; Wang et al., 2005). For example, expression of a heterologous (foreign) DGC from *Caulobacter crescentus* was capable of fully restoring the attachment defects of an *ydeH* mutant in *E. coli* as well as altering the expression of PNAG components (Boehm et al., 2009). Overall, caution is recommended in the interpretation of results from overexpression studies.

Conclusion

Cyclic-di-GMP appears to regulate the physiological functions of the cell in order to coordinate microbial community behaviour and the biofilm state. These turnover proteins also contain putative N-terminus domains that are involved in the sensing of stimuli that allows for the integration of environmental information into the cellular response. Although implicated in many biofilm related activities, including the intrinsic stress response, a link between c-di-GMP, the proteins that make and break it, and the antimicrobial tolerance of bacterial biofilms has not been established. Many hypothesize that c-di-GMP turnover leads to signal transduction cascades within the cell. Subsequently, this could result in modulation of biofilm physiology, altered biofilm cell densities, and changes in stress survival. This rationale is based on the principle that c-di-GMP levels control the switch from motility to sessility in bacteria and that high intracellular levels of second messenger promotes biofilm formation (Hengge 2009; Jenal & Malone, 2006; Romling & Amikam, 2006). High levels of c-di-GMP in the bacterial cell have been associated with increased biofilm formation and induction of genes associated with surface attachment, exopolysaccharide synthesis, and possibly the sensing and activation of the bacterial stress response. Recent work has shown that diverse environmental cues, including antimicrobials, may influence c-di-GMP-mediated signal transduction (Boehm et al., 2009; Hoffman et al., 2005; Price & Raivio, 2009; Yamamoto & Ishihama, 2006; Yamamoto & Ishihama, 2005). An association between c-di-GMP signalling, antibiotic resistance, phenotypic variation, as well as persistence in the host has been demonstrated for *P. aeruginosa* biofilms (Drenkard & Ausubel, 2002; Kirisits et al., 2005; Malone et al., 2010).

Thus, global research has begun to evaluate the idea that the decreased susceptibility of bacterial biofilms to antimicrobials may be mediated through the multitude of cellular and physiological processes regulated by c-di-GMP, and that specific GGDEF/EAL proteins may contribute to biofilm antimicrobial tolerance.

References

Amikam, D., & Galperin, M. Y. (2006). PilZ domain is part of the bacterial c-di-GMP binding protein. *Bioinformatics* 22, 3-6.

Anderl, J. N., Franklin, M. J., & Stewart, P. S. (2000). Role of antibiotic penetration limitation in *Klebsiella pneumoniae* biofilm resistance to ampicillin and ciprofloxacin. *Antimicrob Agents Chemother* 44, 1818-1824.

Antoniani, D., Bocci, P., Maciag, A., Raffaelli, N., & Landini, P. (2010). Monitoring of diguanylate cyclase activity and of cyclic-di-GMP biosynthesis by whole-cell assays suitable for high-throughput screening of biofilm inhibitors. *Appl Microbiol Biotechnol* 85, 1095-1104.

Arnqvist, A., Olsen, A., & Normark, S. (1994). Sigma S-dependent growth-phase induction of the *csgBA* promoter in *Escherichia coli* can be achieved *in vivo* by sigma 70 in the absence of the nucleoid-associated protein H-NS. *Mol Microbiol* 13, 1021-1032.

Barnhart, M. M., & Chapman, M. R. (2006). Curli biogenesis and function. *Annu Rev Microbiol* 60, 131-147.

Boehm, A., Kaiser, M., Li, H., Spangler, C., Kasper, C. A., Ackermann, M., Kaever, V., Sourjik, V., Roth, V., & Jenal, U. (2010). Second messenger-mediated adjustment of bacterial swimming velocity. *Cell* 141, 107-116.

Boehm, A., Steiner, S., Zaehringer, F., Casanova, A., Hamburger, F., Ritz, D., Keck, W., Ackermann, M., Schirmer, T., & Jenal, U. (2009). Second messenger signalling governs *Escherichia coli* biofilm induction upon ribosomal stress. *Mol Microbiol* 72, 1500-1516.

Brombacher, E., Baratto, A., Dorel, C., & Landini P. (2006). Gene expression regulation by the Curli activator CsgD protein: modulation of cellulose biosynthesis and control of negative determinants for microbial adhesion. *J Bacteriol* 188, 2027-2037.

Carter, M. Q., Brandl, M. T., Louie, J. W., Kyle, J. L., Carychao, D. K., Cooley, M. B., Parker, C. T., Bates, A. H., & Mandrell, R. E. (2011). Distinct acid resistance and survival fitness displayed by Curli variants of enterohemorrhagic *Escherichia coli* O157:H7. *Appl Environ Microbiol* 77, 3685-3695.

Cerca, N., & Jefferson, K. K. (2008). Effect of growth conditions on poly-N-acetylglucosamine expression and biofilm formation in *Escherichia coli. FEMS Microbiol Lett* 283, 36-41.

Christen, B., Christen, M., Paul, R., Schmid, F., Folcher, M., Jenoe, P., Meuwly, M., & Jenal, U. (2006). Allosteric control of cyclic di-GMP signaling. *J Biol Chem* 281, 32015-32024.

Christen, M., Christen, B., Folcher, M., Schauerte, A., & Jenal, U. (2005). Identification and characterization of a cyclic di-GMP-specific phosphodiesterase and its allosteric control by GTP. *J Biol Chem* 280, 30829-30837.

Coquet, L., Junter, G. A., & Jouenne, T. (1998). Resistance of artificial biofilms of *Pseudomonas aeruginosa* to imipenem and tobramycin. *J Antimicrob Chemother* 42, 755-760.

Costerton, J. W. (2004). A Short History of the Development of the Biofilm Concept, p. 4-19. *In* M. Ghannoum and G. A. O'Toole (ed.), Microbial Biofilms. ASM Press, Washington, D.C.

Costerton, J. W., Lewandowski, Z., Caldwell, D. E., Korber, D. R., & Lappin-Scott, H. M. (1995). *Microbial biofilms. Annu Rev Microbiol* 49, 711-745.

Costerton, J. W., Stewart, P. S., & Greenberg, E. P. (1999). Bacterial biofilms: a common cause of persistent infections. *Science* 284, 1318-1322.

Cramton, S. E., Gerke, C., Schnell, N. F., Nichols, W. W., & Gotz, F. (1999). The intercellular adhesion (*ica*) locus is present in *Staphylococcus aureus* and is required for biofilm formation. *Infect Immun* 67, 5427-5433.

Danese, P. N., Pratt, L. A., & Kolter, R. (2000). Exopolysaccharide production is required for development of *Escherichia coli* K-12 biofilm architecture. *J Bacteriol* 182, 3593-3596.

Delgado-Nixon, V. M., Gonzalez, G., & Gilles-Gonzalez, M. A. (2000). Dos, a heme-binding PAS protein from *Escherichia coli*, is a direct oxygen sensor. *Biochemistry* 39, 2685-2691.

Drenkard, E., & Ausubel, F. M. (2002). *Pseudomonas* biofilm formation and antibiotic resistance are linked to phenotypic variation. *Nature* 416, 740-743.

Galperin, M. Y. (2004). Bacterial signal transduction network in a genomic perspective. *Environ Microbiol* 6, 552-567.

Galperin, M. Y. (2005). A census of membrane-bound and intracellular signal transduction proteins in bacteria: bacterial IQ, extroverts and introverts. *BMC Microbiol* 5, 35.

Galperin, M. Y., Nikolskaya, A. N., & Koonin, E. V. (2001). Novel domains of the prokaryotic two-component signal transduction systems. *FEMS Microbiol Lett* 203, 11-21.

Garcia, B., Latasa, C., Solano, C., Garcia-del Portillo, F., Gamazo, C., & Lasa, I. (2004). Role of the GGDEF protein family in *Salmonella* cellulose biosynthesis and biofilm formation. *Mol Microbiol* 54, 264-277.

Gerstel, U., Park, C., & Romling, U. (2003). Complex regulation of *csgD* promoter activity by global regulatory proteins. *Mol Microbiol* 49, 639-654.

Gilbert, P., Allison, D. G., & McBain, A. J. (2002). Biofilms *in vitro* and *in vivo*: do singular mechanisms imply cross-resistance? *J Appl Microbiol* 92 Suppl, 98S-110S.

Gilbert, P., Maira-Litran, T., McBain, A. J., Rickard, A. H., & Whyte, F. W. (2002). The physiology and collective recalcitrance of microbial biofilm communities. *Adv Microb Physiol* 46, 202-256.

Goller, C., Wang, X., Itoh, Y., & Romeo, T. (2006). The cation-responsive protein NhaR of *Escherichia coli* activates *pgaABCD* transcription, required for production of the biofilm adhesin poly-beta-1,6-N-acetyl-D-glucosamine. *J Bacteriol* 188, 8022-8032.

Gophna, U., Barlev, M., Seijffers, R., Oelschlager, T. A., Hacker, J., & Ron, E. Z. (2001). Curli fibers mediate internalization of *Escherichia coli* by eukaryotic cells. *Infect Immun* 69, 2659-2665.

Gualdi, L., Tagliabue, L., Bertagnoli, S., Ierano, T., De Castro, C., & Landini, P. (2008). Cellulose modulates biofilm formation by counteracting curli-mediated colonization of solid surfaces in *Escherichia coli*. *Microbiology* 154, 2017-2024.

Hall-Stoodley, L., Costerton, J. W., & Stoodley, P. (2004). Bacterial biofilms: from the natural environment to infectious diseases. *Nat Rev Microbiol* 2, 95-108.

Hammar, M., Arnqvist, A., Bian, Z., Olsen, A., & Normark, S. (1995). Expression of two *csg* operons is required for production of fibronectin- and congo red-binding curli polymers in *Escherichia coli* K-12. *Mol Microbiol* 18, 661-670.

Harrison, J. J., Ceri, H., & Turner, R. J. (2007). Multimetal resistance and tolerance in microbial biofilms. *Nat Rev Microbiol* 5, 928-938.

Harrison, J. J., Turner, R. J., Marques, L., & Ceri, H. (2005). Biofilms: A new understanding of these microbial communities is driving a revolution that may transform the science of microbiology. *American Scientist* 93, 508-515.

Hengge, R. (2009). Principles of c-di-GMP signalling in bacteria. *Nat Rev Microbiol* 7, 263-273.

Hengge, R. (2010). Role of Cyclic Di-GMP in the Regulatory Networks of *Escherichia coli*, p. 230-252. *In* A. J. Wolfe and K. L. Visick (ed.), The Second Messenger Cyclic Di-GMP. ASM Press, Washington, D.C.

Hoffman, L. R., D'Argenio, D. A., MacCoss, M. J., Zhang, Z., Jones, R. A., & Miller, S. I. (2005). Aminoglycoside antibiotics induce bacterial biofilm formation. *Nature* 436, 1171-1175.

Itoh, Y., Rice, J. D., Goller, C., Pannuri, A., Taylor, J., Meisner, J.,. Beveridge, T. J , Preston, 3rd, J. F., & Romeo, T. (2008). Roles of *pgaABCD* genes in synthesis, modification, and export of the *Escherichia coli* biofilm adhesin poly-beta-1,6-N-acetyl-D-glucosamine. *J Bacteriol* 190, 3670-3680.

Itoh, Y., Wang, X., Hinnebusch, B. J., Preston, 3rd, J. F., & Romeo, T. (2005). Depolymerization of beta-1,6-N-acetyl-D-glucosamine disrupts the integrity of diverse bacterial biofilms. *J Bacteriol* 187, 382-387.

Jenal, U., & Malone, J. (2006). Mechanisms of cyclic-di-GMP signaling in bacteria. *Annu Rev Genet* 40, 385-407.

Jonas, K., Edwards, A. N., Simm, R., Romeo, T., Romling, U., & Melefors, O. (2008). The RNA binding protein CsrA controls cyclic di-GMP metabolism by directly regulating the expression of GGDEF proteins. *Mol Microbiol* 70, 236-257.

Jonas, K., Melefors, O., & Romling, U. (2009). Regulation of c-di-GMP metabolism in biofilms. *Future Microbiol* 4, 341-358.

Kader, A., Simm, R., Gerstel, U., Morr, M., & Romling, U. (2006). Hierarchical involvement of various GGDEF domain proteins in rdar morphotype development of *Salmonella enterica* serovar Typhimurium. *Mol Microbiol* 60, 602-616.

Karatan, E., & Watnick, P. (2009). Signals, regulatory networks, and materials that build and break bacterial biofilms. *Microbiol Mol Biol Rev* 73, 310-347.

Kirisits, M. J., Prost, L., Starkey, M., & Parsek, M. R. (2005). Characterization of colony morphology variants isolated from *Pseudomonas aeruginosa* biofilms. *Appl Environ Microbiol* 71, 4809-4821.

Lacey, M. M., Partridge, J. D., & Green, J. (2010). *Escherichia coli* K-12 YfgF is an anaerobic cyclic di-GMP phosphodiesterase with roles in cell surface remodelling and the oxidative stress response. *Microbiology* 156, 2873-2886.

Landini, P. (2009). Cross-talk mechanisms in biofilm formation and responses to environmental and physiological stress in *Escherichia coli*. *Res Microbiol* 160, 259-266.

Lewis, K. (2007). Persister cells, dormancy and infectious disease. *Nat Rev Microbiol* 5, 48-56.

Ma, Q., & Wood, T. K. (2009). OmpA influences *Escherichia coli* biofilm formation by repressing cellulose production through the CpxRA two-component system. *Environ Microbiol* 11, 2735-2746.

Malone, J. G., Jaeger, T., Spangler, C., Ritz, D., Spang, A., Arrieumerlou, C., Kaever, V., Landmann, R., & Jenal, U. (2010). YfiBNR mediates cyclic di-GMP dependent small colony variant formation and persistence in *Pseudomonas aeruginosa*. *PLoS Pathog* 6, e1000804.

Malone, J. G., Williams, R., Christen, M., Jenal, U., Spiers, A. J., & Rainey, P. B. (2007). The structure-function relationship of WspR, a *Pseudomonas fluorescens* response regulator with a GGDEF output domain. *Microbiology* 153, 980-994.

Mendez-Ortiz, M. M., Hyodo, M., Hayakawa, Y., & Membrillo-Hernandez, J. (2006). Genome-wide transcriptional profile of *Escherichia coli* in response to high levels of the second messenger 3',5'-cyclic diguanylic acid. *J Biol Chem* 281, 8090-8099.

Mulcahy, H., Charron-Mazenod, L., & Lewenza, S. (2008). Extracellular DNA chelates cations and induces antibiotic resistance in *Pseudomonas aeruginosa* biofilms. *PLoS Pathog* 4, e1000213.

O'Toole, G., Kaplan, H. B., & Kolter, R. (2000). Biofilm formation as microbial development. *Annu Rev Microbiol* 54, 49-79.

Ogasawara, H., Yamamoto, K., & Ishihama, A. (2011). Role of the biofilm master regulator CsgD in cross-regulation between biofilm formation and flagellar synthesis. *J Bacteriol* 193, 2587-2597.

Patel, R. (2005). Biofilms and antimicrobial resistance. *Clin Orthop Relat Res*, 41-47.

Paul, K., Nieto, V., Carlquist, W. C., Blair, D. F., & Harshey, R. M. (2010). The c-di-GMP binding protein YcgR controls flagellar motor direction and speed to affect chemotaxis by a "backstop brake" mechanism. *Mol Cell* 38, 128-139.

Paul, R., Weiser, S., Amiot, N. C., Chan, C., Schirmer, T., Giese, B., & Jenal, U. (2004). Cell cycle-dependent dynamic localization of a bacterial response regulator with a novel di-guanylate cyclase output domain. *Genes Dev* 18, 715-727.

Pesavento, C., Becker, G., Sommerfeldt, N., Possling, A., Tschowri, N., Mehlis, A., & Hengge, R. (2008). Inverse regulatory coordination of motility and curli-mediated adhesion in *Escherichia coli*. *Genes Dev* 22, 2434-2446.

Povolotsky, T. L., & Hengge, R. (2012). 'Life-style' control networks in *Escherichia coli*: Signaling by the second messenger c-di-GMP. *J Biotechnol* In press.

Price, N. L., & Raivio, T. L. (2009). Characterization of the Cpx regulon in *Escherichia coli* strain MC4100. *J Bacteriol* 191, 1798-1815.

Prigent-Combaret, C., Prensier, G., Le Thi, T. T., Vidal, O., Lejeune, P., & Dorel, C. (2000). Developmental pathway for biofilm formation in curli-producing *Escherichia coli* strains: role of flagella, curli and colanic acid. *Environ Microbiol* 2, 450-464.

Raivio, T. L. (2005). Envelope stress responses and Gram-negative bacterial pathogenesis. *Mol Microbiol* 56, 1119-1128.

Rao, F., Yang, Y., Qi, Y., & Liang, Z. X. (2008). Catalytic mechanism of cyclic di-GMP-specific phosphodiesterase: a study of the EAL domain-containing RocR from *Pseudomonas aeruginosa*. *J Bacteriol* 190, 3622-3631.

Robbe-Saule, V., Jaumouille, V., Prevost, M. C., Guadagnini, S., Talhouarne, C., Mathout, H., Kolb, A., & Norel, F. (2006). Crl activates transcription initiation of RpoS-regulated genes involved in the multicellular behavior of *Salmonella enterica* serovar Typhimurium. *J Bacteriol* 188, 3983-3994.

Romling, U. (2005). Characterization of the rdar morphotype, a multicellular behaviour in Enterobacteriaceae. *Cell Mol Life Sci* 62, 1234-1246.

Romling, U. (2002). Molecular biology of cellulose production in bacteria. *Res Microbiol* 153, 205-212.

Romling, U., & Amikam, D. (2006). Cyclic di-GMP as a second messenger. *Curr Opin Microbiol* 9, 218-228.

Romling, U., Bian, Z., Hammar, M., Sierralta, W. D., & Normark, S. (1998). Curli fibers are highly conserved between *Salmonella typhimurium* and *Escherichia coli* with respect to operon structure and regulation. *J Bacteriol* 180, 722-731.

Romling, U., Rohde, M., Olsen, A., Normark, S., & Reinkoster, J. (2000). AgfD, the checkpoint of multicellular and aggregative behaviour in *Salmonella typhimurium* regulates at least two independent pathways. *Mol Microbiol* 36, 10-23.

Ross, P., Weinhouse, H., Aloni, Y., Michaeli, D., Weinberger-Ohana, P., Mayer, R., Braun, S., de Vroom, E., van der Marel, G. A., van Boom, J. H., & Benziman, M. (1987). Regulation of cellulose synthesis in *Acetobacter xylinum* by cyclic diguanylic acid. *Nature* 325, 279-281.

Ryan, R. P., Fouhy, Y., Lucey, J. F., Crossman, L. C., Spiro, S., He, Y. W., Zhang, L. H., Heeb, S., Camara, M., Williams, P., & Dow, J. M. (2006). Cell-cell signaling in *Xanthomonas campestris* involves an HD-GYP domain protein that functions in cyclic di-GMP turnover. *Proc Natl Acad Sci U S A* 103, 6712-6717.

Ryder, C., Byrd, M., & Wozniak, D. J. (2007). Role of polysaccharides in *Pseudomonas aeruginosa* biofilm development. *Curr Opin Microbiol* 10, 644-648.

Ryjenkov, D. A., Simm, R., Romling, U., & Gomelsky, M. (2006). The PilZ domain is a receptor for the second messenger c-di-GMP: the PilZ domain protein YcgR controls motility in enterobacteria. *J Biol Chem* 281, 30310-30314.

Ryjenkov, D. A., Tarutina, M., Moskvin, O. V., & Gomelsky, M. (2005). Cyclic diguanylate is a ubiquitous signaling molecule in bacteria: insights into biochemistry of the GGDEF protein domain. *J Bacteriol* 187, 1792-1798.

Sanchez-Torres, V., Hu, H., & Wood, T. K. (2011). GGDEF proteins YeaI, YedQ, and YfiN reduce early biofilm formation and swimming motility in *Escherichia coli. Appl Microbiol Biotechnol* 90, 651-658.

Scher, K., Romling, U., & Yaron, S. (2005). Effect of heat, acidification, and chlorination on *Salmonella enterica* serovar Typhimurium cells in a biofilm formed at the air-liquid interface. *Appl Environ Microbiol* 71, 1163-1168.

Schirmer, T., & Jenal, U. (2009). Structural and mechanistic determinants of c-di-GMP signalling. *Nat Rev Microbiol* 7, 724-735.

Schmidt, A. J., Ryjenkov, D. A., & Gomelsky, M. (2005). The ubiquitous protein domain EAL is a cyclic diguanylate-specific phosphodiesterase: enzymatically active and inactive EAL domains. *J Bacteriol* 187, 4774-4781.

Simm, R., Lusch, A., Kader, A., Andersson, M., & Romling, U. (2007). Role of EAL-containing proteins in multicellular behavior of *Salmonella enterica* serovar Typhimurium. *J Bacteriol* 189, 3613-3623.

Simm, R., Morr, M., Kader, A., Nimtz, M., & Romling, U. (2004). GGDEF and EAL domains inversely regulate cyclic di-GMP levels and transition from sessility to motility. *Mol Microbiol* 53, 1123-1134.

Solano, C., Garcia, B., Valle, J., Berasain, C., Ghigo, J. M., Gamazo, C., & Lasa, I. (2002). Genetic analysis of *Salmonella enteritidis* biofilm formation: critical role of cellulose. *Mol Microbiol* 43, 793-808.

Sommerfeldt, N., Possling, A., Becker, G., Pesavento, C., Tschowri, N., & Hengge, R. (2009). Gene expression patterns and differential input into curli fimbriae regulation of all GGDEF/EAL domain proteins in *Escherichia coli. Microbiology* 155, 1318-1331.

Spiers, A. J., & Rainey, P. B. (2005). The *Pseudomonas fluorescens* SBW25 wrinkly spreader biofilm requires attachment factor, cellulose fibre and LPS interactions to maintain strength and integrity. *Microbiology* 151, 2829-2839.

Spoering, A., & Lewis, K. (2001). Biofilm and planktonic cells of *Pseudomonas aeruginosa* have similar resistance to killing by antimicrobials. *Journal of Bacteriology* 183, 6746-6751.

Starkey, M., Hickman, J. H., Ma, L., Zhang, N., De Long, S., Hinz, A., Palacios, S., Manoil, C., Kirisits, M. J., Starner, T. D., Wozniak, D. J., Harwood, C. S., & Parsek, M. R. (2009). *Pseudomonas aeruginosa* rugose small-colony variants have adaptations that likely promote persistence in the cystic fibrosis lung. *J Bacteriol* 191, 3492-3503.

Stevenson, G., Andrianopoulos, K., Hobbs, M., & Reeves, P. R. (1996). Organization of the *Escherichia coli* K-12 gene cluster responsible for production of the extracellular polysaccharide colanic acid. *J Bacteriol* 178, 4885-4893.

Stewart, P. S., & Costerton, J. W. (2001). Antibiotic resistance of bacteria in biofilms. *Lancet* 358, 135-138.

Stoodley, P., Sauer, K., Davies, D. G., & Costerton, J. W. (2002). Biofilms as complex differentiated communities. *Annu Rev Microbiol* 56, 187-209.

Tagliabue, L., Antoniani, D., Maciag, A., Bocci, P., Raffaelli, N., & Landini, P. (2010). The diguanylate cyclase YddV controls production of the exopolysaccharide poly-N-acetylglucosamine (PNAG) through regulation of the PNAG biosynthetic *pgaABCD* operon. *Microbiology* 156, 2901-2911.

Tamayo, R., Tischler, A. D., & Camilli, A. (2005). The EAL domain protein VieA is a cyclic diguanylate phosphodiesterase. *J Biol Chem* 280, 33324-33330.

Tschowri, N., Busse, S., & Hengge, R. (2009). The BLUF-EAL protein YcgF acts as a direct anti-repressor in a blue-light response of *Escherichia coli*. *Genes Dev* 23, 522-534.

Tuckerman, J. R., Gonzalez, G., Sousa, E. H., Wan, X., Saito, J. A., Alam, M., & Gilles-Gonzalez, M. A. (2009). An oxygen-sensing diguanylate cyclase and phosphodiesterase couple for c-di-GMP control. *Biochemistry* 48, 9764-9774.

Ueda, A., & Wood, T. K. (2010). Tyrosine Phosphatase TpbA of *Pseudomonas aeruginosa* Controls Extracellular DNA via Cyclic Diguanylic Acid Concentrations. *Environ Microbiol* 2, 449-455.

Van Houdt, R., & Michiels, C. W. (2005). Role of bacterial cell surface structures in *Escherichia coli* biofilm formation. *Res Microbiol* 156, 626-633.

Vogt, S. L., & Raivio, T. L. (2011). Just scratching the surface: an expanding view of the Cpx envelope stress response. *FEMS Microbiol Lett* 326, 2-11.

Wang, X., Dubey, A. K., Suzuki, K., Baker, C. S., Babitzke, P., & Romeo, T. (2005). CsrA post-transcriptionally represses *pgaABCD*, responsible for synthesis of a biofilm polysaccharide adhesin of *Escherichia coli*. *Mol Microbiol* 56, 1648-1663.

Wang, X., Preston, 3rd, J. F., & Romeo, T. (2004). The *pgaABCD* locus of *Escherichia coli* promotes the synthesis of a polysaccharide adhesin required for biofilm formation. *J Bacteriol* 186, 2724-2734.

Wang, X., Rochon, M., Lamprokostopoulou, A., Lunsdorf, H., Nimtz, M., & Romling, U. (2006). Impact of biofilm matrix components on interaction of commensal *Escherichia coli* with the gastrointestinal cell line HT-29. *Cell Mol Life Sci* 63, 2352-2363.

Waters, C. M., Lu, W., Rabinowitz, J. D., & Bassler, B. L. (2008). Quorum sensing controls biofilm formation in *Vibrio cholerae* through modulation of cyclic di-GMP levels and repression of *vpsT*. *J Bacteriol* 190, 2527-2536.

Weber, H., Pesavento, C.,. Possling, A , Tischendorf, G., & Hengge, R. (2006). Cyclic-di-GMP-mediated signalling within the sigma network of *Escherichia coli*. *Mol Microbiol* 62, 1014-1034.

Weber, H., Polen, T., Heuveling, J., Wendisch, V. F., & Hengge, R. (2005). Genome-wide analysis of the general stress response network in *Escherichia coli*: sigmaS-dependent genes, promoters, and sigma factor selectivity. *J Bacteriol* 187, 1591-1603.

Weinhouse, H., Sapir, S., Amikam, D., Shilo, Y., Volman, G., Ohana, P., & Benziman, M. (1997). c-di-GMP-binding protein, a new factor regulating cellulose synthesis in *Acetobacter xylinum*. *FEBS Lett* 416, 207-211.

White, A. P., Gibson, D. L., Grassl, G. A., Kay, W. W., Finlay, B. B., Vallance, B. A., & Surette, M. G. (2008). Aggregation via the red, dry, and rough morphotype is not a virulence adaptation in *Salmonella enterica* serovar Typhimurium. *Infect Immun* 76, 1048-1058.

White, A. P., Gibson, D. L., Kim, W., Kay, W. W., & Surette, M. G. (2006). Thin aggregative fimbriae and cellulose enhance long-term survival and persistence of *Salmonella*. *J Bacteriol* 188, 3219-3227.

Yamamoto, K., & Ishihama, A. (2006). Characterization of copper-inducible promoters regulated by CpxA/CpxR in *Escherichia coli*. *Biosci Biotechnol Biochem* 70, 1688-1695.

Yamamoto, K., & Ishihama, A. (2005). Transcriptional response of *Escherichia coli* to external copper. *Mol Microbiol* 56, 215-227.

Zahringer, F., Massa, C., & Schirmer, T. (2011). Efficient enzymatic production of the bacterial second messenger c-di-GMP by the diguanylate cyclase YdeH from *E. coli*. *Appl Biochem Biotechnol* 163, 71-79.

Zogaj, X., Nimtz, M., Rohde, M., Bokranz, W., & Romling, U. (2001). The multicellular morphotypes of *Salmonella typhimurium* and *Escherichia coli* produce cellulose as the second component of the extracellular matrix. *Mol Microbiol* 39, 1452-1463.

In: Prokaryotes
Editor: Marina Nisnevitch

ISBN: 978-1-63321-592-4

Chapter 3

Multidrug Transporter Proteins of the Major Facilitator Superfamily: Not Only for Drug Efflux

Kamela O. Alegre, Stephanie Paul and Christopher J. Law*
Queen's University Belfast, School of Biological Sciences, Belfast, United Kingdom

Abstract

Multidrug resistance in prokaryotes is due primarily to efflux of offending antimicrobials from the cell by representatives of several different families of integral membrane transporter proteins. Clearly, in evolutionary terms, these proteins did not arise specifically to pump human-made antimicrobials out of the cell and thereby confer resistance. Despite this, often only their role in antibiotic resistance is characterised and highlighted.

In recent years, however, a transition from the traditional anthropocentric perception of antibiotic resistance mechanisms in microorganisms has occurred, with naturally produced antimicrobials now generally regarded as physiologically important signalling molecules or sources of nutrition for bacteria rather than antimicrobial agents, and bacterial multidrug efflux proteins not merely as a defensive response to antimicrobials but as important players in fundamental physiological processes such as cellular homeostasis.

This emerging perspective supports the notion that a better understanding of the complexities of infection and multidrug resistance in bacteria can be achieved via a more detailed understanding of those physiological processes. In this chapter, we review the 'true' physiological roles of multidrug efflux proteins of the largest non-ATP-hydrolysing family of membrane transporters, the major facilitator superfamily, and explore the evidence for their function in processes such as pH and metal homeostasis, import and export of metabolites and biofilm formation.

Keywords: Multidrug resistance, integral membrane protein, substrate binding, transport

* Corresponding author Email: c.law@qub.ac.uk.

Introduction

The capacity of cells to protect themselves from simultaneous attack by a broad variety of antimicrobial compounds is termed multidrug resistance. The phenomenon of multidrug resistance in prokaryotes derives principally from the efflux activity of representatives of several different families of transporter protein that are integral to the cytoplasmic membrane. The antimicrobial efflux activity of these transporters translates into a major global health challenge, and the associated societal burden is significant; for example, in the USA alone, the economic cost of antimicrobial resistance is estimated at $55 billion per annum (Smith and Coast, 2013). Although it is their role in antimicrobial efflux that gains attention, it is now recognised that multidrug transporters also function in fundamental physiological processes in prokaryotes; it is those functions that likely represent the 'true' or 'natural' roles of the multidrug efflux proteins (Neyfakh, 1997; Piddock, 2006; Krulwich et al., 2005; Fluman and Bibi, 2009).

In bacteria, the multidrug 'effluxome' comprises transporters that belong to one of five families of proteins: the resistance-nodulation-division (RND) family; the ATP-binding cassette (ABC) superfamily; the multidrug and toxic compound extrusion (MATE) family; the small multidrug resistance (SMR) family, and; the major facilitator superfamily (MFS) (Paulsen, 2003). Multidrug transporters of the latter are single component drug/H^+ antiporters that generally consist of 400-600 amino acid residues. Although the sequence conservation between individual MFS transporters is low (Pao et al., 1998; Madej et al., 2013), all appear to share a common fold in which a central translocation pore is straddled by two pseudosymmetrical, transmembrane-spanning α-helical bundle domains, each containing either six or seven helices (Law et al., 2008; Vardy et al., 2004) (Figure 1).

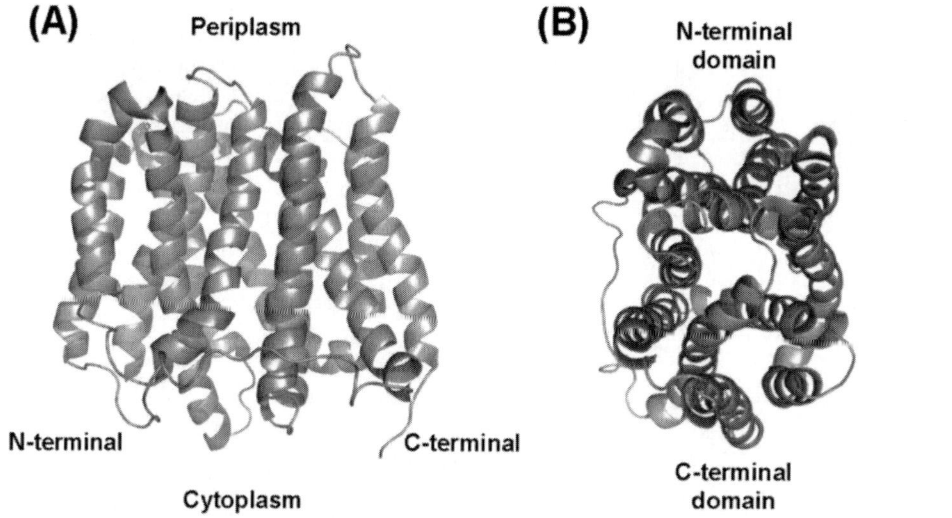

Figure 1. X-ray crystal structure of the 12 TMS multidrug efflux protein EmrD from *Escherichia coli* in occluded conformation. To date, this represents the only MFS multidrug efflux protein to have its structure solved. (A) View of EmrD parallel to the plane of the cytoplasmic membrane. (B) EmrD viewed perpendicular to the membrane from the periplasmic side. Note the two-fold pseudosymmetry of the N- and C-terminal domains. Image was constructed using PyMol (www.pymol.org) from PDB coordinates 2GFP (Yin et al., 2006).

The ability of multidrug efflux members of the MFS to recognise, bind, and subsequently transport a range of chemically and structurally heterogeneous compounds differentiates them from members of the family that transport only a single substrate or group of structurally related substrates (Saier et al., 1998).

Multidrug efflux members of the MFS are placed into one of two sub-families of the superfamily according to the number of times they span the membrane; those that possess 12 transmembrane spanning (TMS) α-helices are classified as belonging to the drug/H^+ antiporter 1 (DHA1) sub-family, and those that possess 14 TMS α-helices are categorised as belonging to the drug/H^+ antiporter 2 (DHA2) sub-family (Reddy et al., 2012). In contrast to the so-called primary transporters that derive their energy via hydrolysis of ATP, the MFS transporters are all energised by the electrochemical gradient generated by the primary transporters, and are therefore termed secondary active transporters (Law et al., 2008). The MFS is the largest superfamily of secondary active transporters with representatives found in all kingdoms of life and in all cell types (Saier et al., 1999; Reddy et al., 2012). Even in single organisms MFS multidrug transporters are abundant; for example, of the 37 proteins that constitute the *Escherichia coli* multidrug 'effluxome', 19 belong to the MFS (Nishino and Yamaguchi, 2001).

In bacterial genomes, the multiplicity of MFS and other family multidrug transporters, often with substantial overlapping substrate specificities (Tal and Schuldiner, 2009; Lewinson et al., 2006; Soo et al., 2011), poses an intriguing question: why does such redundancy exist in organisms that are usually parsimonious with respect to genome organisation and energy expenditure? Although there is some evidence that functional duplication of multidrug transporters may confer a protective advantage to cells in the event of dysfunction of one of them (Holdsworth and Law, 2012), it is likely that the redundancy is superficial and that each transporter actually performs a distinct and specific physiological function. We contend that an understanding of those physiological functions, and their origins, is essential for a fuller understanding of multidrug efflux, and hence, multidrug resistance in prokaryotes. This chapter therefore focuses on physiological or 'natural' roles that have been attributed to specific prokaryotic multidrug transporters of the MFS.

Evolution of Multidrug Efflux in the MFS

Given the ability of individual MFS multidrug proteins to transport a diversity of structurally dissimilar antimicrobials, it is not surprising that the same transporters have the potential to transport substrates of fundamental physiological relevance to the cell. A pertinent question to ask, therefore, is: did multidrug efflux proteins evolve recently as a protective mechanism against chemotherapeutic drugs, or did they arise from substrate-specific transporters that had been encoded in bacterial genomes for perhaps billions of years? The advent of genome sequencing projects over the past decade-and-a-half has illuminated possible answers to this question.

Genomic sequencing has led to the identification and classification of an expansive repertoire of multidrug efflux pumps across various bacterial species (Nikaido and Pages, 2012; Lubelski et al., 2007; Paulsen et al., 1996). The early work by Saier et al., analysed all known putative transport proteins in *E. coli*, *Haemophilus influenza* and *Mycoplasma genitalium*, as well as in the non-pathogenic *Bacillus subtilis*, *Methanococcus janneschii* and *Synechocystis PCC8603*, and revealed their occurrence in four families including the ABC and MFS superfamilies, and the smaller SMR and RND families (Saier et al., 1998). The first members of the MATE family of multidrug efflux proteins, NorM of *Vibrio parahaemolyticus* and its homologue in *E. coli*, were identified in 1998 (Morita et al., 1998), and more recently members have been identified spanning archaea, bacteria and eukaryotes (Kuroda and Tsuchiya, 2009; Omote et al., 2006).

Previous phylogenetic analysis of drug transporters revealed that the ability to confer drug resistance occurred independently, multiple times over an evolutionary time scale, and thus can account for the presence of the these transporters in evolutionarily unrelated families (Saier and Paulsen, 2001; Paulsen, 2003; Saier et al., 1999). It has been suggested that the bacterial multidrug transporters of the MFS were probably derived from a single primordial system, with the 14 TMS proteins being derived from a primordial 12 TMS protein early in evolution, prior to gene duplication and diversifying of the genes to form other members of the superfamily (Paulsen et al., 1996). While substrate specificity is a highly conserved evolutionary trait, multidrug efflux proteins do not seem to exhibit any apparent clustering within their respective superfamilies, either by substrate specificity or by multidrug efflux (Paulsen, 2003), a property exemplified by the MFS. Phylogenetic analysis of 30 bacterial MFS multidrug efflux proteins identified as being involved in various physiological processes including bile salt efflux, metal ion homeostasis, sulphur metabolism and spermidine efflux showed a clear delineation between members of the DHA1 and DHA2 subfamilies, but no coherent grouping within these subfamilies according to the class of compounds transported (Figure 2 and Table 1).

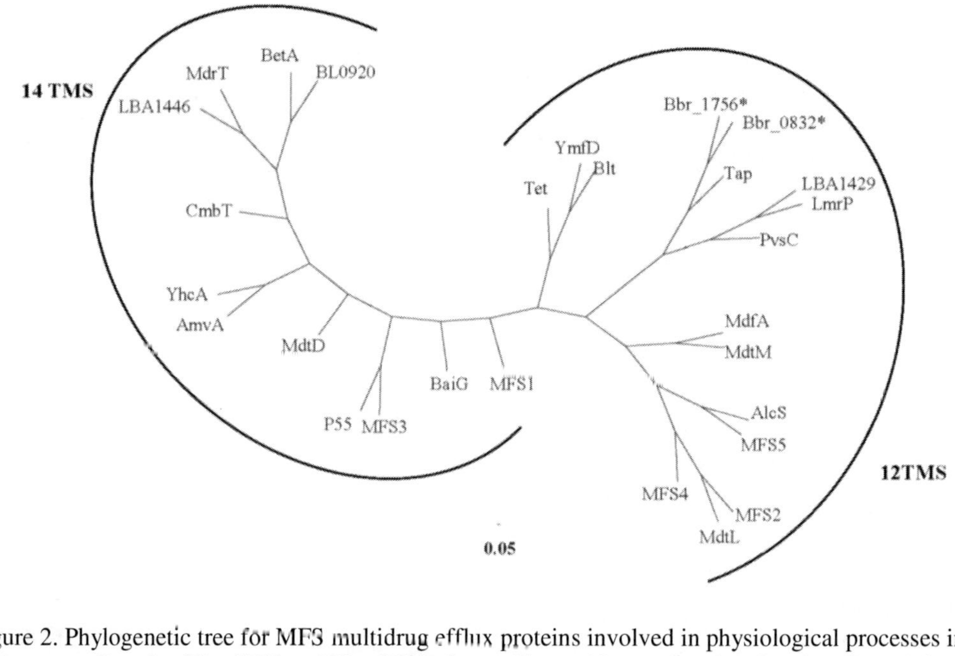

Figure 2. Phylogenetic tree for MFS multidrug efflux proteins involved in physiological processes in prokaryotes. Proteins Bbr_1756 and Bbr_0832, denoted by an (*), are closely related to the members of the 12 TMS (DHA1) family but have 10 and 14 transmembrane spanning regions, respectively. The tree was constructed using Fig Tree Graphical Viewer (http://tree.bio.ed.ac.uk/software/figtree/).

Table 1. Multidrug transporters of the MFS with known or putative roles in physiological processes in prokaryotes

Protein	Organism	TMS	Physiological substrate(s)	Physiological role(s)	Reference
AlcS (Bcr)*	*Bordetella pertussis*	12	Alcaligin	Alcaligin export	Brickman & Armstrong (2005)
AmvA	*Acinetobacter baumannii*	14	DC	Bile salt efflux	Rajamohan et al. (2010)
BaiG	*Eubacterium sp*	14	CA, CDCA	Bile salt import	Mallonee & Hylemon (1996)
Bbr_1756*	*Bifidobacterium breve*	10	CH, DC, GDC, GC, TC	Bile salt efflux	Ruiz et al. (2012)
Bbr_0832*	*Bifidobacterium breve*	14	GC	Bile salt efflux	Gueimonde et al. (2009)
BetA (Bbr_0838)	*Bifidobacterium breve*	14	Bile, CH	Bile and bile salt efflux	Gueimonde et al. (2009)
Blt	*Bacillus subtilis*	12	Spermidine	Spermidine efflux	Woolridge et al. (1997)
BL0920	*Escherichia coli*	14	Bile	Bile efflux	Gueimonde et al. (2009)
Ctr	*Bifidobacterium longum*	9	GC, CH	Bile salt efflux	Price et al. (2006)
CmbT	*Lactococcus lactis*	14	CH	Bile salt efflux, sulphur metabolism	Filipic et al. (2013)
LBA1429	*Lactobacillus acidophilus*	12	GC, TC	Bile salt efflux	Pfeiler & Klaenhammer (2009)
LBA1446	*Lactobacillus acidophilus*	14	GC, TC, GDC, TDC	Bile salt efflux	Pfeiler & Klaenhammer (2009)
LmrP	*Lactococcus lactis*	12	CH	Ca^{2+}/proton anitporter	Mazurkiewicz et al. (2005)
MdfA	*Escherichia coli*	12	Na^{+}, K^{+}	Alkaline pH homeostasis	Lewinson et al. (2004)
MdfA	*Salmonella enterica*	12		Biofilm formation	Baugh et al. (2012)
MdrT	*Listeria monocytogenes*	14	CA	Bile acid efflux	Quillan et al. (2011)
MdtD (IceT)*	*Salmonella typhimurium*	12	Fe^{2+}	Fe^{2+} efflux	Frawley et al. (2013)
MdtK	*Salmonella enterica*	12		Biofilm formation	Baugh et al. (2012)
MdtL	*Escherichia coli*	12	Adenine	pH homeostasis, purine transport	Cho et al. (2011)
MdtM	*Escherichia coli*	12	Na^{+}, K^{+}, Li^{+}, CH, DC	Alkaline pH homeostasis, putative roles in K^{+} and Li^{+} tolerance, bile salt efflux	Holdsworth & Law (2013) Paul et al. (2014)

Table 1. (Continued)

Protein	Organism	TMS	Physiological substrate(s)	Physiological role(s)	Reference
PvsC*	*Vibrio parahaemolyticus*	12	Vibrioferrin	Vibrioferrin secretion	Tanabe et al. (2006)
TetK	*Staphylococcus aureus*	14	Na^+, K^+	K^+ and Na^+ acquisition, alkaline pH homeostasis	Ginn et al. (2000)
TetL	*Escherichia coli*	14	Na^+, K^+	K^+ and Na^+ acquisition, alkaline pH homeostasis	Guffanti & Krulwich (1995)
YhcA*	*Erwinia chrysanthemi*	12	Achromobactin	Achromobactin secretion	Franza et al. (2005)
YmfD*	*Bacillus subtilis*	12	Bacillibactin	Bacillibactin export	Miethke et al. (2008)

Proteins marked by an (*) denote putative multidrug resistance proteins whose proposed functions are based on the presence of conserved amino acid motifs, structural features or limited homologies. *Abbreviations*: CA, cholic acid; CDCA, chenodeoxycholic acid; CH, cholate; DC, deoxycholate; GC, glycocholate; GDC, glycodeoxycholate; TC, taurocholate; TDC, taurodeoxycholate

Furthermore, although the phylogenetic analysis suggested that bacterial MFS multidrug transporters evolved at different times, functional studies have revealed overlapping physiological roles for several of the proteins. For example, MdfA, MdtL and MdtM from *E. coli* and TetL from *B. subtilis* all function in alkaline pH homeostasis (Holdsworth and Law, 2013; Hayes et al., 2006; Lewinson et al., 2004; Cheng et al., 1996a), and BL0920 from *Bifidobacterium longum* NCC2705 and *E. coli* MdtM have been identified as functioning in bile salt efflux (C.J. Law, unpublished data; Gueimonde et al., 2009). It seems, therefore, that at least some MFS multidrug transporters have developed similar physiological functions in order to respond to various environmental stresses, and it is likely these physiological functions emerged prior to multidrug resistance (Saier et al., 1998; Neyfakh, 1997). This contention is contrary to the one that bacteria developed these systems in response to increased exposure to medically relevant drugs. Moreover, comparison of both pathogenic and non-pathogenic bacterial genomes has revealed that the numbers of chromosomally encoded multidrug efflux systems in both systems are comparable (Saier et al., 1998), making dubious any claim that these transporters have arisen recently.

Physiological Roles of MFS Multidrug Transporters

The evolutionary relationship between MFS multidrug transporters and their substrate-specific counterparts provides support that the former play a role in specific physiological processes, and it is those physiological roles that foster the persistence of multidrug transporters in bacterial genomes. However, due to the profound clinical impact of multidrug resistance, it is generally only the transporters' roles in the drug efflux process that gains attention, with the result that physiological functions unrelated to multidrug efflux are often poorly, if at all, characterised (Fluman and Bibi, 2009). This chapter attempts to provide a contemporaneous overview of specific physiological roles that have been attributed to prokaryotic multidrug transporters of the MFS.

Alkaline pH Homeostasis

To survive and grow in alkaline environments, neutralophilic bacteria such as *E. coli* and *B. subtilis* must maintain a cytoplasmic pH - in the narrow range of between 7.4 and 7.8 - that is acidic relative to that of the external environment (Slonczewski et al., 2009; Krulwich et al., 2011). Because cytoplasmic pH is, by definition, inextricably linked to cytoplasmic proton concentration, amongst the strategies employed by neutralophilic bacteria to achieve alkaline pH homeostasis is the upregulation of expression and/or activity of secondary active cation/H^+ antiporters that harvest protons (Krulwich et al., 1998; Krulwich et al., 1994; Krulwich et al., 2009; Booth, 1985). Of the suite of antiporters from various protein families that are identified as functioning in alkaline pH tolerance in bacteria, four are multidrug efflux proteins of the MFS; MdfA, MdtM, TetL and TetK (Lewinson et al., 2004; Holdsworth and Law, 2013; Jin et al., 2002a; Guffanti and Krulwich, 1995; Krulwich et al., 2001).

MdfA (notated in the earlier literature as Cmr) (Nilsen et al., 1996) is a 410 amino acid multidrug transporter from *E. coli* (Lewinson et al., 2004; Bibi et al., 2001) that has orthologues in pathogenic bacteria such as *Shigella flexneri* (99% sequence identity at the amino acid level) (Jin et al., 2002b) and *Salmonella enterica* serovar Typhi (90% sequence identity) (Parkhill et al., 2001). A physiological role for MdfA in alkaline pH resistance was a serendipitous discovery made from observations that an MdfA-dependent *E. coli* phenotype was able to grow at basic pH values normally compatible only with growth of alkaliphilic species (Lewinson et al., 2004). Cultures of *E. coli* with chromosomal *mdfA* deleted failed to grow above pH 8.5, whereas the same mutants expressing the protein from a multicopy plasmid recovered a wild-type phenotype and grow relatively well at pH 10. This alkaline resistance was dependent upon the presence of tens of millimolar concentrations of either sodium or potassium ions in the growth medium, leading to the conclusion that MdfA catalyses a low-affinity $Na^+(K^+)/H^+$ antiport to maintain a stable cytoplasmic pH, acidic relative to outside, when cells are exposed to certain alkaline conditions. MdfA operates under conditions in which previously identified monovalent metal cation/H^+ antiporters known to support alkaline resistance in *E. coli* (such as NhaA, a non-multidrug resistance Na^+/H^+ antiporter) do not operate (Padan et al., 2005; Padan et al., 1999; Padan et al., 2004; Lewinson et al., 2004). The relationship between the multidrug efflux and physiological monovalent cation/H^+ transport activities of MdfA was highlighted by transport assays that revealed that mutant protein defective in antimicrobial transport failed to support growth of *E. coli* cultures at alkaline pH (Lewinson et al., 2004). It should be noted, however, that in contrast to the situation that occurs when MdfA (and MdtM, see below) functions in multidrug efflux and forms a functional interaction with the tripartite RND family transporter AcrAB-TolC to expel cytotoxins from the cell (Lee et al., 2000; Tal and Schuldiner, 2009), the physiological, alkaline resistance function of the single component MFS transporters is independent of any such interaction(s) (Lewinson et al., 2004). A close *E. coli* homologue of MdfA (41% identity at the amino acid level) (Burland et al., 1995; Bibi et al., 2001), originally called YjiO but now known as MdtM, is also a multidrug efflux protein of the MFS that functions in alkaline pH resistance (Holdsworth and Law, 2013). Like MdfA, MdtM catalyses a low-affinity $Na^+(K^+)/H^+$ antiport, operative only within a relatively narrow basic pH window of between pH 9 and pH 10, that is used to harvest protons for buffering cytoplasmic pH (Holdsworth and Law, 2013). The capacity of MdtM to utilise either Na^+ or K^+ as counterions for H^+ uptake enables the bacterial cells to respond efficiently to changes in the monovalent metal cation composition of the growth environment. While the contribution of K^+ concentration to pH homeostasis in *E. coli* is not clearly understood (Booth, 1985), it is feasible that the K^+/H^+ antiport function of MdtM could regulate cytoplasmic pH by utilising the outwardly-directed K^+ gradient to energise proton harvest during growth in an alkaline medium (Plack and Rosen, 1980). The same K^+/H^+ antiport mode may also play a role in metal ion homeostasis in *E. coli* by protecting cells from the cytotoxic effects of high intracellular concentrations of potassium (Holdsworth and Law, 2013).

Although the driving force for MdfA-catalysed $Na^+(K^+)/H^+$ antiport has not been characterised, the same reaction as catalysed by MdtM is electrogenic with > 1 H^+ exchanged per monovalent metal cation (Holdsworth and Law, 2013). Given the close homology between MdfA and MdtM it is perhaps not surprising that both perform a similar physiological function. The similarity of the physiological, alkaline resistance functions of MdfA and MdtM raises questions as to why *E. coli* has such apparent redundancy with respect to these transporters. The answer to this question may lie in the fact that neither chromosomal *mdtM* or *mdfA* can compensate for loss of the other in an alkaline resistance role, suggesting a mechanism by which the cell can utilise the individual transporters to fine tune its response to subtly different environmental challenges (Holdsworth and Law, 2013).

Despite their structural and functional similarities, there are some significant catalytic differences between MdfA and MdtM; for example, in contrast to MdfA, MdtM can also catalyse transport of lithium at alkaline pH via a Li^+/H^+ antiport that may be physiologically important for pH or metal ion homeostasis when Li^+ is present in the environment (Holdsworth and Law, 2013).

The two other MFS transporters confirmed as functioning in pH homeostasis in bacteria are the 14 TMS Tet proteins; TetL from Gram-positive *B. subtilis* (Guffanti and Krulwich, 1995) and TetK from Gram-positive *Staphylococcus aureus* (Ginn et al., 2000). Although in a strict sense these Tet proteins are not true multidrug resistance proteins (their drug substrates are restricted to tetracycline and its derivatives) (Speer et al., 1992), they do display the multi-functionality exhibited by the *bona fide* multidrug transporters; other than tetracycline efflux and alkaline pH resistance, TetL and TetK play significant physiological roles in combating Na^+ stress and in K^+ acquisition (Krulwich et al., 2001). The alkaline resistance roles of TetL and TetK are facilitated by an electrogenic $Na^+(K^+)/H^+$ antiport that moves a net positive charge into the cell during each antiport cycle (Guffanti and Krulwich, 1995; Guffanti et al., 1998; Cheng et al., 1996a; Cheng et al., 1996b).

Metal Ion Homeostasis

Metal ions are pervasive at all levels of bacterial metabolism and play pivotal roles in the structural maintenance of biomolecules, processes involving redox reactions such as respiration, and cell signalling and sensory mechanisms (Silver and Phung, 1996). At the same time, exposure to toxic metals or accumulation of metals within the cell can have deleterious effects on the cell membrane, proteins and DNA (Bruins et al., 2000). As such, metal ion homeostasis in bacteria involves regulation of the uptake, storage and secretion of metal ions. Metal ion transporters belonging to the ABC and MFS superfamilies play a key role in maintaining cellular metal ion concentrations through their involvement in metal acquisition and secretion from bacterial cells (Nelson, 1999).

Metal ion acquisition for bacteria is often limited due to the low solubility and hence low bioavailability of metal ions, such as iron. Moreover, for pathogenic bacteria, innate immune responses to bacterial infections that sequester transition metals ("nutritional immunity") render metal ion acquisition for these microbes all the more challenging (Hood and Skaar, 2012). One of the mechanisms that bacteria have developed to overcome nutrient restriction of iron is through the secretion of siderophores. Siderophores are high affinity, iron-chelating compounds secreted by bacteria to forage for iron by formation of soluble Fe^{3+}-complexes that can subsequently be taken up by bacterial cells (Miethke and Marahiel, 2007). Several MFS multidrug transporters in bacteria have been identified to secrete a wide range of structurally diverse siderophores; these include the alcagin transporter AlcS of *Bordetella pertussis* (Brickman and Armstrong, 2005), the virioferrin transporter PvsC of *Vibrio parahaemolyticus* (Tanabe et al., 2006), the achromobactin transporter YhcA of *Erwinia chrystanthemi* (Franza et al., 2005), and the bacilibactin transporter YmfD of *B. subtilis* (Miethke et al., 2008).

In another metal ion homeostasis role, the bacterial MFS multidrug transporters LmrP (Mazurkiewicz et al., 2005) and IceT, formerly designated MdtD, (Frawley et al., 2013) have been identified as functioning in transport of redox non-active calcium and redox-active iron ions, respectively (Nelson, 1999). As well as extruding a wide range of chemically unrelated compounds, including antibiotics and lipophilic cationic compounds, from the cytosol (Putman et al., 2001; Mazurkiewicz et al., 2002), LmrP of *Lactococcus lactis* is capable also of specifically binding calcium, resulting in an inhibition of multidrug resistance in *L. lactis* (Schaedler and van Veen, 2010; Schaedler et al., 2012). IceT is a recently characterised efflux pump in *S. typhimurium,* thought to link the central metabolism of citrate (an iron-chelating tricarboxylic acid cycle intermediate) to an increased resistance profile to clinically relevant antibiotics and the nitrositive and oxidative stress agents nitric oxide and hydrogen peroxide (Frawley et al., 2013). Given that iron and calcium are both involved in numerous bacterial cellular processes, this evidence suggests physiological roles aside from multidrug resistance for these transporters, thus expanding on the clinical relevance of these transporters in bacterial pathogenicity (Piddock, 2006).

Bile Salt Resistance

While the molecular mechanisms involved in the bacterial genetic response to some cellular stresses, such as pH resistance, in the gastrointestinal tract are fairly well understood, the bacterial response to bile acid resistance is less so. Aside from action on the bacterial cell membrane, bile has various other nefarious effects on the cell, including but not limited to: formation of secondary structures in mRNA, activation of the DNA damage response and of proteins involved in DNA repair, protein misfolding and aggregation (Merritt and Donaldson, 2009; Bernstein et al., 1999). Given the complex nature of this type of stress, it is likely that the bacterial response to bile involves multiple mechanisms and various signalling pathways (Thanassi et al., 1997; Nikaido and Takatsuka, 2009; Begley et al., 2003).

The major function of bile is to emulsify liquids in the digestive tract and component bile salts, which are detergent-like molecules usually found conjugated to Na^+ ions, aid in this process as well as play an important role as a native bactericide within the human gut (Gunn, 2000). Bile salt resistance is essential for bacteria residing in the intestinal tract, with bile salt concentrations of up to 20 mM in the duodenum and between 0.2 and 20% (depending on the individual and the type of food ingested) weight to volume in the small intestine (Love and Dawson, 1998).

As such, enteric bacteria like *E. coli* have developed mechanisms to withstand such conditions while passing through or colonising the gut, and one of these mechanisms is the extrusion of bile salts through multidrug transporters (Piddock, 2006). As previously mentioned, many multidrug resistance pumps of the MFS are involved in the efflux of various structurally and chemically unrelated compounds, and bile salt efflux by multidrug transporters has been measured in both Gram-negative and Gram-positive bacteria (Thanassi et al., 1997; Piddock, 2006). While Gram-positive bacteria are generally accepted as less tolerant to the effects of bile salts than the Gram-negative bacteria, bile salts are often used in enrichment media to inhibit the growth of both Gram-negative and Gram-positive bacterial strains (Kimoto-Nira et al., 2012; Ruiz et al., 2012; Bustos et al., 2012).

Of the 13 MFS multidrug transporters involved in bile salt export and reported here, ten are from Gram-positive bacteria (one from *Eubacterium sp,* three from *Bifidobacterium breve,* two each from *Bifidobacterium longum, L. lactis* and *L. acidophilus*, and one from *Listeria monocytogenes;* refer to Table 1), while only three are from Gram-negative strains (one from *Acinetobacter baumannii* and two from *E. coli*). Within each Gram stain group, it has been observed that bile acid tolerance is not a trait shared among members of the same species, but rather is strain-specific.

For example, of 124 strains of *Bifidobacterium* tested in response to gastrointestinal stresses including bile salt exposure, all but 24 (which exhibited moderate to weak tolerance) were inhibited (Andriantsoanirina et al., 2013). Not surprisingly, *B. breve* and *B. longum,* two of the *Bifido* representatives that showed the highest number of strains moderately tolerant to bile, both contain strains which express MFS transporters involved in bile salt efflux (Bbr_1756, Bbr_0832 and BetA in *B. breve* UCC2003, Ctr in *B. longum* NCIMB 70225 and BL0920 in *B. longum* NCC2705) (Ruiz et al., 2012; Andriantsoanirina et al., 2013; Gueimonde et al., 2009; Price et al., 2006).

In addition to contributing to bile salt efflux, an MFS multidrug transporter from *Eubacterium* sp. strain VPI 12708 has been shown to function in the opposite reaction; that of energy-dependent bile acid uptake driven by the proton motive force across the cell membrane (Mallonee and Hylemon, 1996). This protein, which is part of the bile acid-inducible *bai* operon and is designated as BaiG, shows significant homology to the TetA proteins from Gram-positive bacteria, and, like these proteins, possesses 14 TMS (Mallonee and Hylemon, 1996).

In a reflection of the substrate promiscuity associated with the MFS multidrug transporters (Lewinson et al., 2006; Wong et al., 2014), the BaiG transporter can catalyse uptake of a range of bile acids including cholic acid, chenodeoxycholic acid, cholylglycine, 7-oxocholic acid and deoxycholic acid (Mallonee and Hylemon, 1996).

Sulphur Metabolism

Sulphur is found in all organisms, primarily as a component of amino acids methionine and cysteine and their oxidation products, and in the form of various types of sulphonates (Cook et al., 2006; van der Ploeg et al., 2001). It also forms an integral part of *S*-adenosyl methionine (SAM), a co-substrate that plays a key role in polyamine biosynthesis and is involved in the methylation of over 40 different substrates within the cell including nucleic acids, proteins and lipids (Roje, 2006; Schubert et al., 2003). Despite its abundance in the environment, comprising approximately ~0.1% of the earth's crust, inorganic sulphur is inaccessible for direct biosynthetic incorporation and must first be processed via plant and bacteria assimilation pathways (Muyzer and Stams, 2008; Stipanuk, 1986). The genes involved in sulphur metabolism in Gram-negative *E. coli* and *S. enterica* have been highly scrutinised; the *cys* regulon comprises at least 22 sulphur metabolism genes involved in sulphate and sulphonate assimilation and cysteine biosynthesis (Rocha et al., 2000; Colyer and Kredich, 1996). In Gram-positive *B. subtilis, Clostridium acetobutylicum* and *S. aureus,* the S box regulon regulates genes involved in cysteine and methionine biosynthesis, though the general mechanisms involved in sulphur assimilation are poorly understood in this strain (Grundy and Henkin, 1998).

While there has been much study into how sulphur is integrated into the cell, little work has focused on how sulphate and sulphate-containing metabolic by-products are exported out of the cell. It has been suggested that sulphur transporters would represent members of one of the poorly understood ion channels. One such candidate is the *suzY* gene in the NKNCYSA strain of *Paracoccus pantotrophus*, which shares no similarity to any gene in the Transport Classification Database (www.tcdb.org) (Saier et al., 2006; Saier et al., 2009), and whose gene product is thought to represent a sulphate exporter needed to maintain homeostasis during desulfonation (Rein et al., 2005). A better characterized example is the CmbT transporter from *L. lactis.* The *metC-cysK* operon was characterised by Fernandez *et al.* (2002) and, using random mutagenesis, two genes (*cmbT* and *cmbR*) were found to be involved in the regulation of *metC-cysK* expression (Fernandez et al., 2002). The *cmbR* gene is essential for activation of *metC-cysK* transcription, while *cmbT* is predicted to encode a transmembrane protein.

Further characterisation of the *cmbT* gene product revealed that the CmbT protein is a multidrug transporter of the MFS, with a potential role in sulphur metabolism in *L. lactis* (Filipic et al., 2013). Thiol assimilation and production has been shown to play an important role in sulphur regulation in *L. lactis* and CmbT overproduction was shown to cause an increase in thiol production in media with limited Met/Cys concentrations (Fernandez et al., 2000; Fernandez et al., 2002). Interestingly, CmbT was also shown to cause lactococcal resistance to a wide range of antibiotics including the sulphur-containing lincosamide, antibiotic lincomycin and the sulphonamide antibiotics sulbactam, sulfamethoxazole and sulfadiazine (Filipic et al., 2013). It is possible, therefore, that transport of physiologically relevant sulphur intermediates has led to the indiscriminate efflux of other compounds, leading to the multidrug resistance profile seen in *L. lactis,* although the mechanism or evolutionary pressure for such an action is unclear. Sulphur exporters must be present at all phylogenetic levels due to the cellular requirement of sulphate excretion during organosulphate and sulphur ester processing, but whether multidrug transport is a recurring theme remains to be seen (Rein et al., 2005).

Polyamine Transport

Polyamines (putrescine, spermidine and spermine) are organic compounds, synthesised by all living cells and essential for cell growth and differentiation (Tabor and Tabor, 1985; Pegg, 1986). Spermidine and spermine are synthesised from putrescine and are the predominant polyamines present in eukaryotic cells, while putrescine and spermidine predominate in prokaryotic cells (Igarashi and Kashiwagi, 2000). The cellular content and metabolism of polyamines is tightly regulated by biosynthesis, degradation and transport (Igarashi and Kashiwagi, 1999) and their concentrations are cell growth and cell-cycle dependent in bacteria and eukaryotes, respectively (Pegg, 1986; Ishii et al., 1994). Spermidine toxicity has been shown to inhibit protein synthesis and cell growth in *E. coli* thus polyamine catabolism, as well as anabolism, must be tightly regulated within bacterial cells (Higashi et al., 2008; Limsuwun and Jones, 2000). Members of the ABC transporter superfamily implicated in spermidine and spermidine derivative export have been identified in *E. coli* (Furuchi et al., 1991; Kashiwagi et al., 1990) and *S. pneumonia* (Shaw et al., 2010), while less is known about spermidine efflux pumps of the RND and major facilitator superfamilies (Chan and Chua, 2010; Woolridge et al., 1997).

Blt, or Bmr-like transporter, from *B. subtilis* is the only single-component member of the MFS family of multidrug transporters with known involvement in spermidine efflux (Woolridge et al., 1997). It shares 51% sequence similarity with *B. subtilis* multidrug resistance protein Bmr, and though both proteins display resistance to the same spectrum of drugs, their expression patterns are dissimilar (Ahmed et al., 1993; Neyfakh, 1992; Neyfakh et al., 1991). While Bmr is transcribed as a monocistronic mRNA, *blt* forms a single operon with *bltD*, whose gene product is a spermidine/spermine acetyltransferase involved in spermidine metabolism (Woolridge et al., 1997). The close proximity of the genes encoding for the spermidine efflux transporter and the spermidine acetyltransferase suggests a physiological role for the *blt-bltD* operon in regulating the amount of spermidine within the cell, whereby spermidine content is down-regulated by both expression of an efflux protein and chemical modification of spermidine into downstream metabolic by-products (Woolridge et al., 1997).

Spermine and spermidine have been shown to increase the susceptibility to β-lactam antibiotics in Gram-negative *P. aeruginosa*, *E. coli*, and *S. enterica* serovar Typhimurium and in Gram-positive *S. aureus* (Kwon and Lu, 2007). Furthermore, *E. coli* cells lacking spermidine were found to be less susceptible to the effects of novobiocin and certain aminoglycocides (Dhamdhere et al., 2010). In *B. subtilis*, overproduction of the Blt transporter results in increased spermidine efflux, as well as a multidrug resistant phenotype (Woolridge et al., 1997). It seems spermidine can synergistically modulate the susceptibility to certain antibiotics, and this modulation likely occurs at antibiotic-specific binding sites. On the other hand, where multidrug resistant profiles can be seen as in the case of Blt, multidrug efflux is likely opportunistic, involving non-specific substrate recognition.

Purine Transport

Purine and pyrimidine nucleotides not only constitute part of the nucleic acids but they also play critical roles in enzymatic reactions and cell signaling processes as well as acting as the principal carriers of cellular energy (Rolfes, 2006) . In enterobacteria, purine metabolism is regulated by the PurR transcription factor which acts directly on the *de novo* biosynthesis of pyrimidine and the transport of purine nucleotides (Cho et al., 2011). In response to exogenous adenine, PurR binds to upstream regions of the *xanP*, *mdtL* and *yieG* genes but does not bind to *xanP* or *mdtL* in the absence of exogenous adenine (Cho et al., 2011). These results indicated that *yieG* encodes a high-affinity transport system for adenine, which is redundant in the presence of an excess of the substrate.

Furthermore, it was suggested that the *mdtL* gene, the product of which is a multidrug transporter of the MFS, might be responsible for the low-affinity adenine transport (Cho et al., 2011). Thus, the transport system of the PurR regulon probably consists of two nucleobase: cation symporter-2 (NCS2) family proteins (encoded by *xanP* and *yieG*) that function as high affinity purine transporters and a single MFS transporter (MdtL) that functions as a low affinity purine transporter.

Interestingly, the *mdtL* gene product may also function physiologically in pH homeostasis; although in aerated *E. coli* cultures there appears to be no pH dependence of *mdtL* up-regulation, it is dependent upon acidic pH under anaerobic conditions (Hayes et al., 2006). However, further studies are required to confirm this physiological role for MdtL.

Biofilm Formation

Biofilms are complex communities of microbially-originated, matrix-enclosed, three-dimensional accretions that can adhere to both biotic and abiotic surfaces (Branda et al., 2005; Costerton et al., 1995). Sessile biofilm bacteria predominate in nature and, due to their much greater resistance to antimicrobial chemotherapies compared to their planktonic counterparts, are of intense clinical significance (Costerton et al., 1987; Gomez and Prince, 2007; Diederen, 2008; Lovegrove, 2004). Biofilm cells also differ fundamentally from free-living planktonic cells in that they display a different phenotype due to derepression of expression of a large number of genes (An and Parsek, 2007; Lazazzera, 2005).

Transporters of all five main families (ABC, MFS, MATE, RND and SMR) of multidrug efflux proteins have been recognised as playing a role in biofilm formation (Kvist et al., 2008; Baugh et al., 2012). In uropathogenic *E. coli* strains 83972 and VR50, expression of three MFS multidrug efflux genes (*mdtL, mdtG* and *mdfA*) is upregulated during biofilm growth (Kvist et al., 2008), although the specific contribution of these particular gene products to production of biofilm is not known.

However, studies on biofilm formation by *Salmonella* Typhimurium revealed that deletion or dysfunction of any one of the bacterium's complement of multidrug transporters abrogated biofilm formation and resulted in a common phenotype (Baugh et al., 2012). At least two multidrug efflux members of the MFS (MdtK and MdfA) have been implicated as having a physiological role in biofilm formation. Although, as in the case of *E. coli* described above, the mechanistic role played by these transporters in biofilm formation in *Salmonella* is not understood, it may involve interplay between regulation of efflux and the biosynthesis of curli, the latter being essential proteinaceous components of the *Salmonella* biofilm extracellular matrix (Baugh et al., 2012; Garcia et al., 2004; Jonas et al., 2007; Cogan et al., 2004).

Conclusion

We hope that this chapter has provided some insight into the astonishingly broad spectrum of physiologically relevant roles played by prokaryotic multidrug transporters of the MFS. Given the clinical, industrial and environmental importance of these structurally conserved but extremely functionally diverse superfamily of transporter proteins, much more investigative work into their molecular biology, biochemistry and biophysics is clearly warranted. Recent advances in sequencing technology that has resulted in disclosure of many more genomes, combined with advances in membrane protein structural biology, means that we are now poised to gain that fuller understanding of these proteins.

Acknowledgments

This work was made possible by financial support (grant reference BB/K014226/1) from the Biotechnology and Biological Sciences Research Council (BBSRC).

References

Ahmed, M., Borsch, C.M., Neyfakh, A.A., Schuldiner, S. (1993). Mutants of the Bacillus subtilis multidrug transporter Bmr with altered sensitivity to the antihypertensive alkaloid reserpine. *J. Biol. Chem.* 268, 11086-11089.

An, D., Parsek, M.R. (2007). The promise and peril of transcriptional profiling in biofilm communities. *Curr. Opin. Microbiol.* 10, 292-296.

Andriantsoanirina, V., Allano, S., Butel, M.J., Aires, J. (2013). Tolerance of Bifidobacterium human isolates to bile, acid and oxygen. *Anaerobe.* 21, 39-42.

Baugh, S., Ekanayaka, A.S., Piddock, L.J., Webber, M.A. (2012). Loss of or inhibition of all multidrug resistance efflux pumps of Salmonella enterica serovar Typhimurium results in impaired ability to form a biofilm. *J. Antimicrob. Chemother.* 67, 2409-2417.

Begley, M., Hill, C., Gahan, C.G. (2003). Identification and disruption of btlA, a locus involved in bile tolerance and general stress resistance in Listeria monocytogenes. *FEMS Microbiol. Lett.* (218), 31-38.

Bernstein, C., Bernstein, H., Payne, C.M., Beard, S.E., Schneider, J. (1999). Bile salt activation of stress response promoters in Escherichia coli. *Curr. Microbiol.* 39, 68-72.

Bibi, E., Adler, J., Lewinson, O., Edgar, R. (2001). MdfA, an interesting model protein for studying multidrug transport. *J. Mol. Microbiol. Biotechnol.* 3, 171-177.

Booth, I.R. (1985). Regulation of cytoplasmic pH in bacteria. *Microbiol Revs.* 49, 359-78.

Branda, S.S., Vik, S., Friedman, L., Kolter, R. (2005). Biofilms: the matrix revisited. *Trends Microbiol.* 13, 20-26.

Brickman, T.J., Armstrong, S.K. (2005). Bordetella AlcS transporter functions in alcaligin siderophore export and is central to inducer sensing in positive regulation of alcaligin system gene expression. *J. Bacteriol.* 187, 3650-3661.

Bruins, M.R., Kapil, S., Oehme, F.W. (2000). Microbial resistance to metals in the environment. *Ecotox. Env. Saf.* 45, 198-207.

Burland, V., Plunkett, G., 3rd, Sofia, H.J., Daniels, D.L., Blattner, F.R. (1995). Analysis of the Escherichia coli genome VI: DNA sequence of the region from 92.8 through 100 minutes. *Nuc. Acid Res.* 23, 2105-2119.

Bustos, A.Y., Saavedra, L., De Valdez, G.F., Raya, R.R., Taranto, M.P. (2012). Relationship between bile salt hydrolase activity, changes in the internal pH and tolerance to bile acids in lactic acid bacteria. *Biotech. Letts.* 34, 1511-1518.

Chan, Y.Y., Chua, K.L. (2010). Growth-related changes in intracellular spermidine and its effect on efflux pump expression and quorum sensing in Burkholderia pseudomallei. *Microbiology.* 156, 1144-1154.

Cheng, J., Guffanti, A.A., Wang, W., Krulwich, T.A., Bechhofer, D.H. (1996a). Chromosomal tetA(L) gene of Bacillus subtilis: regulation of expression and physiology of a tetA(L) deletion strain. *J. Bacteriol.* 178, 2853-2860.

Cheng, J., Hicks, D.B., Krulwich, T.A. (1996b). The purified Bacillus subtilis tetracycline efflux protein TetA(L) reconstitutes both tetracycline-cobalt/H^+ and $Na^+(K^+)/H^+$ exchange. *Proc. Natl. Acad. Sci. USA.* 93, 14446-14451.

Cho, B.K., Federowicz, S.A., Embree, M., Park, Y.S., Kim, D., Palsson, B.O. (2011). The PurR regulon in Escherichia coli K-12 MG1655. *Nuc. Acid Res.* 39, 6456-6464.

Cogan, T.A., Jorgensen, F., Lappin-Scott, H.M., Benson, C.E., Woodward, M.J., Humphrey, T.J. (2004). Flagella and curli fimbriae are important for the growth of Salmonella enterica serovars in hen eggs. *Microbiology.* 150, 1063-1071.

Colyer, T.E., Kredich, N.M. (1996). In vitro characterization of constitutive CysB proteins from Salmonella typhimurium. *Mol. Micro.* 21, 247-256.

Cook, A.M., Denger, K., Smits, T.H. (2006). Dissimilation of C3-sulfonates. *Arch. Microbiol.* 185, 83-90.

Costerton, J.W., Cheng, K.J., Geesey, G.G., Ladd, T.I., Nickel, J.C., Dasgupta, M., Marrie, T.J. (1987). Bacterial biofilms in nature and disease. *Annu. Rev. Micro.* 41, 435-64.

Costerton, J.W., Lewandowski, Z., Caldwell, D.E., Korber, D.R., Lappin-Scott, H.M. (1995). Microbial biofilms. *Annu. Rev. Micro.* 49, 711-745.

Dhamdhere, G., Krishnamoorthy, G., Zgurskaya, H.I. (2010). Interplay between drug efflux and antioxidants in Escherichia coli resistance to antibiotics. *Antimicrob. Agents Chemother.* 54, 5366-5368.

Diederen, B.M. (2008). Legionella spp. and Legionnaires' disease. *J. Infect.* 56, 1-12.

Fernandez, M., Kleerebezem, M., Kuipers, O.P., Siezen, R.J., Van Kranenburg, R. (2002). Regulation of the metC-cysK operon, involved in sulfur metabolism in Lactococcus lactis. *J. Bacteriol.* 184, 82-90.

Fernandez, M., Van Doesburg, W., Rutten, G.A., Marugg, J.D., Alting, A.C., Van Kranenburg, R., Kuipers, O.P. (2000). Molecular and functional analyses of the metC gene of Lactococcus lactis, encoding cystathionine beta-lyase. *App. Env. Microbiol.* 66, 42-48.

Filipic, B., Golic, N., Jovcic, B., Tolinacki, M., Bay, D.C., Turner, R.J., Antic-Stankovic, J., Kojic, M., Topisirovic, L. (2013). The cmbT gene encodes a novel major facilitator multidrug resistance transporter in Lactococcus lactis. *Res. Microbiol.* 164, 46-54.

Fluman, N., Bibi, E. (2009). Bacterial multidrug transport through the lens of the major facilitator superfamily. *Biochim. Biophys. Acta.* 1794, 738-747.

Franza, T., Mahe, B., Expert, D. (2005). Erwinia chrysanthemi requires a second iron transport route dependent of the siderophore achromobactin for extracellular growth and plant infection. *Mol. Micro.* 55, 261-275.

Frawley, E.R., Crouch, M.L., Bingham-Ramos, L.K., Robbins, H.F., Wang, W., Wright, G.D., Fang, F.C. (2013). Iron and citrate export by a major facilitator superfamily pump regulates metabolism and stress resistance in Salmonella Typhimurium. *Proc. Natl. Acad. Sci. USA.* 110, 12054-12059.

Furuchi, T., Kashiwagi, K., Kobayashi, H., Igarashi, K. (1991). Characteristics of the gene for a spermidine and putrescine transport system that maps at 15 min on the Escherichia coli chromosome. *J. Biol. Chem.* 266, 20928-20933.

Garcia, B., Latasa, C., Solano, C., Garcia-Del Portillo, F., Gamazo, C., Lasa, I. (2004). Role of the GGDEF protein family in Salmonella cellulose biosynthesis and biofilm formation. *Mol. Micro.* 54, 264-277.

Ginn, S.L., Brown, M.H., Skurray, R.A. (2000). The TetA(K) tetracycline/H(+) antiporter from Staphylococcus aureus: mutagenesis and functional analysis of motif C. *J. Bacteriol.* (182), 1492-1498.

Gomez, M.I., Prince, A. (2007). Opportunistic infections in lung disease: Pseudomonas infections in cystic fibrosis. *Curr. Opin. Pharma.* 7, 244-251.

Grundy, F.J., Henkin, T.M. (1998). The S box regulon: a new global transcription termination control system for methionine and cysteine biosynthesis genes in gram-positive bacteria. *Mol. Micro.* 30, 737-749.

Gueimonde, M., Garrigues, C., Van Sinderen, D., De Los Reyes-Gavilan, C.G., Margolles, A. (2009). Bile-inducible efflux transporter from Bifidobacterium longum NCC2705, conferring bile resistance. *App. Env. Micro.* 75, 3153-3160.

Guffanti, A.A., Cheng, J., Krulwich, T.A. (1998). Electrogenic antiport activities of the Gram-positive Tet proteins include a $Na^{+}(K^{+})/K^{+}$ mode that mediates net K^{+} uptake. *The J. Biol. Chem.* 273, 26447-26454.

Guffanti, A.A., Krulwich, T.A. (1995). Tetracycline/H^{+} antiport and Na^{+}/H^{+} antiport catalyzed by the Bacillus subtilis TetA(L) transporter expressed in Escherichia coli. *J. Bacteriol.* 177, 4557-4561.

Gunn, J.S. (2000). Mechanisms of bacterial resistance and response to bile. *Microbe. Infect.* 2, 907-913.

Hayes, E.T., Wilks, J.C., Sanfilippo, P., Yohannes, E., Tate, D.P., Jones, B.D., Radmacher, M.D., Bondurant, S.S., Slonczewski, J.L. (2006). Oxygen limitation modulates pH regulation of catabolism and hydrogenases, multidrug transporters, and envelope composition in Escherichia coli K-12. *BMC Micro.* 6, 89.

Higashi, K., Terui, Y., Suganami, A., Tamura, Y., Nishimura, K., Kashiwagi, K., Igarashi, K. (2008). Selective structural change by spermidine in the bulged-out region of double-stranded RNA and its effect on RNA function. *J. Biol. Chem.* 283, 32989-32994.

Holdsworth, S.R., Law, C.J. (2012). Functional and biochemical characterisation of the Escherichia coli major facilitator superfamily multidrug transporter MdtM. *Biochimie.* 94, 1334-1346.

Holdsworth, S.R., Law, C.J. (2013). Multidrug resistance protein MdtM adds to the repertoire of antiporters involved in alkaline pH homeostasis in Escherichia coli. *BMC Micro.* 13, 113-132.

Hood, M.I., Skaar, E.P. (2012). Nutritional immunity: transition metals at the pathogen-host interface. *Nature reviews. Microbiology.* 10, 525-537.

Igarashi, K., Kashiwagi, K. (1999). Polyamine transport in bacteria and yeast. *Biochem. J.* 344, 633-642.

Igarashi, K., Kashiwagi, K. (2000). Polyamines: mysterious modulators of cellular functions. *Biochem. Biophys. Res. Comm.* 271, 559-564.

Ishii, I., Takada, H., Terao, K., Kakegawa, T., Igarashi, K., Hirose, S. (1994). Decrease in spermidine content during logarithmic phase of cell growth delays spore formation of Bacillus subtilis. *Cell. Mol. Biol.* 40, 925-931.

Jin, J., Guffanti, A.A., Bechhofer, D.H., Krulwich, T.A. (2002a). Tet(L) and tet(K) tetracycline-divalent metal/H+ antiporters: characterization of multiple catalytic modes and a mutagenesis approach to differences in their efflux substrate and coupling ion preferences. *J. Bacteriol.* 184, 4722-4732.

Jin, Q., Yuan, Z., Xu, J., Wang, Y., Shen, Y., Lu, W., Wang, J., Liu, H., Yang, J., Yang, F., Zhang, X., Zhang, J., Yang, G., Wu, H., Qu, D., Dong, J., Sun, L., Xue, Y., Zhao, A., Gao, Y., Zhu, J., Kan, B., Ding, K., Chen, S., Cheng, H., Yao, Z., He, B., Chen, R., Ma, D., Qiang, B., Wen, Y., Hou, Y., Yu, J. (2002b). Genome sequence of Shigella flexneri 2a: insights into pathogenicity through comparison with genomes of Escherichia coli K12 and O157. *Nuc. Acid Res.*. 30, 4432-4441.

Jonas, K., Tomenius, H., Kader, A., Normark, S., Romling, U., Belova, L.M., Melefors, O. (2007). Roles of curli, cellulose and BapA in Salmonella biofilm morphology studied by atomic force microscopy. *BMC Microbiol.* 7, 70-77.

Kashiwagi, K., Hosokawa, N., Furuchi, T., Kobayashi, H., Sasakawa, C., Yoshikawa, M., Igarashi, K. (1990). Isolation of polyamine transport-deficient mutants of Escherichia coli and cloning of the genes for polyamine transport proteins. *J. Biol. Chem.* 265, 20893-20897.

Kimoto-Nira, H., Suzuki, S., Yakabe, T., Suzuki, C. (2012). Relationships between fatty acid composition and bile tolerance in lactobacillus isolates from plants and from non-plant materials. *Canad. J. Microbiol.* 58, 1396-1404.

Krulwich, T.A., Cheng, J., Guffanti, A.A. (1994). The role of monovalent cation/proton antiporters in Na(+)-resistance and pH homeostasis in Bacillus: an alkaliphile versus a neutralophile. *J. Exp. Biol.* 196, 457-470.

Krulwich, T.A., Hicks, D.B., Ito, M. (2009). Cation/proton antiporter complements of bacteria: why so large and diverse? *Mol. Micro.* 74, 257-260.

Krulwich, T.A., Ito, M., Hicks, D.B., Gilmour, R., Guffanti, A.A. (1998). pH homeostasis and ATP synthesis: studies of two processes that necessitate inward proton translocation in extremely alkaliphilic Bacillus species. *Extremophiles.* 2, 217-222.

Krulwich, T.A., Jin, J., Guffanti, A.A., Bechhofer, H. (2001). Functions of tetracycline efflux proteins that do not involve tetracycline. *J. Mol. Microbiol. Biotechnol.* 3, 237-246.

Krulwich, T.A., Lewinson, O., Padan, E., Bibi, E. (2005). Do physiological roles foster persistence of drug/multidrug-efflux transporters? A case study. *Nat. Rev. Microbiol.* 3, 566-572.

Krulwich, T.A., Sachs, G., Padan, E. (2011). Molecular aspects of bacterial pH sensing and homeostasis. *Nat. Rev. Microbiol.* 9, 330-343.

Kuroda, T., Tsuchiya, T. (2009). Multidrug efflux transporters in the MATE family. *Biochim. Biophys. Acta.* 1794, 763-768.

Kvist, M., Hancock, V., Klemm, P. (2008). Inactivation of efflux pumps abolishes bacterial biofilm formation. *App. Env. Microbiol.* 74, 7376-7382.

Kwon, D.H., Lu, C.D. (2007). Polyamine effects on antibiotic susceptibility in bacteria. *Antimicrob. Agents Chemother.* 51, 2070-2077.

Law, C.J., Maloney, P.C., Wang, D.N. (2008). Ins and outs of major facilitator superfamily antiporters. *Annu. Rev. Microbiol.* 62, 289-305.

Lazazzera, B.A. (2005). Lessons from DNA microarray analysis: the gene expression profile of biofilms. *Curr. Opin. Microbiol.* 8, 222-227.

Lee, A., Mao, W., Warren, M.S., Mistry, A., Hoshino, K., Okumura, R., Ishida, H., Lomovskaya, O. (2000). Interplay between efflux pumps may provide either additive or multiplicative effects on drug resistance. *J. Bacteriol.* 182, 3142-3150.

Lewinson, O., Adler, J., Sigal, N., Bibi, E. (2006). Promiscuity in multidrug recognition and transport: the bacterial MFS Mdr transporters. *Mol. Micro.* 61, 277-284.

Lewinson, O., Padan, E., Bibi, E. (2004). Alkalitolerance: a biological function for a multidrug transporter in pH homeostasis. *Proc. Natl. Acad. Sci. USA.* 101, 14073-14078.

Limsuwun, K., Jones, P.G. (2000). Spermidine acetyltransferase is required to prevent spermidine toxicity at low temperatures in Escherichia coli. *J. Bacteriol.* 182, 5373-5380.

Love, M.W., Dawson, P.A. (1998). New insights into bile acid transport. *Curr. Opin. Lipid.* 9, 225-229.

Lovegrove, J.M. (2004). Dental plaque revisited: bacteria associated with periodontal disease. *J. New Zealand Soc.Periodont.* 7-21.

Lubelski, J., Konings, W.N., Driessen, A.J. (2007). Distribution and physiology of ABC-type transporters contributing to multidrug resistance in bacteria. *Micro. Mol. Biol. Revs.* 71, 463-476.

Madej, M.G., Dang, S., Yan, N., Kaback, H.R. (2013). Evolutionary mix-and-match with MFS transporters. *Proc. Natl. Acad. Sci. USA* 110, 5870-5874.

Mallonee, D.H., Hylemon, P.B. (1996). Sequencing and expression of a gene encoding a bile acid transporter from Eubacterium sp. strain VPI 12708. *J. Bacteriol.* 178, 7053-7058.

Mazurkiewicz, P., Konings, W.N., Poelarends, G.J. (2002). Acidic residues in the lactococcal multidrug efflux pump LmrP play critical roles in transport of lipophilic cationic compounds. *J. Biol. Chem.* 277, 26081-26088.

Mazurkiewicz, P., Sakamoto, K., Poelarends, G.J., Konings, W.N. (2005). Multidrug transporters in lactic acid bacteria. *Mini Rev. Med. Chem.* 5, 173-181.

Merritt, M.E., Donaldson, J.R. (2009.) Effect of bile salts on the DNA and membrane integrity of enteric bacteria. *J. Med. Microbiol.* 58, 1533-1541.

Miethke, M., Marahiel, M.A. (2007). Siderophore-based iron acquisition and pathogen control. *Microbiol. Mol. Biol. Revs.* 71, 413-451.

Miethke, M., Schmidt, S., Marahiel, M.A. (2008). The major facilitator superfamily-type transporter YmfE and the multidrug-efflux activator Mta mediate bacillibactin secretion in Bacillus subtilis. *J. Bacteriol.* 190, 5143-5152.

Morita, Y., Kodama, K., Shiota, S., Mine, T., Kataoka, A., Mizushima, T., Tsuchiya, T. (1998). NorM, a putative multidrug efflux protein, of Vibrio parahaemolyticus and its homolog in Escherichia coli. *Antimicrob. Agents Chemother.* 42, 1778-1782.

Muyzer, G., Stams, A.J. (2008). The ecology and biotechnology of sulphate-reducing bacteria. *Nat. Revs. Microbiol.* 6, 441-454.

Nelson, N. (1999). Metal ion transporters and homeostasis. *EMBO J.* 18, 4361-4371.

Neyfakh, A.A. (1992). The multidrug efflux transporter of Bacillus subtilis is a structural and functional homolog of the Staphylococcus NorA protein. *Antimicrob. Agents Chemother.* 36, 484-485.

Neyfakh, A.A. (1997). Natural functions of bacterial multidrug transporters. *Trends Microbiol.* 5, 309-313.

Neyfakh, A.A., Bidnenko, V.E., Chen, L.B. (1991). Efflux-mediated multidrug resistance in Bacillus subtilis: similarities and dissimilarities with the mammalian system. *Proc. Natl. Acad. Sci. USA.* 88, 4781-4785.

Nikaido, H., Pages, J.M. (2012). Broad-specificity efflux pumps and their role in multidrug resistance of Gram-negative bacteria. *FEMS Microbiol. Revs.* 36, 340-363.

Nikaido, H., Takatsuka, Y. (2009). Mechanisms of RND multidrug efflux pumps. *Biochim. Biophys. Acta.* 1794, 769-781.

Nilsen, I.W., Bakke, I., Vader, A., Olsvik, O., El-Gewely, M.R. (1996). Isolation of cmr, a novel Escherichia coli chloramphenicol resistance gene encoding a putative efflux pump. *J. Bacteriol.* 178, 3188-3193.

Nishino, K., Yamaguchi, A. (2001). Analysis of a complete library of putative drug transporter genes in Escherichia coli. *J. Bacteriol.* 183, 5803-5812.

Omote, H., Hiasa, M., Matsumoto, T., Otsuka, M., Moriyama, Y. (2006). The MATE proteins as fundamental transporters of metabolic and xenobiotic organic cations. *Trends Pharma. Sci.* 27, 587-593.

Padan, E., Bibi, E., Ito, M., Krulwich, T.A. (2005). Alkaline pH homeostasis in bacteria: new insights. *Biochim. Biophys. Acta.* 1717, 67-88.

Padan, E., Gerchman, Y., Rimon, A., Rothman, A., Dover, N., Carmel-Harel, O. (1999). The molecular mechanism of regulation of the NhaA Na^+/H^+ antiporter of Escherichia coli, a key transporter in the adaptation to Na^+ and H^+. *Novartis Found. Symp.* 221, 183-196; discussion 196-199.

Padan, E., Tzubery, T., Herz, K., Kozachkov, L., Rimon, A., Galili, L. (2004). NhaA of Escherichia coli, as a model of a pH-regulated Na^+/H^+antiporter. *Biochim. Biophys. Acta.* 1658, 2-13.

Paul, S., Alegre, K.O., Holdsworth, S.R., Rice, M., Brown, J.A., McVeigh, P., Kelly, S.M., Law, C.J. (2014). A single-component multidrug transporter of the major facilitator superfamily is part of a network that protects Escherichia coli from bile salt stress. *Mol. Micro.* 92, 872-884.

Pao, S.S., Paulsen, I.T., Saier, M.H., Jr. (1998). Major facilitator superfamily. *Microbiol. Mol. Biol. Rev.* 62, 1-34.

Parkhill, J., Dougan, G., James, K.D., Thomson, N.R., Pickard, D., Wain, J., Churcher, C., Mungall, K.L., Bentley, S.D., Holden, M.T., Sebaihia, M., Baker, S., Basham, D., Brooks, K., Chillingworth, T., Connerton, P., Cronin, A., Davis, P., Davies, R.M., Dowd, L., White, N., Farrar, J., Feltwell, T., Hamlin, N., Haque, A., Hien, T.T., Holroyd, S., Jagels, K., Krogh, A., Larsen, T.S., Leather, S., Moule, S., O'gaora, P., Parry, C., Quail, M., Rutherford, K., Simmonds, M., Skelton, J., Stevens, K., Whitehead, S., Barrell, B.G. (2001). Complete genome sequence of a multiple drug resistant Salmonella enterica serovar Typhi CT18. *Nature.* 413, 848-852.

Paulsen, I.T. (2003). Multidrug efflux pumps and resistance: regulation and evolution. *Curr. Opin. Microbiol.* 6, 446-51.

Paulsen, I.T., Brown, M.H., Skurray, R.A. (1996). Proton-dependent multidrug efflux systems. *Microbiol. Rev.* 60, 575-608.

Pegg, A.E. (1986). Recent advances in the biochemistry of polyamines in eukaryotes. *Biochem. J.* 234, 249-262.

Pfeiler, E.A., Klaenhammer, T.R. (2009). Role of transporter proteins in bile tolerance of Lactobacillus acidophilus. . *App. Env. Microbiol.* 75, 6013-16.

Piddock, L.J. (2006). Multidrug-resistance efflux pumps - not just for resistance. *Nat. Rev. Microbiol.* 4, 629-636.

Plack, R.H., Jr., Rosen, B.P. (1980). Cation/proton antiport systems in Escherichia coli. Absence of potassium/proton antiporter activity in a pH-sensitive mutant. *J. Biol. Chem.* 255, 3824-3825.

Price, C.E., Reid, S.J., Driessen, A.J., Abratt, V.R. (2006). The Bifidobacterium longum NCIMB 702259T ctr gene codes for a novel cholate transporter. *App. Env. Microbiol.* 72, 923-926.

Putman, M., Van Veen, H.W., Degener, J.E., Konings, W.N. (2001). The lactococcal secondary multidrug transporter LmrP confers resistance to lincosamides, macrolides, streptogramins and tetracyclines. *Microbiology.* 147, 2873-2880.

Quillin, S.J., Schwartz, K.T., Leber, J.H. (2011). The novel Listeria monocytogenes bile sensor BrtA controls expression of the cholic acid efflux pump MdrT. *Mol. Micro.* 81, 129-42.

Rajamohan, G., Srinivasan, V.B., Gebreyes, W.A. (2010). Molecular and functional characterization of a novel efflux pump, AmvA, mediating antimicrobial and disinfectant resistance in Acinetobacter baumannii. *J. Antimicrob. Chemother.* 65, 1919-1925.

Reddy, V.S., Shlykov, M.A., Castillo, R., Sun, E.I., Saier, M.H., Jr. (2012). The major facilitator superfamily (MFS) revisited. *FEBS J.* 279, 2022-2035.

Rein, U., Gueta, R., Denger, K., Ruff, J., Hollemeyer, K., Cook, A.M. (2005). Dissimilation of cysteate via 3-sulfolactate sulfo-lyase and a sulfate exporter in Paracoccus pantotrophus NKNCYSA. *Microbiology.* 151, 737-747.

Rocha, E.P., Sekowska, A., Danchin, A. (2000). Sulphur islands in the Escherichia coli genome: markers of the cell's architecture? *FEBS Letts.* 476, 8-11.

Roje, S. (2006). S-Adenosyl-L-methionine: beyond the universal methyl group donor. *Phytochemistry.* 67, 1686-1698.

Rolfes, R.J. (2006). Regulation of purine nucleotide biosynthesis: in yeast and beyond. *Biochem. Soc. Trans.* 34, 786-790.

Ruiz, L., Zomer, A., O'connell-Motherway, M., Van Sinderen, D., Margolles, A. (2012). Discovering novel bile protection systems in Bifidobacterium breve UCC2003 through functional genomics. *App. Env. Microbiol.* 78, 1123-1131.

Saier, M.H., Jr., Beatty, J.T., Goffeau, A., Harley, K.T., Heijne, W.H., Huang, S.C., Jack, D.L., Jahn, P.S., Lew, K., Liu, J., Pao, S.S., Paulsen, I.T., Tseng, T.T., Virk, P.S. (1999). The major facilitator superfamily. *J. Mol. Microbiol. Biotechnol.* 1, 257-279.

Saier, M.H., Jr., Paulsen, I.T. (2001). Phylogeny of multidrug transporters. *Sem. Cell Dev. Biol.* 12, 205-213.

Saier, M.H., Jr., Paulsen, I.T., Sliwinski, M.K., Pao, S.S., Skurray, R.A., Nikaido, H. (1998). Evolutionary origins of multidrug and drug-specific efflux pumps in bacteria. *FASEB J.* 12, 265-274.

Saier, M.H., Jr., Tran, C.V.,Barabote, R.D. (2006). TCDB: the Transporter Classification Database for membrane transport protein analyses and information. *Nuc. Acid Res.* 34, D181-186.

Saier, M.H., Jr., Yen, M.R., Noto, K., Tamang, D.G., Elkan, C. (2009). The Transporter Classification Database: recent advances. *Nuc. Acid Res.* 37, D274-278.

Schaedler, T.A., Tong, Z., Van Veen, H.W. (2012). The multidrug transporter LmrP protein mediates selective calcium efflux. *J. Biol. Chem.* 287, 27682-27690.

Schaedler, T.A., Van Veen, H.W. 2010. A flexible cation binding site in the multidrug major facilitator superfamily transporter LmrP is associated with variable proton coupling. *FASEB J.* 24, 3653-3661.

Schubert, H.L., Blumenthal, R.M., Cheng, X. (2003). Many paths to methyltransfer: a chronicle of convergence. *Trends Biochem. Sci.* 28, 329-335.

Shaw, F.L., Elliott, K.A., Kinch, L.N., Fuell, C., Phillips, M.A., Michael, A.J. (2010). Evolution and multifarious horizontal transfer of an alternative biosynthetic pathway for the alternative polyamine sym-homospermidine. *J. Biol. Chem.* 285, 14711-14723.

Silver, S., Phung, L.T. (1996). Bacterial heavy metal resistance: new surprises. *Annu. Rev. Microbiol.* 50, 753-789.

Slonczewski, J.L., Fujisawa, M., Dopson, M., Krulwich, T.A. (2009). Cytoplasmic pH measurement and homeostasis in bacteria and archaea. *Adv. Micro. Physiol.* 55, 1-79, 317.

Smith, R., Coast, J. (2013). The true cost of antimicrobial resistance. *BMJ.* 346, f1493.

Soo, V.W., Hanson-Manful, P., Patrick, W.M. (2011). Artificial gene amplification reveals an abundance of promiscuous resistance determinants in Escherichia coli. *Proc. Natl. Acad. Sci. USA* 108, 1484-1489.

Speer, B.S., Shoemaker, N.B., Salyers, A.A. (1992). Bacterial resistance to tetracycline: mechanisms, transfer, and clinical significance. *Clin. Microbiol. Revs.* 5, 387-399.

Stipanuk, M.H. (1986). Metabolism of sulfur-containing amino acids. *Ann. Rev. Nutr.* 6, 179-209.

Tabor, C.W., Tabor, H. (1985). Polyamines in microorganisms. *Microbiol. Revs.* 49, 81-99.

Tal, N., Schuldiner, S. (2009). A coordinated network of transporters with overlapping specificities provides a robust survival strategy. *Proc. Natl. Acad. Sci. USA.* 106, 9051-9056.

Tanabe, T., Nakao, H., Kuroda, T., Tsuchiya, T., Yamamoto, S. (2006). Involvement of the Vibrio parahaemolyticus pvsC gene in export of the siderophore vibrioferrin. *Microbiol. Immunol.* 50, 871-876.

Thanassi, D.G., Cheng, L.W., Nikaido, H. (1997). Active efflux of bile salts by Escherichia coli. *J. Bacteriol.* 179, 2512-2518.

Van Der Ploeg, J.R., Barone, M., Leisinger, T. (2001). Expression of the Bacillus subtilis sulphonate-sulphur utilization genes is regulated at the levels of transcription initiation and termination. *Mol. Micro.* 39, 1356-1365.

Vardy, E., Arkin, I.T., Gottschalk, K.E., Kaback, H.R., Schuldiner, S. (2004). Structural conservation in the major facilitator superfamily as revealed by comparative modeling. *Protein Sci.* 13, 1832-1840.

Wong, K., Ma, J., Rothnie, A., Biggin, P.C., Kerr, I.D. (2014). Towards understanding promiscuity in multidrug efflux pumps. *Trends Biochem. Sci.* 39, 8-16.

Woolridge, D.P., Vazquez-Laslop, N., Markham, P.N., Chevalier, M.S., Gerner, E.W., Neyfakh, A.A. (1997). Efflux of the natural polyamine spermidine facilitated by the Bacillus subtilis multidrug transporter Blt. *J. Biol. Chem.* 272, 8864-8866.

Yin, Y., He, X., Szewczyk, P., Nguyen, T., Chang, G. (2006). Structure of the multidrug transporter EmrD from Escherichia coli. *Science*. 312, 741-744.

In: Prokaryotes
Editor: Marina Nisnevitch

ISBN: 978-1-63321-592-4

Chapter 4

Adaptation of Bacterial Cells to Toxic Aromatic Hydrocarbons

Rivka Cahan[1,*], Esti Michael[1,2] and Nissim Swissa[1,2]
[1]Department of Chemical Engineering, Ariel University, Ariel, Israel
[2]The Mina and Everard Goodman Faculty of Life Sciences, Bar-Ilan University, Ramat-Gan, Israel

Abstract

Aromatic hydrocarbons affect the environment, and in particular the fauna and flora. Organic solvents are among the most toxic chemicals. Biodegradation of these molecules has been found under both aerobic and anaerobic conditions. Bacteria utilize these compounds as an energy and carbon source. Microorganisms have evolved various biochemical and physiological adaptation mechanisms which protect them from the lethal effects of organic solvents. The adaptation mechanisms include: alteration of the properties of the bacterial cell membrane, morphological and bacterial proteome changes.

Keywords: Aromatic hydrocarbons, outer membrane vesicles, bacterial proteome, porins and efflux pumps

Introduction

Aromatic hydrocarbons affect the environment, and in particular the fauna and flora. Organic solvents are among the most toxic chemicals. These solvents include phenol, toluene, benzene, xylene and styrene, which dissolve the cell membrane, disorganize it, cause the loss of lipids and proteins, and eventually lead to cell death (Agarry et al., 2008). The toxicity of aromatic compounds generally is correlated with their hydrophobicity, which can be defined

[*] Corresponding author: Email: rivkac@ariel.ac.il.

as the logarithm of the partition coefficient P of a specific solvent between a standardized 1:1 mixture of *n*-octanol and water. Chemical compounds with log P values ranging between 1 and 4 are regarded as being highly toxic for microorganisms. The hydrophobicity of compounds with a log P less than 1 is too low to enable their penetration through the membrane, whereas above log P 4, the substances are too poorly water-soluble and thus too poorly bioavailable to cause any toxic effect. The higher the hydrophobicity within the log P ranges of 1 to 4, the higher is the compound's toxic effect on bacteria, demonstrating a systematic nearly linear correlation (Inoue & Horikoshi, 1989; Weber & de Bont, 1996).

Bacteria that degrade toxic aromatic hydrocarbon compounds have been isolated from extremely hot as well as from subzero environments (Aislabie et al., 2000). Biodegradation of these molecules has been found under both aerobic and anaerobic conditions (Chakraborty et al., 2005; Cafaro et al., 2004). Bacteria utilize these compounds as an energy and carbon source (Chakraborty et al., 2005).

Microorganisms have evolved various biochemical and physiological adaptation mechanisms which protect them from the lethal effects of organic solvents.

The adaptation mechanisms include: (1) alteration of the properties of the bacterial cell membrane; (2) morphological changes such as bacterial cell size and formation of vesicles; (3) changes in the bacterial proteome which include proteins involved in energy metabolism, stress proteins, efflux pumps and porins.

Alteration of the Properties of the Bacterial Cell Membrane in the Presence of Organic Solvents

Gram-negative bacteria are surrounded by inner and outer membranes. The inner lipid bilayer contains mainly phosphatidylethanolamine. The outer membrane consists of an asymmetric bilayer, where the inner leaflet is enriched with phosphatidylethanolamine and the outer leaflet is enriched with lipopolysaccharides (LPS). The inner and outer leaflets are separated by a periplasmic space that contains a hydrophilic layer, the peptidoglycan, which provides additional stability (Geldera et al., 2000; Schirmer, 1998). The outer membrane of Gram-negative bacteria is a hydrophobic barrier which functions as a molecular sieve. Molecules with a molecular mass of greater than 600-1,000 Da cannot penetrate through the outer membrane into the periplasm. The low permeability of the outer membrane toward hydrophobic molecules is ascribed to the presence of the LPS (Sikkema et al., 1995). The cell wall of Gram-positive bacteria, in contradistinction to that of Gram-negative bacteria, lacks the outer membrane but is surrounded by several layers of peptidoglycan.

The cell membrane is the main and first target for the toxic effects of solvents. Changes in the membrane composition therefore play a crucial role in the mechanisms of bacterial solvent tolerance (Inoue & Horikoshi, 1989). The temperature required to induce a change in the lipid physical state from the ordered gel phase (where the fatty acid hydrocarbon chains are fully extended and closely packed) to the disordered liquid-crystalline phase (where the hydrocarbon chains are randomly oriented and "fluid") is called the phase transition temperature. Several factors directly affect the phase transition temperature of membrane phospholipids: the hydrocarbon length, degree of unsaturation, charge, and headgroup species. Fatty acids exhibit different transition temperatures (melting points) at which they alter their gel phase, depending on their length, degree of saturation, and configuration (*cis, trans, cyclo, iso, anteiso*).

The changes in the cytoplasmic membrane in response to a solvent prevent the influx of solvents by decreasing the membrane permeability and fluidity (Weber & de Bont, 1996). The membrane lipid composition of *Pseudomonas oleovorans* was studied during two-phase fermentations. More unsaturated and *trans*-fatty acids were formed in the presence of octane and the fatty acids showed an increase in acyl chain length leading to decreased membrane lipid fluidity as a counteracting adaptation to the alkane solvent (Chen et al., 1995a,b). Heipieper et al. (1992) reported that isomerization of the *cis* bond of unsaturated esterified fatty acids into the corresponding *trans*-configuration occurred mainly in palmitoleic acid (C16:1Δ9) and *cis*-vaccenic acid (C18:1Δ11) (Heipieper et al. 1992). *Cis-trans* isomerization was proven to be a reaction to the presence of solvents such as toluene (Weber et al., 1994; Heipieper et al., 1994). The content of branched (10-methyl) fatty acids in Rhodococcus opacus increased to up to 34% of the total fatty acids when the cells were grown in the presence of several organic solvents, compared with cells which were grown on fructose. Furthermore, the content of trans-hexadecenoic acid increased up to 18% in the presence of phenol or chlorophenol (Tsitko et al., 1999). An increased content of 10-methylheptanoic and 10-methyloctanoic acids was observed in Pimelobacter sp. that grew on pyridine (Rhee et al., 1996). Growing *P. putida* P8 cells in the presence of sub-lethal phenol concentrations in the medium led to a change in the fatty acid composition of the lipids, thereby increasing the degree of saturation. At phenol concentrations which completely inhibited the growth of *P. putida*, the cells were still able to increase the content of the *trans* fatty acids and simultaneously decrease the proportion of the corresponding 9-*cis*-hexadecenoic acid. Conversion of fatty acids was also induced by 4-chlorophenol in non-growing cells in which *de novo* synthesis of lipids had ceased. Isomerization of the double bond in the presence of chloramphenicol indicates a constitutively operating enzyme system (Heipieper et al., 1992). The increased *trans/cis* ratio enhances the ordering, density, and packing of the membrane, which lead to a decrease in the membrane fluidity (Chen et al., 1995b; Keweloh & Heipieper, 1996). The *cis-trans* isomerization mechanism was shown to be independent of energy, cell growth and *de novo* biosynthesis of lipids. This allows the bacteria to adapt to organic solvents even when environmental conditions are not suitable for growth and synthesis of new membrane fatty acids (Heipieper et al., 1992; Heipieper & de Bont, 1994). It was shown that the partitioning of the organic solvent into the lipid part of the membrane prevents the solvent molecules from entering the membrane (Heipieper et al., 1996; Heipieper et al., 2003). The enzyme responsible for this conversion, *cis-trans* isomerase (Cti), is located in the periplasmic space and is considered to be a constitutive enzyme. It does not require any energy or co-factors (von Wallbrunn et al., 2003; Kiran et al., 2004). Changes in the *cis/trans* ratio due to activity of the enzyme Cti are considered short-term solvent adaptation mechanisms (de Bont, 1998), whereas the change in the degree of saturation is a long-term adaptation mechanism (Isken & Heipieper, 2002). A direct correlation was observed between the hydrophobicity of organic compounds (log P), their concentration-dependent growth inhibition and the *trans/cis* ratio of unsaturated membrane fatty acids. It was shown that *cis-trans* isomerization alone does not provide the necessary protection against organic solvents, since there are microorganisms that are able to perform the isomerization but are still solvent-sensitive (Ramos et al. 1997).

The bacterial cell hydrophobic envelope has great influence on the cells' ability to adhere to surfaces or substrates. A change in the surface hydrophobicity of Gram-negative bacteria exposed to toxic organic solvents was already observed in detail (Neumann et al., 2005). Bacteria harboring quite hydrophobic envelopes, such as solvent-tolerant *Mycobacterium* sp., showed cell aggregates in the presence of butan-1-ol. In this case, the hydrophobic character promotes cellular aggregation as a way to diminish the binding of toxic hydrophilic compounds (de Carvalho et al., 2004).

In conclusion, the major adaptations on the level of the bacterial cell membrane are modifications in the saturated-to-unsaturated fatty acids ratio of hydrocarbons, *cis*-to-*trans* isomerization of unsaturated fatty acids and production of changes in phospholipid headgroups. These changes are interpreted as a way for bacteria to maintain membrane fluidity and impermeability or to restore membrane integrity, thus reducing partitioning of hydrocarbons in the membrane.

Morphological Changes That Leading to Bacterial Cells' Adaptation to Organic Solvents

Outer membrane vesicles (OMVs) are bilayered spherical membranous structures with a diameter range of 50 to 250 nm, which are constantly discharged from the surface of several Gram-negative bacteria during bacterial growth (Beveridge, 1999; Zhou et al., 1998). The OMVs possess LPS, outer membrane proteins, phospholipids, and periplasmic constituents. OMVs were found in the planktonic or biofilm bacterial cell mode of growth (Beveridge et al., 1997). The OMVs are a secretion and delivery system which is distinct from the well-studied soluble protein secretion system. In the OMVs secretion system, soluble molecules are released in a complex with insoluble material (Kulp & Kuehn, 2010). OMVs may aid in bacterial survival by removing a variety of harmful, intrinsic or extrinsic, agents from the bacterial population (McBroom & Kuehn, 2007). High growth temperatures can increase OMV production. It was found that the Tol/Pal system is a master regulator of vesiculation (Deatherage et al., 2009). *P. aeruginosa* produces 11 autolysins which are important for the synthesis of the peptidoglycan layer (Bernadsky et al., 1994; Beveridge, 1999). The major autolysin is a 26 kDa peptidoglycan hydrolase (PGase) packed into OMVs. It was found that OMVs can lyse other bacteria when these bacteria suffer from poor nutrition and poor growth. OMVs can attack both Gram-negative and Gram-positive bacteria. The OMVs adhere to the surface of Gram-positive bacteria and hydrolyze the peptidoglycan. The same mechanism occurs for bacteria covered by a S-layer on top of the peptidoglycan. The OMVs attack the bacterial cells through a single-hit route which produces a single large lesion in the cell wall (Li et al., 1996; 1998). The OMVs of *P. aeruginosa*, *Proteus mirabilis* and *Serratia marcescens* were found to packed with phospholipase C, proteases, proelastase, and hemolysins which provide a route for delivery of these virulence factors (Kadurugamuwa & Beveridge, 1995).

Pseudomonas putida IH-2000 is a toluene-tolerant microorganism that was found to release OMVs in the range of about 80 to 100 nm, as a response to the addition of toluene to the culture medium. The toluene molecules which adhere to the outer membrane were eliminated by the shedding of the OMVs. These OMVs contained a higher concentration of toluene molecules than the concentration found in the cell membrane (Kobayashi et al., 2000). The OMVs were detected only when IH-2000 cells were grown in the presence of toluene. The number of OMVs associated with toluene molecules decreased over time. The maximal amount of toluene associated with the OMVs was 0.625 mol/mol of lipid at 0.5 h after the addition of toluene. It then decreased to 0.166 mol/mol at 2.0 h, which was almost the same as the 0.172 mol/mol at 6.0 h after the addition of toluene (Kobayashi et al., 2000). The adherence of toluene to OMVs is an ideal model of organic solvent tolerance. This model shows that the outer membrane of Gram-negative bacteria is not a simple barrier, but rather an active eliminator of a toxic organic solvent.

Staphylococcus aureus cells treated with the aromatic hydrocarbons displayed irregular rod shapes with wrinkled surfaces (Pinkart et al., 1996). When *Pseudomonas aeruginosa* IBBML1 cells were grown in the presence of a mixture of aromatic hydrocarbons composed of benzene, toluene and cyclohexane, some lesions were observed on the outer membrane, exopolysaccharides were found to surround the bacterial cells and the cells' shape was modified (Lazaroaie, 2008). Gram-positive *Bacilus licheniformis* S-86 bacterial cells showed extensive filamentation and an increase in cell volume (4.7-fold) when growing in the presence of 0.6% 3-methylbutan-1-ol. Furthermore, the cytoplasmic membrane division was not accompanied by cell wall partition, which appears extremely thick at this site, in the presence of 3-methylbutan-1-ol (Torres et al., 2009). It was suggested that cell filamentation is a possible physiological mechanism caused by interference of cell division gene expression as a response to exposure to stress environments (Mattick et al., 2000). Another kind of morphological change was observed in *Staphylococcus* sp. ZZ1 in response to toluene. These cells produced an unusual hydrophilic carbohydrate extracellular capsule which prevents organic solvents from reaching the cell membrane (Zahir et al., 2006). However, increased production of the capsular polymer emulsan was also observed in cultures of the Gram-negative bacterium *Acinetobacter venetianus* Rag-1 when it was grown with ethanol as the sole carbon source (Panilaitis et al., 2007).

Pseudomonas putida P8 changed its cell size when grown in the presence of aromatic compounds such as phenol and 4-chlorophenol. The same effect was found for an alkanol-degrading bacterium, *Enterobacter* sp. VKGH12, in the presence of n-butanol. The size of the bacteria increased with increasing concentrations of n-butanol. This was observed only after the addition of non-lethal concentrations. The decrease in the ratio between surface area and cell volume as a response to the presence of organic solvents leads to a reduction in the cell surface area, and especially the cytoplasmic membrane which is the major target for the toxic effects. This reduction of the relative surface area represents an adaptive response to the presence of such compounds (Neumann et al., 2005). On the other hand, a decrease in cell size was observed in *Enterobacter* sp. cells as an adaptive response to the solvent 1-butanol when grown on it as the sole carbon and energy source. The same phenomenon was observed with *Bacillus subtilis* which reduced their cell size when treated with aromatic hydrocarbons (Pinkart et al., 1996). In this case, the increase in the cell surface area can provide an enhanced possibility for the uptake of potential growth substrates. The addition of butanol led to the opposite effect in cells grown on glucose, which were not solvent-adapted: a decrease in the surface area-volume ratio. *Mycobacterium* species which were grown in a two-phase aqueous-organic and organic media tended to shrink and decrease their surface roughness, indicating that they decreased their surface area (de Carvalho et al., 2004). A similar study on biphasic systems containing different organic solvents observed morphological changes in very different bacterial species upon solvent exposure and concomitant incorporation of solvent molecules into the cellular membrane which resulted in a loss of cell viability (de Carvalho & da Fonseca, 2004).

In conclusion, morphological changes in bacterial cells that may lead to bacterial adaptation to solvents include formation of OMVs, changes in cell surface area and in cell size.

Changes in the Bacterial Proteome

In order to understand the importance of proteins in the bacterial cells' adaptive mechanism, *Pseudomonas putida* KT2440 were exposed to sub-lethal inhibitory concentrations of aromatic hydrocarbons and variations in protein expression were analyzed using two-dimensional gel electrophoresis and mass spectrometry (MS) analysis. The up-regulated proteins include proteins involved in the: (1) Oxidative stress response (AhpC, SodB, Tpx and Dsb).

These proteins were previously described as being involved in bacterial protection against oxidative stress and other forms of stress such as the formation of reactive oxygen species (ROS) (Santos et al., 2004; Winn, 2003); (2) General stress response (UspA, HtpG, GrpE and Tig) which enhanced the efficiency of cell repair mechanisms against phenol-induced damage; (3) Energy metabolism (AcnB, AtpH, Fpr, AceA, NuoE, and MmsA-1) (Santos et al., 2004). Many environmental stress factors can cause a transient decrease in the energy charge of the cells, resulting in the activation of ATP-regenerating reactions and the inhibition of ATP-utilizing reactions (Schaechter, 2001); (4) Fatty acid biosynthesis (FabB, AccC-1 and FabBx1) (Santos et al., 2004). Regulation of membrane fluidity as a response to aromatic hydrocarbons in bacterial cells may be exerted by the modification of the relative levels of saturated/unsaturated fatty acids present in membrane phospholipids (Ramos et al., 2002) and as a recovery mechanism for damaged membrane phospholipids resulting from phenol-induced oxidative stress (Rodriguesa et al., 2010); (5) Inhibition of cell division (MinD).

The inhibition of growth may act as a protective mechanism to prevent cells from spending energy in cell division or to avoid multiplication of mutated cells resulting from DNA oxidative damage caused by phenol-induced ROS (Winn, 2003); (6) Cell envelope biosynthesis (LpxC, VacJ, and MurA); (7) Transcription regulation (OmpR and Fur); and (8) Transport of small molecules (TolC, BraC, AotJ, AapJ, FbpA and OprQ). Among the down-regulated proteins are those involved in nucleotide biosynthesis (PurM, PurL, PyrH and Dcd) and cell motility (FliC) (Santos et al., 2004). Another study on the proteome level of *Pseudomonas putida* DOT-T1E which were grown in the presence of toluene showed drastic changes in the protein pattern observed in 2D gel electrophoresis.

The identified proteins which were increased in the presence of toluene can be classified into four groups: (1) proteins involved in the catabolism of toluene; (2) proteins involved in the channeling of metabolic intermediates to the Krebs cycle and activation of purine biosynthesis; (3) proteins involved in sugar transport; (4) stress-related proteins. The high energy demand which can be seen by the elevation of the expressed proteins in groups 2 and

3 shows that the bacterial cells' tolerance to organic solvents maybe achieved via activation of the cell metabolism (Segura et al., 2005).

Transport Proteins: Efflux Pumps and Porins

Efflux Pumps

The efflux pumps can be divided in two major groups, primary and secondary, based on structural features and bio-energy needed for their activation. Primary efflux pumps hydrolyze ATP as a source of energy and are also called ATP-binding cassette (ABC) transporters. Secondary efflux pumps utilize the proton gradient provided by the proton motive force resulting from electron transport. Primary efflux pumps (ABC transporters) are ubiquitous membrane systems involved in different transport functions such as the efflux of toxins, metabolites and drugs (Putman et al., 2000; Marquez, 2005). The secondary efflux pumps are classified into four groups: (1) The facilitator superfamily, a ubiquitous system that ensures transport of sugars, intermediate metabolites and drugs (Saier et al., 1999); (2) The resistance nodulation and cell division (RND) family that is involved in the transport of lipophilic or amphiphilic molecules or toxic divalent cations and is also responsible for the solvent-tolerance of some bacterial strains (Fernandes et al., 2003). All RND efflux pumps consist of three proteins: the membrane fusion protein located in the periplasmic space of the cell envelope, the transporter protein which performs the actual transport of the agent from the periplasm/cytoplasm, and the TolC protein which is an outer membrane protein creating the channel for transporting the agent (Amaral et al., 2007); (3) The small multi-drug resistance SMR family which is involved in the efflux of lipophilic cationic drugs in bacteria; (4) The multidrug and toxic compound extrusion (MATE) family (Marquez, 2005).

The first RND efflux system for toluene was purified by Kieboom and was shown to be responsible for *P. putida* S12's tolerance to organic solvents (Kieboom *et al.*, 1998). The *ttgABC*, *ttgDEF* genes from *P. putida* DOT-T1E were shown to be involved with the efflux systems in the bacterial solvent-tolerance (Mosqueda & Ramos, 2000). The AcrAB-TolC efflux pump, which belongs to the RND family, enhances the release of solvents which accumulate intracellularly in *E. coli* bacterial cells (Tsukagoshi & Aono, 2000). AcrAB and AolC are marA/soxS/rob regulon genes MarA and SoxS proteins are transcriptional activators which belong to the AraC/XylS family (Barbosa & Levy, 2000).

Phenol exposure led to an increase (up to 35-fold) in protein subunits of known solvent efflux pump systems (e.g., TtgA, TtgC, Ttg2A, Ttg2C) (Rodriguesa et al., 2010). TtgC and TtgA are protein subunits of the efflux pump TtgABC, which is a RND efflux transporter associated with the tolerance of *P. putida* to toluene, styrene, xylene, ethlylbenzene and propylbenzene. Ttg2A and Ttg2C of an ABC-type transport system are known to be involved in the resistance of *P. putida* to organic solvents (Rodriguesa et al., 2010; Mosqueda et al., 1999). In P. putida DOT-T1E, the RND efflux pumps TtgABC are expressed constitutively, but TtgDEF is induced by aromatic hydrocarbons. The efflux pump TtgGHI in P. putida DOT-T1E was described by Rojas et al. (2001). The ttgGHI genes form an operon that is expressed at a higher level in the presence of toluene. Survival analysis of induced and non-induced cells revealed that the TtgABC and TtgGHI pumps extruded toluene, styrene, m-xylene, ethylbenzene, and propylbenzene, whereas the TtgDEF pump removed only toluene and styrene (Rojas et al., 2001; Duque et al., 2001). It was reported that P. putida S12 cells that survived a toluene shock carried an extra copy of the 2.6-kb insertion sequence ISS12 within the srpS gene which blocks srpS expression and thereby enhances the expression of the SrpABC efflux pump. It was proposed that P. putida S12 employs ISS12 as a specific mutator element in order to generate mutations that allow the bacterium to swiftly adapt to extreme solvent conditions (Wery et al., 2001).

Porins

Nearly half of the mass of the outer membrane consists of proteins. Most of these proteins are transmembranal pore-forming proteins called porins, which have a specific structure capable of producing water-filled channels for diffusion of molecules (Nikaido, 2003). The porins are important to bacterial physiology, as established by the use of porin-deficient mutants. Porins can be found as a trimeric structure or as monomers (Sugawara et al., 2006; Sugawara & Nikaido, 1992). Porins that form a trimeric structure exhibit high resistance to degradation by proteolytic enzymes, detergents and high temperatures (Avidan et al., 2008). The molecular weights of the monomers of the trimeric structure porins range from 30 to 50 kDa.

The two major classes of porins, the general and the specific, differ primarily in their substrate selectivity. Many Gram-negative bacterial cells possess a high number of general porins that provide an aqueous environment for the diffusion of small structurally hydrophilic molecules. The kinetics of diffusion through the general porins is affected by physicochemical properties such as size, shape, electrical charge, polarity and the concentration gradient of the molecule across the membrane. In contradistinction to general porins, the specific porins possess specific substrate binding sites for substrates such as carbohydrates or nucleosides (Maier et al., 1988).

This facilitated uptake is related to nutrient-deficient conditions when the simple diffusion of nutrients through the general porins becomes rate-limiting for growth. Many bacterial cells, for example *Escherichia coli*, possess a large number of general porins and few specific ones, while *Pseudomonas* sp. and other closely related soil bacterial cells use specific porins almost exclusively for uptake through the outer membrane. The specific porins of *P. aeruginosa* include the glucose-specific porin OprB, the phosphate- and polyphosphate-specific porins OprP and OprO, respectively, and the basic amino acid-specific porin OprD. The OprD family (consisting of 18 homologous species) was first described for *P. aeruginosa*. Since then, this family was found in over 100 metabolically versatile soil bacterial strains (Mampel et al., 2004). A third class of porins consists of the TolC family, which link an outer membrane channel with an inner membrane export system and thus bridge the periplasmic space (Koronakis et al., 2000). In *E. coli*, two major cation-selective porins, OmpC and OmpF, are synthesized under standard laboratory conditions, while the anion-selective porin PhoE is produced under phosphate limitation (Geldera et al., 2000).

Several studies showed a change in the porin type as well as their concentration as a function of exposure to aromatic hydrocarbons. The exposure of *P. putida* KT2440 cells to 800 mg/L phenol resulted in a decrease in the content of the porins OprB, OprF, OprG and OprQ. The porin OprB was previously related to phenol influx in *P. putida* KT2440 cells

(Kivistik et al., 2006). The decrease in porins as a response to phenol exposure may be partially mediated by post-translational modifications, as suggested by the relative content of the multiple forms of identified porins. Therefore, exposure to phenol may trigger an adaptive decrease in outer membrane permeability by decreasing the porin concentrations. This cellular response was previously hypothesized and may act in coordination with the activation of efflux pump systems to reduce the intracellular concentration of phenol (Isken & de Bont, 1998).

The abundance of the highly conserved porin OprF form with a higher molecular weight was found to increase following exposure to phenol, whereas the lower molecular weight form was drastically reduced. The OprF in *P. aeruginosa*, appears in two different conformations, the closed and open channels. The closed conformer constitutes an N-terminal β-barrel domain and a C-terminal periplasmic α-helix domain and the open channel conformer constitutes a single β-barrel domain that possesses a larger number of β-strands compared with the closed channel conformer. The closed channel conformer, which comprises 95% of the OprF population, plays an important function in the stabilization of the bacterial cell envelope structure (Nestorovich et al., 2006; Rawling et al., 1998), while the open channel conformer, which is only approximately 5% of the OprF population, is responsible for the protein activity (Nestorovich et al., 2006; Rodrigues et al., 2010). The abundance of the outer membrane porin OmpC increased 32.4, 21.9, and 23.4-fold in succinate-utilizing cells of strain EbN1 shocked with *p*-cresol, phenol, and the solvent mixture, respectively (Trautwein et al., 2008). Similarly, solvent-tolerant *P. putida* S12 up-regulated the OprH porin (12-fold) upon toluene stress, probably in order to stabilize the outer membrane (Volkers et al., 2006).

In conclusion, the response to phenol involves an increase in different solvent efflux pump systems that export toxic organic solvents to the external medium, together with a decreased content of porins, presumably leading to decreased cell permeability. These coordinated responses may also involve post-translational modifications that contribute to the decrease in the solvent concentration inside the cells. There is also induction of the proteins of the protective mechanisms in order to ensure cell surface and cell structure integrity as well as to increase the activity of the respiratory chain to supply the ATP demands.

References

Agarry, S.E., Durojaiye, A.O., Solomon, B.O. (2008). Microbial degradation of phenols: a review. *Int. J. Environ. Pollut.* 32, 12-28.

Aislabi, J., Foght, J., Saul, D. (2000). Aromatic hydrocarbon-degrading bacteria from soil near Scott Base, Antarctica. *Polar Biol.* 23, 183-188.

Amaral, L., Engi, H., Viveiros, M., Molnar, J. (2007). Comparison of multidrug resistant efflux pumps of cancer and bacterial cells with respect to the same inhibitory agents. *In Vivo* 21, 237-244.

Avidan, O., Kaltageser, E., Pechatnikov, I., Wexler, H.M., Shainskaya, A., Nitzan, Y. (2008). Isolation and characterization of porins from *Desulfovibrio piger* and *Bilophila wadsworthia*: structure and gene sequencing. *Arch. Microbiol.* 190, 641–650.

Barbosa, T.M., Levy, S.B. (2000). Differential expression of over 60 chromosomal genes in *Escherichia coli* by constitutive expression of MarA. *J. Bacteriol.* 182, 3467–3474.

Bernadsky, G., Beveridge, T.J. Clarke. A.J. (1994). Analysis of the sodium dodecyl sulfate-stable peptidoglycan autolysins of select gram-negative pathogens by using renaturing polyacrylamide gel electrophoresis. *J. Bacteriol.* 176, 5225-5232.

Beveridge, T.J. (1999). Structures of Gram-negative cell walls and their derived membrane vesicles. *J. Bacteriol.* 181, 4725-4733.

Beveridge, T.J., Makin, S.A., Kadurugamuwa, J.L. Li, Z. (1997). Interactions between biofilms and the environment. *FEMS Microbiol.* Rev. 20, 291-303.

Cafaro, V., Izzo, V., Scognamiglio, R., Notomista, E., Capasso, P., Casbarra, A., Pucci, P., Donato, A. (2004). Phenol Hydroxylase and Toluene/*o*-Xylene Monooxygenase from *Pseudomonas stutzeri* OX1: Interplay between two enzymes. *Appl. Environ. Microbiol.* 70, 2211-2219.

Chakraborty, R., O'Connor, S.M., Chan, E., Coates, J.D. (2005). Anaerobic degradation of benzene, toluene, ethylbenzene, and xylene compounds by *Dechloromonas* strain RCB. *Appl. Environ. Microbiol.* 71, 8649-8655.

Chen, Q., Janssen, D.B., Witholt, B. (1995a) Growth on octane alters the membrane lipid fatty acids of *Pseudomonas oleovorans* due to the induction of *alkB* and synthesis of octanol. *J. Bacteriol.* 177, 6894-6901.

Chen, Q., Nijenhuis, A., Preusting, H., Dolfing, J., Janssen, D.B., Witholt, B. (1995b). Effects of octane on the fatty acid composition and transition temperature of *Pseudomonas oleovorans* membrane lipids during growth in two-liquid phase continuous cultures. *Enz. Microb. Technol.* 17, 647-652.

De Bont, J.A.M. (1998). Solvent-tolerant bacteria in biocatalysis. *Trends Biotechnol.* 16, 493-499.

De Carvalho, C.C.C.R., Cruz, A., Angelova, B., Fernandes, P., Pons, M.N., Pinheiro, H.M., Cabral, J.M.S., da Fonseca, M.M.R. (2004). Behaviour of *Mycobacterium* sp. NRRL B-3805 whole cells in aqueous, organic-aqueous and organic media studied by fluorescence microscopy. *Appl. Microbiol. Biotechnol.* 64, 695-701.

De Carvalho, C.C.C.R., da Fonseca, M.M.R. (2004). Principal component analysis applied to bacterial cell behavior in the presence of organic solvents. *Biocatalysis and Biotransformation.* 22, 203-214.

Duque, E., Segura, A., Mosqueda, G., Ramos, J.L. (2001). Global and cognate regulators control the expression of the organic solvent efflux pumps TtgABC and TtgDEF of Pseudomonas putida. *Mol. Microbiol.* 39, 1100–1106.

Fernandes, P., Ferreira, B.S., Cabral, J.M.S. (2003). Solvent tolerance in bacteria: role of efflux pumps and cross-resistance with antibiotics. *Int. J. Antimicrob. Agents.* 22, 211-216.

Geldera, P.V., Dumasa, F., Winterhalter, M. (2000). Understanding the function of bacterial outer membrane channels by reconstitution into black lipid. *Biophys. Chem.* 85, 153-167.

Heipieper, H.J., de Bont, J.A.M. (1994). Adaptation of *Pseudomonas putida* S12 to ethanol and toluene at the level of fatty acid composition of membranes. *Appl. Environ. Microbiol.* 60, 4440-4444.

Heipieper, H.J., Diefenbach, R., Keweloh, H. (1992). Conversion of *cis* unsaturated fatty acids to *trans*, a possible mechanism for the protection of phenol-degrading

Pseudomonas putida P8 from substrate toxicity. *Appl. Environ. Microbiol.* 58, 1847-1852.

Heipieper, H.J., Meinhardt, F., Segura, A. (2003). The *cis-trans* isomerase of unsaturated fatty acids in *Pseudomonas* and *Vibrio*: biochemistry, molecular biology and physiological function of a unique stress adaptive mechanism. *FEMS Microbiol. Lett.* 229, 1-7.

Heipieper, H.J., Meulenbeld, G., van Oirschot, Q., de Bont, J.A.M. (1996). Effect of environmental factors on the *trans/cis* ratio of unsaturated fatty acids in *Pseudomonas putida* S12. *Appl. Environ. Microbiol.* 62, 2773-2777.

Inoue, A., Horikoshi, K. (1989). A Pseudomonas thrives in high concentrations of toluene. *Nature.* 338, 264–266.

Isken, S., de Bont, J.A.M. (1998). Bacteria tolerant to organic solvents. *Extremophiles.* 2, 229-238.

Isken, S., Heipieper, H.J. (2002). Toxicity of organic solvents to microorganisms. In: encyclopedia of environmental microbiology. Bitton G. (Ed.) John Wiley, New York. Vol. 6, 3147-3155.

Kadurugamuwa, J. L., Beveridge, T. J. (1995). Virulence factors are released from *Pseudomonas aeruginosa* in association with membrane vesicles during normal growth and exposure to gentamicin: a novel mechanism of enzyme secretion. *J. Bacteriol.* 177, 3998-4008.

Keweloh, H., Heipieper, H.J. (1996). *Trans* unsaturated fatty acids in bacteria. *Lipids* 31, 129-137.

Kieboom, J., Dennis, J.J., de Bont, J. A. M., Zylstra, G. J. (1998). Identification and molecular characterization of an efflux pump involved in *Pseudomonas putida* S12 solvent tolerance. *J. Biol. Chem.* 273, 85-91.

Kiran, M.D., Prakash, J.S., Annapoorni, S., Dube, S., Kusano, T., Okuyama, H., Murata, N., Shivaji, S. (2004). Psychrophilic *Pseudomonas syringae* requires *trans*-monounsaturated fatty acid for growth at higher temperature. *Extremophiles* 8: 401-410.

Kivistik, P.A., Putrins, M., Puvi, K., Ilves, H., Kivisaar, M., Horak, R. (2006). The ColRS two-component system regulates membrane functions and protects *Pseudomonas putida* against phenol. *J. Bacteriol.* 188, 8109-8117.

Kobayashi, H., Hirayama, K.H., Horikoshi, K. (2000). Novel toluene elimination system in a toluene-tolerant microorganism. *J. Bacteriol.* 182, 6451-6455.

Koronakis, V., Sharff, A., Koronakis, E., Luisi, B., Hughes, C. (2000). Crystal structure of the bacterial membrane protein TolC central to multidrug efflux and protein export. *Nature.* 405, 914-919.

Kulp, A., Kuehn, M.J. (2010). Biological functions and biogenesis of secreted bacterial outer membrane vesicles. *Annu. Rev. Microbiol.* 64, 163–184.

Lazaroaie, M.M. (2008). *Pseudomonas aeruginosa* IBBML1 Adaptation to high concentrations of hydrocarbons. *Electronic. J. Biology.* 4, 17-26.

Li, Z., Clarke, A. J., Beveridge, T. J. (1996). A major autolysin of *Pseudomonas aeruginosa*, its subcellular distribution, its potential role in cell growth and division, and its secretion in surface membrane vesicles. *J. Bacteriol.* 178, 2479-2488.

Li, Z., Clarke, A. J., Beveridge, T. J. (1998). Gram-negative bacteria produce membrane vesicles which are capable of killing other bacteria. *J. Bacteriol.* 180, 5478-5483.

Maier, C., Bremer, E., Schmid, A., Benz, R. (1988). Pore-forming activity of the Tsx protein from the outer membrane of Escherichia coli: demonstration of a nucleoside-specific binding site. *J. Biol. Chem.* 263, 2493-2499.

Mampel, J., Maier, E., Tralau, T., Ruff, J., Benz, R., Cook, A.M. (2004). A novel outer-membrane anion channel (porin) as part of a putatively two-component transport system for 4-toluenesulphonate in *Comamonas testosteroni* T-2. *Biochem J.* 383, 91–99.

Marquez, B. (2005). Bacterial efflux systems and efflux pumps inhibitors. *Biochimie.* 87, 1137-1147.

Mattick, K.L., Jørgensen, F., Legan, J.D., Cole, M.B., Porter, J., Lappin-Scott, H.M., Humphrey, T.J. (2000). Survival and filamentation of *Salmonella enterica* Serovar *Enteritidis* PT4 and *Salmonella enteric* Serovar *Typhimurium* DT104 at low water activity. *Appl. Environ. Microbiol.* 66, 1274-1279.

McBroom, A.J., Kuehn, M.J. (2007). Release of outer membrane vesicles by Gram-negative bacteria is a novel envelope stress response. *Mol. Microbiol.* 63, 545-558.

Mosqueda, G., Ramos, G.M.I., Ramos, J.L. (1999). Toluene metabolism by the solvent-tolerant Pseudomonas putida DOT-T1 strain, and its role in solvent impermeabilization. *Gene.* 232, 69-76.

Mosqueda, G., Ramos, J.L. (2000). A set of genes encoding a second toluene efflux system in *Pseudomonas putida* DOT-T1E is linked to the *tod* genes for toluene metabolism. *J. Bacteriol.* 182, 937-943.

Nestorovich, E.M., Sugawara, E., Nikaido, H., Bezrukov. S.M. (2006). *Pseudomonas aeruginosa* porin OprF: properties of the channel. *J. Biol. Chem.* 281, 16230–16237.

Neumann, G., Veeranagouda, Y., Karegoudar, T.B., Sahin, O., Mäusezahl, I., Kabelitz, N., Kappelmeyer, U., Heipieper, H.J. (2005). Cells of *Pseudomonas putida* and *Enterobacter* sp. adapt to toxic organic compounds by increasing their size. *Extremophiles.* 9, 163-168.

Nikaido, H. (2003). Molecular basis of bacterial outer membrane permeability revisited. *Microbiol. Mol. Biol. Rev.* 67, 593- 656.

Panilaitis, B., Castro, G.R., Solaiman, D., Kaplan, D.L. (2007). Biosynthesis of emulsan biopolymers from agro-based feedstocks. *J. Appl. Microbiol.* 102, 531–537.

Pinkart, H.C., Wolfram, J.W., Rogers, R., White, D.C. (1996). Cell envelope changes in solvent-tolerant and solvent-sensitive *Pseudomonas putida* strains following exposure to *o*-xylene. *Appl. Environ. Microbiol.* 62, 1129–1132.

Putman, M., Van Veen, H.W., Konings, W.N. (2000). Molecular properties of bacterial multidrug transporters. *Microbiol. Mol. Biol. Rev.* 64, 672–693.

Ramos, J.L., Duque, E., Gallegos, M.T., Godoy, P., Ramos-Gonzalez, M.I., Rojas A., Teran, W., Segura, A. (2002). Mechanisms of solvent tolerance in Gram-negative bacteria. *Annu. Rev. Microbiol.* 56, 743-768.

Ramos, J.L., Duque, E., Rodriguez-Herva, J.J., Godoy, P., Haidour, A., Reyes, F., Fernandez-Barrero, A. (1997). Mechanisms for solvent tolerance in bacteria. *J. Biol. Chem.* 272, 3887-3890.

Rawling, E.G., Brinkman, F.S., Hancock, R.E. (1998). Roles of the carboxy-terminal half of *Pseudomonas aeruginosa* major outer membrane protein OprF in cell shape, growth in low-osmolarity medium, and peptidoglycan association. *J. Bacteriol.* 180, 3556-3562.

Rhee, S.K., Lee, G.M., Kim, Y.B., Lee, S.T. (1996). Effect of pyridine on the fatty acid composition of *Pimelobacter* sp. *FEMS Microbiol. Lett.* 141, 139-143.

Rodrigues, C.R., Santos, P.M., Benndorf, D., Rapp, E., Sá-Correiaa, I. (2010). Response of *Pseudomonas putida* KT2440 to phenol at the level of membrane proteome. *J. Proteomics*. 73, 1461-1478.

Rojas, A., Duque, E., Mosqueda, G., Golden, G., Hurtado, A., Ramos, J.L., Segura, A. (2001). Three efflux pumps are required to provide efficient tolerance to toluene in *Pseudomonas putida* DOT-T1E. *J. Bacteriol*. 183, 3967-3973.

Saier, M.H., Beatty, J.T., Goffeau, A., Harley, K.T., Heijne, W.H., Huang, S.C., Jack, D.L., Jahn, P.S., Lew, K., Liu, J., Pao, S.S., Paulsen, I.T., Tseng, T.T., Virk, P.S. (1999). The major facilitator superfamily. *J. Mol. Microbiol. Biotechnol*. 2, 257–279.

Santos, P.M., Benndorf, D., Correia, I. (2004). Insights into *Pseudomonas putida* KT2440 response to phenol-induced stress by quantitative proteomics. *Proteomics*. 4, 2640–2652.

Segura, A., Godoy, P., van Dillewijn, P., Hurtado, A., Arroyo, N., Santacruz, S., Ramos, J.L. (2005). Proteomic analysis reveals the participation of energy- and stress-related proteins in the response of *Pseudomonas putida* DOT-T1E to toluene. *J. Bacteriol.* 187, 5937–5945.

Schaechter, M. (2001). *Escherichia coli* and *Salmonella* 2000: the view from here, *Microbiol. Mol. Biol. Rev*. 65, 119–130.

Schirmer, T. (1998). General and specific porins from bacterial outer membranes. *J. Struct. Biol.* 121, 101-109.

Sikkema, J., De Bont, J.A. M., Poolman, B. (1995). Mechanisms of membrane toxicity of hydrocarbons. *Microbiol. Rev*. 59, 201–222.

Sugawara, E., Nestorovich, E. M., Bezrukov, S. M., Nikaido, H. (2006). *Pseudomonas aeruginosa* porin OprF exists in two different conformations. *J. Biol. Chem.* 281, 16220–16229.

Sugawara, E., Nikaido, H. (1992). Pore-forming activity of OmpA protein of *Escherichia coli*. *J. Biol. Chem*. 267, 2507–2511.

Torres, S., Martínez, M.A., Pandey, A., Castro, G.R. (2009). An organic–solvent-tolerant esterase from thermophilic Bacillus licheniformis S-86. *Bioresour. Technol.* 100, 896–902.

Trautwein, K., Kühner, S., Wöhlbrand, L., Halder, T., Kuchta, K., Steinbüchel, A., Rabus, R. (2008). Solvent stress response of the denitrifying bacterium "Aromatoleum aromaticum" Strain EbN1. *Appl. Environ. Microbiol.* 74, 2267–2274.

Tsitko, I.V., Zaitsev, G.M., Lobanok, A.G., Salkinoja-Salonen, M.S. (1999). Effect of aromatic compounds on cellular fatty acid composition of *Rhodococcus opacus. Appl. Environ. Microbiol.* 65, 853-855.

Tsukagoshi, N., Aono, R. (2000). Entry into and release of solvents by *Escherichia coli* in an organic-aqueous two-liquid-phase system and substrate specificity of the AcrAB-TolC solvent-extruding pump. *J. Bacteriol*. 182, 4803–4810.

Volkers, R. J.M., de Jong, A.L. Hulst, A. G. van Baar, B.L.M., de Bont, J.A.M., Wery, J. (2006). Chemostat-based proteomic analysis of toluene-affected *Pseudomonas putida* S12. *Environ. Microbiol*. 8, 1674-1679.

von Wallbrunn, A., Richnow, H.H., Neumann, G., Meinhardt, F., Heipieper, H.J. (2003). Mechanism of *cis-trans* isomerization of unsaturated fatty acids in *Pseudomonas putida*. *J. Bacteriol*. 185, 1730-1733.

Weber, F. J., de Bont, J.A.M. (1996). Adaptation mechanisms of microorganisms to the toxic effects of organic solvent on membranes. *Biochim. Biophys. Acta.*1286, 225–245.

Weber, F.J., Isken, S., de Bont, J.A.M. (1994). *Cis/trans* isomerization of fatty acids as a defense mechanism of *Pseudomonas putida* strains to toxic concentrations of toluene. *Microbiology*. 140, 2013-2017.

Wery, J., Hidayat, B., Kieboom, J., de Bont, J.A. (2001). *An insertion sequence prepares* Pseudomonas putida *S12 for severe solvent stress. J. Biol. Chem.* 276, 5700-*5706.*

Winn, L. M. (2003). Homologous recombination initiated by benzene metabolites: A potential role of oxidative stress. *Toxicol. Sci.* 72, 143–149.

Zahir, Z., Seed, K.D., Dennis, J.J. (2006). Isolation and characterization of novel organic solvent-tolerant bacteria. *Extremophiles* 10, 129–138.

Zhou, L., Srisatjaluk, R., Justus, D.E., Doyle, R.J. (1998). On the origin of membrane vesicles in Gram-negative bacteria. *FEMS Microbiol. Lett.* 163, 223–228.

In: Prokaryotes
Editor: Marina Nisnevitch

ISBN: 978-1-63321-592-4

Chapter 5

Hfq-Like RNA Chaperones and Small Non-Coding Regulatory RNAs: A New Paradigm of Bacterial Gene Regulation

Jorge H. Leitão[1,2*], Christian G. Ramos[2], Joana R. Feliciano[2], Sílvia A. Sousa[2], André M. Grilo[2] and Paulo J. P. da Costa[2]
[1]Department of Bioengineering
[2]Institute for Biotechnology and Bioengineering.
Instituto Superior Técnico, Universidade de Lisboa, Lisboa, Portugal

Abstract

The Hfq protein was discovered almost forty years ago as an *Escherichia coli* host factor required for the replication of the RNA genome of the bacteriophage Qβ. However, the exact role of the protein as a global regulator of bacterial physiology and metabolism emerged much later, through the characterization of the broad phenotypes exhibited by *hfq* mutants in *E. coli* and in the zoonotic pathogen *Brucella abortus*. Recent studies have shown that the protein also plays a critical role on the pathogenicity of several Gram-negative and Gram-positive pathogens. The understanding of the pleiotropic phenotypes exhibited by *hfq* mutants emerged recently with evidence that Hfq plays a major role in the post-transcriptional regulation of gene expression, by facilitating the interactions between small non-coding regulatory RNAs and their target mRNAs. This knowledge added a new layer of complexity to our understanding of prokaryotic gene regulation. This chapter is focused on the protein Hfq and on its roles on pleiotropic and virulence phenotypes in several bacteria, including recent results from our research group on the roles of Hfq-like proteins on stress resistance and virulence of bacteria of the *Burkholderia cepacia* complex (Bcc). The role of Hfq as a mediator of interactions of sRNAs with their target mRNAs will also be presented.

Keywords: RNA chaperone Hfq, small non-coding RNAs, *Burkholderia cepacia* complex

* Corresponding author: Email: jorgeleitao@tecnico.ulisboa.pt.

Introduction

Most of our current knowledge on the structure and functions of Hfq derives from the *E. coli* protein, the first to be described (Kajitani and Ishihama, 1991). In this bacterium, Hfq is an abundant protein, estimated by Kajitani et al. (Kajitani et al., 1994) to be present in numbers as high as 30,000 to 60,000 molecules per cell, largely associated with ribosomes and accounting for 0.04 to 0.06 % of the total protein content of the cell during the exponential phase of growth.

The *E. coli* Hfq protein, together with other prokaryotic Hfq-like proteins and proteins of archaeal and eukaryotic origin, form the large Sm/LSm protein superfamily (Vogel and Luisi, 2011; Tan et al., 1988; Séraphin, 1995). This superfamily was named Sm/LSm after the discovery that autoimmune antibodies from patients with systemic lupus erythematous recognize an antigen, named Sm, in a group of proteins from these patients (Tan et al., 1988), and the identification, by sensitive database searches, of a group of proteins with no clear counterpart with the human Sm protein complex, therefore named Sm-like proteins (LSm) (Séraphin, 1995). LSm proteins are also found in single-cell eukaryotes and in archaea. Members of this superfamily are involved in a variety of RNA processing events, including pre-mRNA splicing, telomere replication, histone-mRNA processing, and mRNA degradation (Wilusz and Wilusz, 2005; Valentin-Hansen et al., 2004).

The *E. coli* Hfq protein is composed of 102 amino acid residues. Like other members of the Sm/LSm superfamily, Hfq possesses a conserved bipartite motif consisting of two regions named Sm1 and Sm2 and known to be involved in RNA binding, interrupted by a spacer region of variable length in other Sm/LSm members (Figure 1). In the *E. coli* protein, this conserved core spans amino acid residues 7 to 66. The conserved core of this protein, as well as that of Hfqs from other bacteria, forms a α-helix in the N-terminus, followed by five-stranded antiparallel β-sheets (Figure 1).

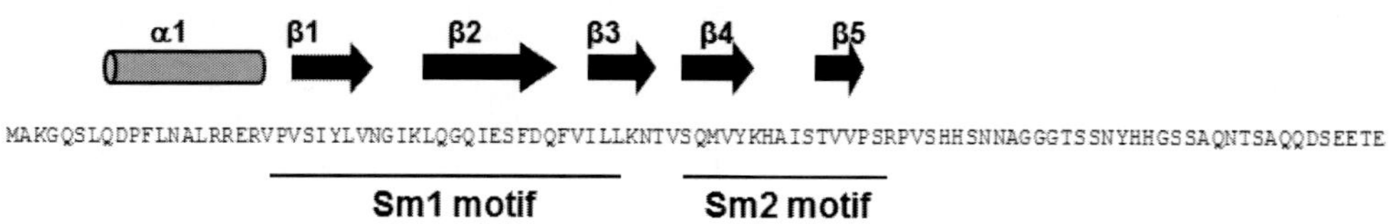

Figure 1. Schematic representation of the amino acid sequence of the *E. coli* Hfq protein, the secondary structure (upper part), and the conserved Sm1 and Sm2 motifs (lower part).

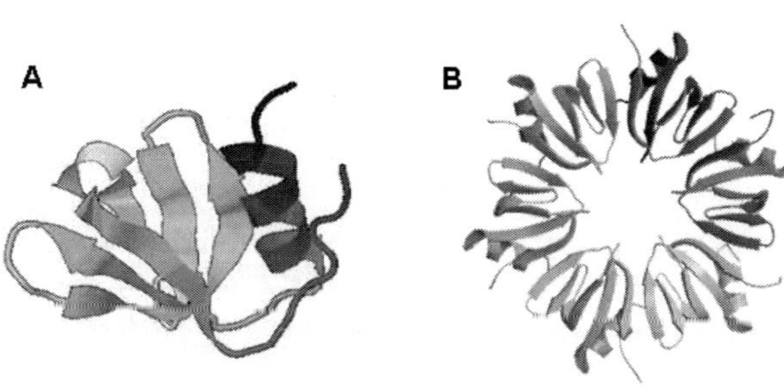

Figure 2. Monomeric (A) and hexameric (B) structures of the Hfq protein. Structures were modelled using the RasWin freeware software.

This structure enables the protein to assemble into a quaternary ring-shaped homo-hexameric structure, with the histidine 57 residue playing a crucial role in this assembling (Figure 2). Evidence from structural studies suggests that the C-terminus region of the protein adopts a disordered structure which extends outward from the hexameric ring (Figure 2). Evidence also shows that when the protein is in the hexameric ring-like structure, the inner proximal face of the Sm domain preferentially binds A/U-rich RNAs, while the distal face exhibits a higher affinity for A-rich RNA, namely polyadenylated RNAs (Fender et al., 2011).

Mutation of *Hfq* Results in Pleiotropic Phenotypes

The initial descriptions of the roles played by Hfq on bacterial physiology were made by Tsui et al. (1994). These authors described pronounced pleiotropic phenotypes in association with mutations in the *hfq* gene in *E. coli*. The phenotypes included decreased growth rates and biomass yields, decreased negative supercoiling of plasmids in stationary phase, increased cell size, increased susceptibility to osmotic stress, increased oxidation of carbon sources, and increased sensitivity to ultraviolet light (Tsui et al., 1994). More recent work with uropathogenic and enterohemorrhagic *E. coli* strains showed that Hfq also plays a critical role on the virulence and pathogenicity (Kulesus et al., 2008). For instance, the protein was found as critical for bladder and kidneys effective colonization in a murine urinary tract infection model by the uropathogenic *E. coli* isolate UTI89 (Kulesus et al., 2008). Although in this clinical isolate a non-functional *hfq* did not affect the adherence or the invasiveness, the development of intracellular micro-colonies was impaired when the bacterium located within the terminally differentiated epithelial cells lining the lumen of the bladder (Kulesus et al., 2008). The enterohemorrhagic *E. coli* O157:H7 strain EDL933 contains a pathogenicity island named the Locus of Enterocyte Effacement (LEE). The LEE locus encodes most of the proteins of a type III secretion system, as well as proteins involved in attachment to erythrocytes and effacing lesions. Kendall et al. (2011) found that in the hfq deletion mutant the levels of transcripts originating from the LEE pathogenicity island and of those corresponding to the *qseBC* two-component regulatory system were down-regulated, while those corresponding to the *stx2*AB system encoding the deadly shiga toxin were up-regulated.

Hfq Is Required for Virulence in Several Human Pathogens

The comparison of virulence of *hfq* mutants and their respective wild-type strains of various bacterial pathogens using several models of infection, have evidenced the roles played by Hfq on the regulation of expression of a wide range of virulence-related traits.

Initial evidence on the involvement of Hfq on virulence came from research performed with the zoonotic pathogen *Brucella abortus*, an intracellular pathogen that causes abortion among domestic animals, and the severe persistent febrile disease brucellosis in humans. In this pathogen, an *hfq* mutant strain exhibited increased sensitivity to hydrogen peroxide, and a decreased ability to survive to acidic conditions and to prolonged starvation conditions

(Robertson and Roop, 1999). In addition, the mutant failed to replicate within cultured macrophages, and was rapidly cleared from the spleens and livers of infected mice (Robertson and Roop, 1999).

In the pneumonia causative agent *Klebsiella pneumoniae*, mutation of the *hfq* gene resulted in an attenuated virulence phenotype in a mouse model of infection, in altered capsular polysaccharide production, and in tolerance to stresses and homeostasis of the cell envelope (Chiang et al., 2011). In addition, microarray data obtained by Chiang et al. (2011) indicated that about one fifth of the genes of *K. pneumoniae* were differentially expressed in an *hfq* mutant. The authors suggested that in this bacterial species, Hfq directly regulates individual genes expression by regulating the sigma factors RpoS and RpoE, being positioned at a hierarchical upper level that allows the control of both the physiological fitness and virulence potential (Chiang et al., 2011).

In the intracellular pathogen *Legionella*, an *hfq* deletion mutant exhibited defects in growth and pigmentation, as well as slight defects in the ability of the pathogen to invade and multiply intracellularly in the amoeba and the macrophage infection models (McNealy et al., 2005). The Hfq regulatory networks in *Legionella* were proposed to be quite distinct from those in other gram-negative bacteria (McNealy et al., 2005).

Hfq was also found to play an important role in another intracellular pathogen, *Francisellla tularensis*, responsible for tularemia in humans and in a large number of animal species. Deletion of the *hfq* gene in this intracellular pathogen was shown by Meibom et al. (2009) to lead to pleiotropic phenotypes, such as reduced growth rate, increased susceptibility to cellular stresses and severely attenuated virulence in a mice model of infection.

Further evidence of the impact of Hfq on the microbial cell physiology and virulence come from the plague disease causing agent *Yersinia pestis*. Hfq was recently found to play an important role on this microorganism´s virulence by regulating the intracellular levels of the second messenger c-di-GMP levels and biofilm formation (Bellows et al., 2012). The protein was shown to promote the accumulation of the mRNA corresponding to the phosphodiesterase HmsP and to destabilize the mRNA transcript corresponding to the diguanylate cyclase HmsT (Bellows et al., 2012). Hfq was therefore proposed to be used by *Y. pestis* to link c-di-GMP levels to environmental conditions. In addition, the lack of a functional Hfq resulted in the severely attenuated virulence of *Y. pestis* in animal models of plague (Bellows et al., 2012). In a more recent study by Rempe et al. (2012), the protein was found as essential for the infection of the flea gut by *Y. pestis*.

The protein Hfq was also found to play important roles in the bacterial species *Pseudomonas aeruginosa*, the *Burkholderia cepacia* complex (Bcc), and *Stenotrophomonas maltophilia*, which are known pathogens to patients suffering from the genetic disease cystic fibrosis (CF). Mutation of the *hfq* gene in *Pseudomonas aeruginosa* was found by Sonnleitner et al. (2003) to lead to pleiotropic phenotypes and reduced virulence in the *Galleria melonella* and the mice infection models. Interestingly, the phenotypes exhibited by the mutant include impaired production of the exopolysaccharide alginate. Alginate is one of the virulence factors secreted by the bacterium when infecting the respiratory tract of patients suffering from the genetic disease cystic fibrosis, known to allow the bacterium to evade the host immune system, and to contribute to its persistence in the CF lung (Govan and Deretic, 1996). Other observed phenotypes included reduced growth specific rate and yield upon batch growth in different media, reduced production of the exotoxins elastase and pyocyanin, and impaired swimming, twitching, and swarming motilities (Sonnleitner et al., 2003).

In the case of the opportunistic human pathogen *Stenotrophomonas maltophilia*, studies using a deletion mutant showed that the *hfq* gene is required for biofilm formation and cell motility (Roscetto et al., 2012). The *hfq* mutant presented an increased susceptibility to tobramycin and amikacin, and an increased resistance to tetracycline, chloramphenicol and ciprofloxacin, compared to the wild-type (Roscetto et al., 2012). Remarkably, and opposed to the findings described in other opportunistic pathogens, no differences were detected in the resistance of the *hfq* deletion mutant and the wild-type strain to osmotic or oxidative stresses (Roscetto et al., 2012).

The *Burkholderia cepacia* complex (Bcc) is a group of plant, animal and human opportunistic pathogens, distributed by at least seventeen closely related bacterial species of the β-proteobacteria (Mahenthiralingam et al., 2005), widely distributed by both natural and man-made habitats. These bacteria emerged in the early 1980s as important pathogens, especially to patients with CF. Infections by Bcc are particularly feared by CF patients and caregivers due to the unpredictable clinical outcome of infections, the well-documented easy patient-to-patient transmission of particularly epidemic isolates, and the difficult eradication of infections due to the intrinsic resistance of these bacteria to clinically-relevant antibiotics (Leitão et al., 2010; Leitão et al., 2008; Cunha et al., 2004; Govan et al., 1993).

These bacteria are singular for encoding in their genomes 2 distinct and functional Hfq-like proteins, a smaller protein named Hfq and composed of 79 amino acid residues, and a larger protein, named Hfq2 and composed of 188 amino acid residues (Sousa et al., 2010; Ramos et al., 2011). Despite the existence of 2 *hfq*-encoding genes in its genome, infection experiments with mutants derived from the highly transmissible clinical isolate *B. cenocepacia* J2315 showed that the mutants in either the *hfq* or the *hfq2* genes were impaired in their ability to infect the nematode *Caenorhabditis elegans* (Figure 3).

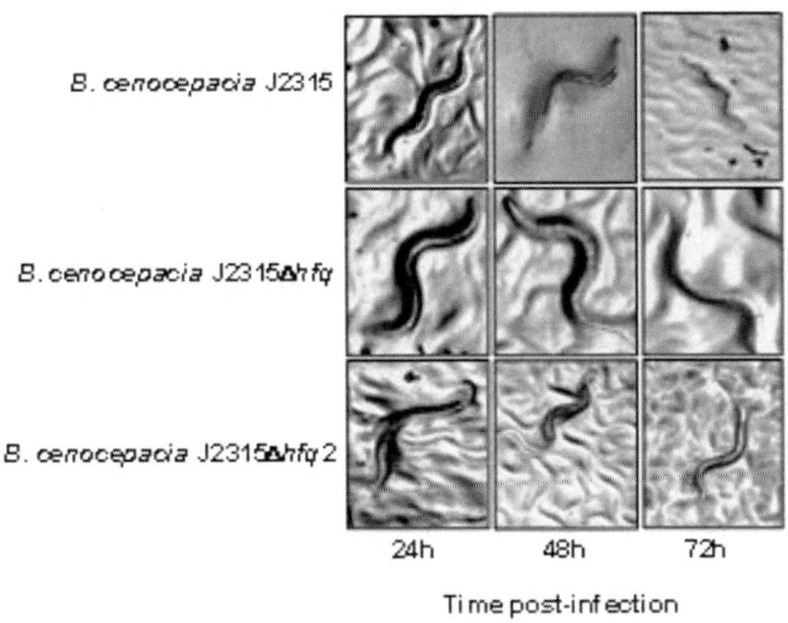

Figure 3. Visual appearance of *Caenorhabditis elegans* worms, after infection for the indicated time, by *B. cenocepacia* J2315 wild-type strain or *B. cenocepacia* J2315 mutants with a non-functional *hfq* (Δ*hfq*), or *hfq2* (Δ*hfq2*). Slimmer worms present a poor health status and a shorter life span.

Besides bacteria of the Bcc and some species of the *Burkholderia* genus, only a few other prokaryotes are known to encode 2 *hfq*-like genes within their genome sequences, namely the archaeal species *Methanobacterium thermoautotrophicum* and *Archaeoglobus fulgidus*, the bacterial species *Bacillus anthracis*, *Magnetospirillum magnetotacticum*, and *Novosphingobium aromativcivorans*. Nevertheless, it is worth to mention that so far, Bcc bacteria remain as the unique microbes bearing in their genomes 2 *hfq*-like genes that have been functionally characterized.

Hfq Proteins from Bacteria of the *Burkholderia Cepacia* Complex

The *B. cepacia hfq* gene was found after the screening of a mutant library prepared by plasposon random mutagenesis for exopolysaccharide (EPS)-deficient mutants. Upon isolation of one EPS-deficient mutant, the interrupted gene was rescued after total DNA extraction, restriction with EcoRI, self-ligation, and transformation of *E. coli* with the ligation mixture. Nucleotide sequence analysis revealed that the plasposon disrupted the promoter region of a gene encoding a protein 81% identical to the *E. coli* Hfq protein (Sousa et al., 2010). The *B. cepacia* Hfq protein is shorter than the *E. coli* protein, being composed of 79 amino acid residues, with a molecular weight and pI of 8.85 kDa and 8.1, respectively (Sousa et al., 2010). Sequence analysis showed that the protein contain the conserved motifs Sm1 and Sm2 (Figure 4). The *B. cepacia* Hfq protein was predicted to fold into a secondary structure remarkably similar to that of the *E. coli* protein, with a N-terminus secondary structure composed of an α-helix followed by five β-strands (Figure 4), and a C-terminus shorter than that of the *E. coli* protein, extending for only 14 amino acid residues, in a predicted disordered structure.

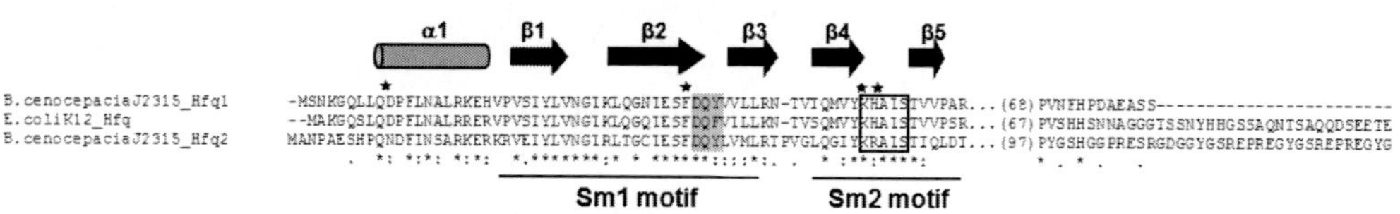

Figure 4. Sequence alignment of the *E. coli* Hfq protein and the *B. cenocepacia* J2315 Hfq and Hfq2 proteins, showing the conserved motifs Sm1 and Sm2, as well as a conserved secondary structure composed of one α-helix followed by five β-strands. The non-conserved C-terminal regions of the proteins were not included in this alignment. ★ indicate conserved residues that are involved in RNA binding by the *E. coli* Hfq. Loops 3 and 5 involved in RNA binding, are boxed in grey and enclosed in a rectangle, respectively. Identical amino acid residues are marked with an *, conserved or semi-conserved substitutions are marked with double dots or single dots, respectively.

Searches for homologues to the *B. cenocepacia hfq* gene within the genome sequences of other Bcc strains available through the Integrated Microbial Genomes database (IMG) (Markowitz et al., 2006) revealed that the gene sequence is highly conserved, with the encoded Hfq proteins of Bcc organisms exhibiting a level of 95% identity at the amino acid residue level. In addition, the genetic organization *engA*-***hfq***-*hflX*-*hflK*-*hflC* (Figure 5) is also conserved among the sequenced genomes of Bcc bacteria (Sousa et al., 2010; Ramos et al., 2011).

The *BCAL1538* gene, encoding Hfq2, was initially found by Blast searches to encode an *hfq*-like protein within the genome sequence of the *B. cenocepacia* J2315 strain. Homologues to *BCAL1538* were identified in all the genomes of Bcc strains. Subsequent work confirmed that *BCAL1538* encodes an Hfq-like protein, and the gene was named *hfq2* (Sousa et al., 2010; Ramos et al., 2011). The gene also locates in chromosome 1, approximately 370 kb apart from *hfq*. The *hfq2* gene is flanked by genes putatively encoding membrane proteins, exported proteins, and an acyl co-enzyme A synthase, an arrangement completely distinct from that found for the *hfq* genes of *E. coli* and *B. cenocepacia* (Figure 5). The comparison of the genetic context of the *hfq* genes of *B. cepacia* and *E. coli* revealed a semi-conserved arrangement. In fact, the *hflC*, *hflK* and *hflX* genes appear downstream the *hfq* gene in both bacterial genomes. However, in *E. coli* the genes *miaA*, *mutL*, and *amiB* are immediately located upstream *hfq*, while in *B. cenocepacia engA* is the gene immediately upstream *hfq* (Figure 5).

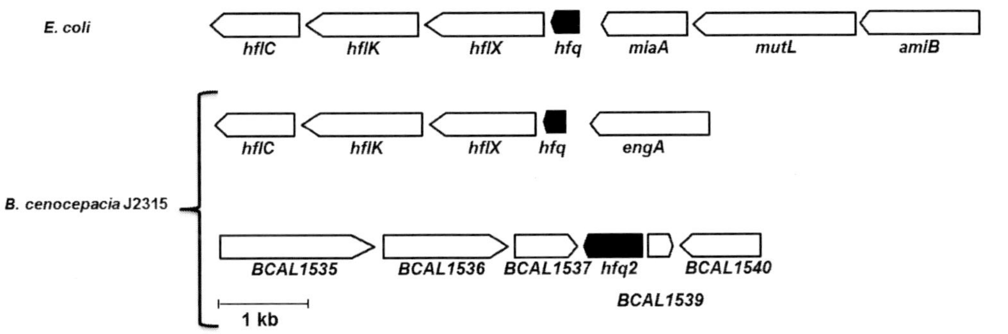

Figure 5. Genetic organization of the genome regions flanking the *E. coli hfq* gene and the *B. cenocepacia* J2315 *hfq* and *hfq2* genes. See text for details on the encoded genes.

In *E. coli*, HflX is a GTPase found to bind to the 50S subunit of the ribosome, while HflC and HflK are membrane-associated proteases of poorly understood physiological function (Dutta et al., 2009). EngA is a GTPase that participates in ribosome biogenesis (Tomar et al., 2009), while the *miaA* gene product is involved in the formation of Δ2-isopentenyl adenosine in tRNA (Caillet and Droogmans, 1988).

The gene *hfq2* encodes an unusually long protein, composed of 188 amino acid residues. Hfq2 conserves the Sm1 and Sm2 motifs typical of other bacterial Hfq proteins, but possess an unusually large C-terminus, being 109 amino acid residues longer than the other *B. cenocepacia* Hfq protein. Interestingly, in the case of the Bcc strains *B. cenocepacia* J2315 and MCO-3, *B. vietnamiensis* G4 and *B. multivorans* ATCC17616, the C- terminus extension is composed of a glycine-rich region (^{70}G to ^{173}G), a DNA-binding domain (^{89}G to ^{129}E), and the repetitive motif RE(P/S)RRXX(E/G)(G/S) that appears five times in the Hfq2 protein of *B. cenocepacia* J2315 strain and a few other Bcc bacteria. Although the biological significance of this extension remains mostly unknown, an implication is that it prevents the protein to assemble in the hexameric form found for the Hfq proteins of *E. coli*, *B. cenocepacia*, and other bacterial species. Furthermore, this C-terminus extension was shown to bind DNA, a common feature among Hfq proteins, except for the *B. cenocepacia* J2315

Hfq protein (Ramos et al., 2011). Results from discontinuous native polyacrylamide gel electrophoresis confirmed that the protein mostly assembles in a trimeric structure instead of the hexameric form observed for the *B. cenocepacia* Hfq (Ramos et al., 2011). Altogether, these findings corroborated the bioinformatics predictions indicating that *B. cenocepacia* J2315 encodes 2 distinct Hfq-like genes in their genomes.

The roles of Hfq have also been investigated in gram-positive bacteria, as is the case of the food-born pathogen *Listeria monocytogenes*, responsible for serious and life-threatening infections in humans, with clinical symptoms like febrile gastroenteritis, abortion, septicaemia, and meningitis (Vazquez-Boland et al., 2001). In this gram-positive bacterium, deletion of the *hfq* gene resulted in decreased survival during long term amino acid starvation, and decreased total colony forming units in the spleen and liver of the mice model of tissue colonization after three days of infection (Christiansen et al., 2004).

Regulation of *Hfq* Expression

Despite the pleiotropic phenotypes described for several bacterial species due to mutations in *hfq*, knowledge on the regulation of expression of the gene remains poorly characterized. In *E. coli*, the *hfq* gene is part of the superoperon *amiB-mutL-miaA-**hfq**-hflX-hflK-hflC* (Figure 5). All adjacent genes in this superoperon were shown to be co-transcribed in cells growing exponentially under non-stress conditions (Tsui et al., 1994). Tsui et al. (1996) also showed that the superoperon contains three σ^{32}-dependent heat shock (HS) promoters [P(*mutL*)HS, P(*miaA*)HS, and P1(*hfq*)HS] and four σ^{70}–dependent promoters [P(*mutL*), P(*miaA*), P2(*hfq*), and P3(*hfq*)], being the three heat shock promoters actively transcribed at 37ºC in vitro by σ^{32} (Tsui et al., 1996). In addition, the amounts of Hfq were maintained under heat shock conditions at 44 or 50 ºC. While these multiple promoters originate a complex transcription pattern, it is thought that this feature enables the cell to be functional and survive at both moderate and high temperatures (Tsui et al., 1996). The protein was found to regulate its own expression by means of destabilization of the corresponding transcript (Tsui et al., 1997). This autoregulation is achieved by means of binding of the protein to 2 distinct sites of the 5´untranslated region (UTR) of the *hfq* mRNA, inhibiting the formation of the translation initiation complex. These events expose RNase E binding sites on the *hfq* mRNA and the message degradation occurs, at least, by the action of RNase E (Vecerek et al., 2005). Studies in *Shigella flexneri*, which causes bacillary dysentery in human populations worldwide, showed that the pleiotropic regulator DksA is required for maximal transcription of *hfq* during exponential growth and also for growth under stringent conditions (Sharma and Payne, 2006).

Sobrero and Valverde found that the Hfq protein of the legume symbiotic *Sinorhizobium meliloti* strain 2011 autocontrols its own expression at the translational level by using chromosomal *hfq-lacZ* fusions (Sobrero and Valverde, 2011).

In the *B. cenocepacia* J2315, we found that the two *hfq*-like genes, *hfq* and *hfq2*, are differentially expressed (Ramos et al., 2011). In fact, the levels of transcripts corresponding to *hfq* are maximal in cells at the early exponential phase of growth and decreased by about 2.5 fold in cells in the stationary phase. Contrastingly, the levels of transcripts corresponding to *hfq2* were minimal in cells in the early exponential phase of growth and increased by about 2 fold, reaching maximal levels in cells in the stationary phase of growth (Figure 6).

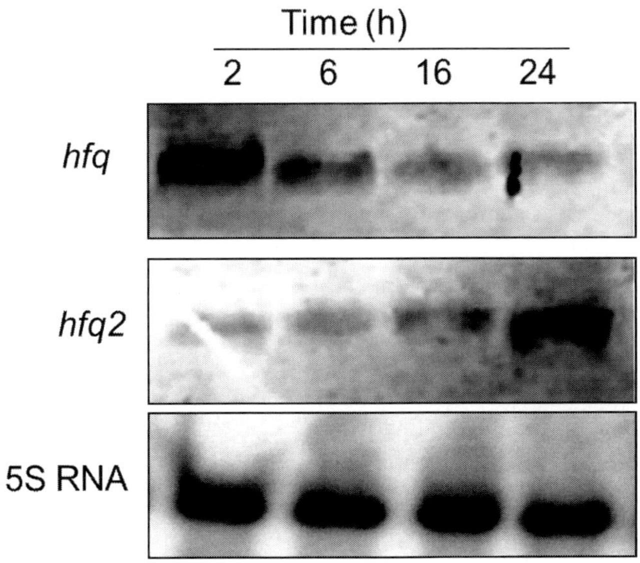

Figure 6. Representative Northern blots showing the relative abundance of *hfq* or *hfq2* mRNAs, in cells of *B. cenocepacia* J2315 grown for the indicated time. Results were obtained by Northern blot analysis using 2 µg of total RNA per lane, and adequate probes for *hfq*, *hfq2* and 5S RNA (used as loading control in Northern blots).

The relative abundance of transcripts corresponding to hfq2 throughout the growth curves was in good agreement with a quorum-sensing promoter sequence found upstream the hfq2 gene (Ramos et al., 2011). The distinct patterns of regulation of *hfq* and *hfq2* genes suggest that the levels of the corresponding proteins must be under strict regulation, to fulfil the correct balance between the two RNA chaperones, required by the cell to survive stress conditions and for efficient virulence (Ramos et al., 2011). In fact, similar levels of mRNA corresponding to *hfq* and *hfq2* were observed in cells of *B. cenocepacia* J2315 infecting the nematode *C.* elegans (Ramos et al., 2011).

Hfq Mediates RNA Interactions

Most of the pleiotropic phenotypes associated with the deletion of the *hfq* gene are presently known to be the result of the role of Hfq as a mediator of anti-sense interactions between small non-coding regulatory RNAs (hereafter named sRNAs) and mRNAs. sRNAs belong to a class of RNA molecules that are transcribed but generally not translated into proteins, as is the case of transfer RNAs (tRNA), ribosomal RNAs (rRNA), and other minor RNA species. sRNAs typically range 50-300 nucleotides in length, and can be classified as cis- or trans-encoded sRNAs, depending whether they are encoded within the same genomic region of their target mRNA, or in a distinct genetic region, respectively (Chao and Vogel., 2010; Waters and Storz, 2009; Storz, 2002). sRNAs are widespread in bacteria and form gene regulatory networks that control the cellular response to stresses such as cold shock, oxidative stress, iron concentration, sugar metabolism, and quorum sensing, among others (Beisel and Storz, 2010). sRNAs are post transcriptional regulators, with most of them acting by base-pairing with specific mRNAs (the target mRNAs). As a consequence of this base-pairing, the ribosome binding site of the mRNA can either become covered or exposed, leading, respectively to the inhibition or activation of the translation initiation (Papenfort and Vogel, 2009). This base-pairing is also thought to recruit the primary endoribonuclease RNase E, which apparently has a little sequence specificity but exhibits a strong preference to hydrolyze A/U-rich single stranded regions (Carpousis, 2007). Once RNase E initiates cleavage, the coupled degradation of both the sRNA and its mRNA target proceed rapidly (Massé et al., 2003).

The best known class of sRNAs are the trans-acting sRNAs which encompass a short segment of 7 to 12 nucleotides that base pair with the cognate region of the target mRNA (the seed region) by partial or total complementarity (Bandyra et al., 2012). The interactions of these trans-acting sRNAs with their targets require the action of Hfq. In its hexameric form, Hfq has two faces, the proximal and the distal face. Interactions with U-rich RNA sequences seem to occur in the proximal face, while binding to A-rich sequences occur preferentially in the distal face (Mikulecky et al., 2004).

Insights on the molecular details of the role of Hfq as a mediator of RNA interactions were obtained by Soper et al. (2001) in an elegant study performed with the DsrA sRNA and the *rpoS* mRNA leader sequence. The study demonstrated that both RNAs bind to Hfq with similar affinity, but interact with opposite faces of the Hfq hexamer. Hfq acts by binding and restructuring the rpoS mRNA. Binding of DsrA to Hfq is required for maximal annealing in vitro, indicating that transient interactions with both RNAs contribute to the regulatory mechanism (Soper et al., 2001).

In a recent study, Fender et al. (2011) showed that an active cycling of RNAs occur at the surface of Hfq, during which RNAs displace each other on Hfq, on a short time scale, in an active RNA concentration-driven cycling. The authors proposed a mechanism in which a competitor RNA transiently associates to the RNA-HFQ complexes, RNAs exchange binding sites, and then one of the RNAs eventually dissociates from the complex (Fender et al., 2011).

sRNAs Are Involved in Post-Transcription Gene Regulation

Over the last years, an ever-increasing number of sRNAs have been described in bacteria. In an organism like *E. coli*, at least 83 sRNAs are known to be transcribed ((Waters and Storz, 2009; Gottesman and Storz, 2011).

Although sRNAs have been initially found almost serendipitously, high-throughput methods have been recently used to identify sRNAs. This is the case of RNA deep-sequencing strategies, which allowed the identification of sRNAs from several bacterial pathogens (Chao and Vogel, 2010). Bioinformatics approaches have also been used to predict the occurrence of sRNAs in sequenced genomes, although it is the opinion of various authors that these predictions often fail. Other methodologies have been used to experimentally identify bacterial sRNAs (Sharma and Vogel, 2009). These approaches led to the identification of hundreds of sRNAs, most of them with unknown function. A short list of *E. coli* sRNAs that have been functionally characterized is presented in Table 1.

Table 1. Examples of sRNAs from *E. coli* for which the function and at least one target mRNA are known

Name	Target (s)	Function	Reference
omrA	*ompT, cirA, fecA, fepA, ompR, csgD*	Regulates the protein composition of the outer membrane	Guiller and Gottesman, 2008; Tjaden et al., 2006; Wassarman et al., 2001; Holmqvist et al., 2010
rprA	*rpoS, csgD, ydaM*	Response to osmotic shock	Majdalani et al., 2002; Jorgensen et al., 2012; Mika et al., 2012
rydB	*rpoS*	Unknown	Wassarman et al., 2001
ryfB (OhsC)	*shoB*	Regulates the expression of the toxic peptide ShoB	Fozo et al., 2008
ryhA (ArcZ)	*arcB, rpoS*	Regulation of *rpoS*	Mandin and Gottesman, 2010; Papenfort et al., 2009; Soper et al., 2010
sgrS	*ptsG, manXYZ, sgrR*	Enables transport of glucose into the cytoplasm	Vanderpool and Gottesman, 2004; Kawamoto et al., 2005; Rice and Vanderpool, 2011
gadY	*gadX* and *gadW* mRNA	Low pH response	Opdyke et al., 2004, Hayes et al., 2006;
isrA (McaS)	*csgD, flhD, pgaA*	Regulates flagellar motility and biofilm formation	Jorgensen et al., 2012; Thomason et al., 2012
micA	*ompA, phoPQ, ompX*	Envelope stress response	Udekwu and Wagner, 2007; Coornaert et al., 2010; Johansen et al., 2008
omrB	*ompT, cirA, fecA, gntP, ompR, csgD*	Composition of the outer membrane	Guillier and Gottesman, 2008; Tjaden et al., 2006; Holmqvist et al., 2010; Guillier et al., 2006
rseX	*ompA, ompC*	Survival in the absence of the protease RseP	Douchin et al., 2006
dicF	FtsZ	Inhibits cell division	Bouché and Bouché, 1989; Tétart and Bouché, 1992
rybB	*rpoS, ompC, ompW, rpoE*	Cell envelope stress	Vogel et al., 2003; Thompson et al., 2007; Johansen et al., 2006

This strategy already allowed the identification of 24 sRNAs from *B. cenocepacia* J2315, although their targets remain unknown (Ramos et al., 2013). We have also recently functionally characterized the sRNAs h2cR (Ramos et al., 2012). Interestingly, the sRNA h2cR was found as being encoded in the DNA strand complementary to that where the *hfq2* gene promoter is encoded, being classified as a cis-encoded sRNA, and suggesting that it might regulate the *hfq2* mRNA. Studies performed confirmed that in fact h2cR negatively regulate *hfq2* mRNA, by a mechanism involving the direct binding of the sRNA to part of the 5′-UTR region of the mRNA. Furthermore, our findings using the nematode *C. elegans* as an infection model and bacterial strains in which the h2cR sRNA was either overexpressed or silenced also suggest that this sRNA plays a role on bacterial persistence in the host (Ramos et al., 2012).

Work in progress at our research group is focused on the functional and molecular characterization of the biological roles played by sRNAs on the biology of the opportunistic pathogens of the *Burkholderia cepacia* complex and their contributions to the host-pathogen relationship.

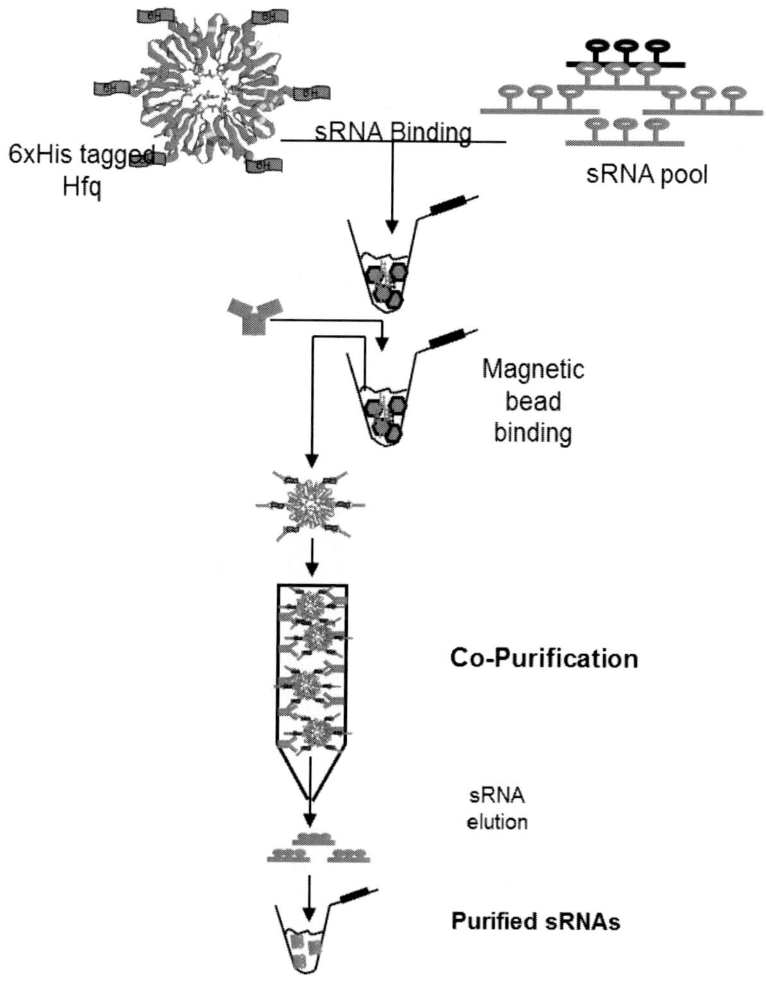

Figure 7. Experimental strategy used to isolate and purify sRNAs from *B. cenocepacia* J2315.

Our research group has recently conducted experiments envisaging the identification of sRNAs from bacteria of the Bcc, by using an experimental strategy based on the co-purification of sRNA-Hfq complexes followed by cloning, sequencing and functional analysis of the identified sRNAs. A schematic representation of this strategy is shown in Figure 7.

Conclusion

Since its discovery almost four decades ago until the discovery of its role as a mediator of RNA interactions, Hfq proteins have attracted the attention of many research groups. Recent developments have disclosed its role as a riboregulator, and Hfq is now being used to find bacterial sRNAs, the vast majority of which playing still unknown functions. Future work, either focused on the protein itself, or on the exploitation of its ability to bind RNAs, will undoubtedly further our knowledge on the molecular mechanisms and regulatory networks orchestrated by this singular protein.

Acknowledgments

This work was partially funded by Fundação para a Ciência e a Tecnologia (Portugal) through contract PTDC/BIA-MIC 119091/2010, and scholarships to S. A. Sousa, C. G. Ramos, A. M. Grilo, J. R. Feliciano, and P. J. P. da Costa.

References

Bandyra, K.J., Said, N., Pfeiffer, V., Górna, M.W., Vogel, J., Luisi, B.F., (2012). The seed region of a small RNA drives the controlled destruction of the target mRNA by the endoribonuclease RNase E. *Mol. Cell* 47, 943-953.

Beisel, C.L., Storz, G., (2010). Base pairing small RNAs and their roles in global regulatory networks. F*EMS Microbiol. Rev.* 34, 866-882.

Bellows, L.E., Koestler, B.J., Karaba, S.M., Waters, C.M., Lathem, W.W., (2012). Hfq-dependent, co-ordinate control of cyclic diguanylate synthesis and catabolism in the plague pathogen *Yersinia pestis*. *Mol. Microbiol.* 86, 661-674.

Bouché, F., Bouché, J.P., (1989). Genetic evidence that DicF, a second division inhibitor encoded by the *Escherichia coli dicB* operon, is probably RNA. *Mol. Microbiol.* 3, 991-994.

Caillet, J., Droogmans, L., (1988). Molecular cloning of the *Escherichia coli miaA* gene involved in the formation of delta 2-isopentenyl adenosine in tRNA. *J. Bacteriol.* 170, 4147-4152.

Carpousis, A.J., (2007). The RNA Degradosome of *Escherichia coli*: an mRNA-degrading machine assembled on RNase E. *Ann. Rev. Microbiol.* 61, 71-87.

Chao, Y., Vogel, J., (2010). The role of Hfq in bacterial pathogens. *Cur. Opin. Microbiol.* 13 24-33.Chiang, M.-K., Lu, M.-C., Liu, L.-C., Lin, C.-T., Lai, Y.-C., (2011). Impact of Hfq on global gene expression and virulence in *Klebsiella pneumoniae*. *PLoS ONE* 6, e22248.

Christiansen, J.K., Larsen, M.H., Ingmer, H., Sogaard-Andersen, L., Kallipolitis, B.H., (2004). The RNA-binding protein Hfq of *Listeria monocytogenes*: role in stress tolerance and virulence. *J. Bacteriol.* 186, 3355-3362.

Coornaert, A., Lu, A., Mandin, P., Springer, M., Gottesman, S., Guillier, M., (2010). MicA sRNA links the PhoP regulon to cell envelope stress. *Mol. Microbiol.* 76, 467-479.

Cunha, M.V., Sousa, S.A., Leitão, J.H., Moreira, L.M., Videira, P.A., Sá-Correia, I., (2004). Studies on the involvement of the exopolysaccharide produced by cystic fibrosis-associated isolates of the *Burkholderia cepacia* complex in biofilm formation and in persistence of respiratory infections. *J. Clin. Microbiol.* 42, 3052-3058.

Douchin, V., Bohn, C., Bouloc, P., (2006). Down-regulation of porins by a small RNA bypasses the essentiality of the regulated intramembrane proteolysis protease RseP in *Escherichia coli*. *J. Biol. Chem.* 281, 12253-12259.

Dutta, D., Bandyopadhyay, K., Datta, A.B., Sardesai, A.A., Parrack, P., (2009). Properties of HflX, an enigmatic protein from *Escherichia coli*. *J. Bacteriol.* 191, 2307-2314.

Fender, A., Elf, J., Hampel, K., Zimmermann, B., Wagner, E.G.H., (2011). RNAs actively cycle on the Sm-like protein Hfq. *Genes Dev.* 24, 2621-2626.

Fozo, E.M., Kawano, M., Fontaine, F., Kaya, Y., Mendieta, K.S., Jones, K.L., Ocampo, A., Rudd, K.E., Storz, G., (2008). Repression of small toxic protein synthesis by the Sib and OhsC small RNAs. Mol. Microbiol. 70, 1076-1093.

Gottesman, S., Storz, G., (2011). Bacterial small RNA regulators: versatile roles and rapidly evolving variations. *Cold Spring Harb. Perspect. Biol.* 3. pii: a003798.

Govan, J.R., Deretic, V., (1996). Microbial pathogenesis in cystic fibrosis: mucoid *Pseudomonas aeruginosa* and *Burkholderia cepacia*. *Microbiol. Rev.* 60, 539-574.

Govan, J.R.W., Doherty, C.J., Nelson, J.W., Brown, P.H., Greening, A.P., Maddison, J., Dodd, M., Webb, A.K., (1993). Evidence for transmission of *Pseudomonas cepacia* by social contact in cystic fibrosis. *Lancet* 342, 15-19.

Guillier, M., Gottesman, S., (2008). The 5' end of two redundant sRNAs is involved in the regulation of multiple targets, including their own regulator. *Nucleic Acids Res.* 36, 6781-6794.

Guillier, M., Gottesman, S., Storz, G., (2006). Modulating the outer membrane with small RNAs. *Genes Dev.* 20, 2338-2348.

Hayes, E., Wilks, J., Sanfilippo, P., Yohannes, E., Tate, D., Jones, B., Radmacher, M., BonDurant, S., Slonczewski, J., (2006). Oxygen limitation modulates pH regulation of catabolism and hydrogenases, multidrug transporters, and envelope composition in *Escherichia coli* K-12. *BMC Microbiol.* 6, 89.

Holmqvist, E., Reimegard, J., Sterk, M., Grantcharova, N., Römling, U., Wagner, E.G.H., (2010). Two antisense RNAs target the transcriptional regulator CsgD to inhibit curli synthesis. *EMBO J.* 29, 1840-1850.

Johansen, J., Eriksen, M., Kallipolitis, B., Valentin-Hansen, P., (2008). Down-regulation of outer membrane proteins by noncoding RNAs: unraveling the cAMP-CRP- and sigmaE-dependent CyaR-*ompX* regulatory case. *J. Mol. Biol.* 383, 1-9.

Johansen, J., Rasmussen, A.A., Overgaard, M., Valentin-Hansen, P., (2006). Conserved small non-coding RNAs that belong to the sigmaE regulon: role in down-regulation of outer membrane proteins. *J. Mol. Biol.* 364, 1-8.

Jørgensen, M.G., Nielsen, J.S., Boysen, A., Franch, T., Møller-Jensen, J., Valentin-Hansen, P., (2012). Small regulatory RNAs control the multi-cellular adhesive lifestyle of *Escherichia coli*. *Mol. Microbiol.* 84, 36-50.

Kajitani, M., Ishihama, A., (1991). Identification and sequence determination of the host factor gene for bacteriophage Qβ. *Nucleic Acids Res.* 19, 1063-1066.

Kajitani, M., Kato, A., Wada, A., Inokuchi, Y., Ishihama, A., (1994). Regulation of the *Escherichia coli hfq* gene encoding the host factor for phage Q beta. *J. Bacteriol.* 176, 531-534.

Kawamoto, H., Morita, T., Shimizu, A., Inada, T., Aiba, H., (2005). Implication of membrane localization of target mRNA in the action of a small RNA: mechanism of post-transcriptional regulation of glucose transporter in *Escherichia coli*. *Genes Dev.* 19, 328-338.

Kendall, M.M., Gruber, C.C., Rasko, D.A., Hughes, D.T., Sperandio, V., (2011). Hfq virulence regulation in enterohemorrhagic *Escherichia coli* O157:H7 Strain 86-24. *J. Bacteriol.* 193, 6843-6851.

Kulesus, R.R., Diaz-Perez, K., Slechta, E.S., Eto, D.S., Mulvey, M.A., (2008). Impact of the RNA chaperone Hfq on the fitness and virulence potential of uropathogenic *Escherichia coli. Infect. Immun.* 76, 3019-3026.

Leitão, J.H., Sousa, S.A., Ferreira, A.S., Ramos, C.G., Silva, I.N., Moreira, L.M., (2010). Pathogenicity, virulence factors, and strategies to fight against *Burkholderia cepacia* complex pathogens and related species. *Appl. Microbiol. Biotechnol.* 87, 31-40.

Leitão, J.H., Sousa, S.A., Cunha, M.V., Salgado, M.J., Melo-Cristino, J., Barreto, M.C., Sá-Correia, I., (2008). Variation of the antimicrobial susceptibility profiles of *Burkholderia cepacia* complex clonal isolates obtained from chronically infected cystic fibrosis patients: a five-year survey in the major Portuguese treatment center. *Eur. J. Clin. Microbiol. Infect. Dis.* 27, 1101-1111.

Mahenthiralingam, E., Urban, T.A., Goldberg, J.B., (2005). The multifarious, multireplicon *Burkholderia cepacia* complex. *Nat. Rev. Microbiol.* 3, 144-156.

Majdalani, N., Hernandez, D., Gottesman, S., (2002). Regulation and mode of action of the second small RNA activator of RpoS translation, RprA. *Mol. Microbiol.* 46, 813-826.

Mandin, P., Gottesman, S., (2010). Integrating anaerobic/aerobic sensing and the general stress response through the ArcZ small RNA. *EMBO J.* 29, 3094-3107.

Markowitz, V.M., Korzeniewski, F., Palaniappan, K., Szeto, E., Werner, G., Padki, A., Zhao, X., Dubchak, I., Hugenholtz, P., Anderson, I., Lykidis, A., Mavromatis, K., Ivanova, N., Kyrpides, N.C., (2006). The integrated microbial genomes (IMG) system. *Nucleic Acids Res.* 34 (suppl_1), D344-348.

Massé, E., Escorcia, F.E., Gottesman, S., (2003). Coupled degradation of a small regulatory RNA and its mRNA targets in *Escherichia coli. Genes Dev.* 17, 2374-2383.

McNealy, T.L., Forsbach-Birk, V., Shi, C., Marre, R., (2005). The Hfq homolog in *Legionella pneumophila* demonstrates regulation by LetA and RpoS and interacts with the global regulator CsrA. *J. Bacteriol.* 187, 1527-1532.

Meibom, K.L., Forslund, A.-L., Kuoppa, K., Alkhuder, K., Dubail, I., Dupuis, M., Forsberg, A., Charbit, A., (2009). Hfq, a novel pleiotropic regulator of virulence-associated genes in *Francisella tularensis. Infect. Immun.* 77, 1866-1880.

Mika, F., Busse, S., Possling, A., Berkholz, J., Tschowri, N., Sommerfeldt, N., Pruteanu, M., Hengge, R., (2012). Targeting of csgD by the small regulatory RNA RprA links stationary phase, biofilm formation and cell envelope stress in *Escherichia coli. Mol. Microbiol.* 84, 51-65.

Mikulecky, P.J., Kaw, M.K., Brescia, C.C., Takach, J.C., Sledjeski, D.D., Feig, A.L. (2004). *Escherichia coli* Hfq has distinct interaction surfaces for DsrA, rpoS and poly(A) RNAs. *Nat. Struct. Mol. Biol.* 11, 1206-1214.

Opdyke, J.A., Fozo, E.M., Hemm, M.R., Storz, G., (2011). RNase III participates in GadY-dependent cleavage of the gadX-gadW mRNA. *J. Mol. Biol.* 406, 29-43.

Opdyke, J.A., Kang J.-G., Storz, G., (2004). GadY, a small-RNA regulator of acid response genes in *Escherichia coli. J. Bacteriol.* 186, 6698-6705.

Papenfort, K., Said, N., Welsink, T., Lucchini, S., Hinton, J.C.D., Vogel, J., (2009). Specific and pleiotropic patterns of mRNA regulation by ArcZ, a conserved, Hfq dependent small RNA. *Mol. Microbiol.* 74, 139-158.

Papenfort, K., Vogel, J., (2009). Multiple target regulation by small noncoding RNAs rewires gene expression at the post-transcriptional level. *Res. Microbiol.* 160, 278-287.

Ramos, C.G., da Costa, P.J.P., Döring, G., Leitão, J.H., (2012). The novel cis-encoded small RNA h2cR Is a negative regulator of *hfq2* in *Burkholderia cenocepacia*. *PLoS One* 7, e47896.

Ramos, C.G., Grilo, A.M., da Costa, P.J.P., Leitão, J.H., (2013). Experimental identification of small non-coding regulatory RNAs in the opportunistic human pathogen *Burkholderia cenocepacia* J2315. *Genomics* 101, 139-148.

Ramos, C.G., Sousa, S.A., Grilo, A.M., Feliciano, J.R., Leitão, J.H., (2011). The second RNA chaperone Hfq2, is also required for survival to stress and the full virulence of *Burkholderia cenocepacia* J2315. *J. Bacteriol.* 193, 1515-1526.

Rempe, K.A., Hinz, A.K., Vadyvaloo, V., (2012). Hfq regulates biofilm gut blockage that facilitates flea-borne transmission of *Yersinia pestis*. *J. Bacteriol.* 194, 2036-2040.

Rice, J.B., Vanderpool, C.K., (2011). The small RNA SgrS controls sugar-phosphate accumulation by regulating multiple PTS genes. *Nucleic Acids Res.* 39, 3806-3819.

Robertson, G.T., Roop, R.M., (1999). The *Brucella abortus* host factor I (HF-I) protein contributes to stress resistance during stationary phase and is a major determinant of virulence in mice. *Mol. Microbiol.* 34, 690-700.

Roscetto, E., Angrisano, T., Costa, V., Casalino, M., Förstner, K.U., Sharma, C.M., Di Nocera, P.P., De Gregorio, E., (2012). Functional characterization of the RNA chaperone Hfq in the opportunistic human pathogen *Stenotrophomonas maltophilia*. *J. Bacteriol.* 194, 5864-5874.

Séraphin, B., (1995). Sm and Sm-like proteins belong to a large family: identification of proteins of the U6 as well as the U1, U2, U4 and U5 snRNPs. *EMBO J.* 14, 2089-2098.

Sharma, A.K., Payne, S.M., (2006). Induction of expression of *hfq* by DksA is essential for *Shigella flexneri* virulence. *Mol. Microbiol.* 62, 469-479.

Sharma, C.M., Vogel, J., (2009). Experimental approaches for the discovery and characterization of regulatory small RNA. *Curr. Opin. Microbiol.* 12, 536-546.

Sobrero, P., Valverde, C., (2011). Evidences of autoregulation of *hfq* expression in *Sinorhizobium meliloti* strain 2011. *Arch. Microbiol.* 193, 629-639.

Sonnleitner, E., Hagens, S., Rosenau, F., Wilhelm, S., Habel, A., Jager, K.-E., Blasi, U., (2003). Reduced virulence of a *hfq* mutant of *Pseudomonas aeruginosa* O1. *Microb. Pathog.* 35, 217-228.

Soper, T.J., Doxzen, K., Woodson, S.A., (2001). Major role for mRNA binding and restructuring in sRNA recruitment by Hfq. *RNA* 17, 1544-1550.

Soper, T., Mandin, P., Majdalani, N., Gottesman, S., Woodson, S.A., (2010). Positive regulation by small RNAs and the role of Hfq. *Proc. Natl. Acad. Sci. USA* 107, 9602-9607.

Sousa, S.A., Ramos, C.G., Moreira, L.M., Leitão, J.H., (2010). The *hfq* gene is required for stress resistance and full virulence of *Burkholderia cepacia* to the nematode *Caenorhabditis elegans*. *Microbiology* 56, 896-908

Storz, G., (2002). An Expanding Universe of Noncoding RNAs. *Science* 296, 1260-1263.

Tan, E.M., Shan, E.K., Sullivan, K.F., Rubin, R.L., (1988). Antinuclear antibodies (ANAs): diagnostically specific immune markers and clues toward the understanding of systemic autoimmunity. *Clin. Immunol. Immunopathol.* 47, 121-141.

Tétart, F., Bouché, J.-P., (1992). Regulation of the expression of the cell-cycle gene *ftsZ* by DicF antisense RNA. Division does not require a fixed number of FtsZ molecules. *Mol. Microbiol.* 6, 615-620.

Thomason, M.K., Fontaine, F., De Lay, N., Storz, G., (2012). A small RNA that regulates motility and biofilm formation in response to changes in nutrient availability in *Escherichia coli. Mol. Microbiol.* 84, 17-35.

Thompson, K.M., Rhodius, V.A., Gottesman, S., (2007). Sigma-E regulates and is regulated by a small RNA in *Escherichia coli. J. Bacteriol.* 189, 4243-4256.

Tjaden, B., Goodwin, S.S., Opdyke, J.A., Guillier, M., Fu, D.X., Gottesman, S., Storz, G., (2006). Target prediction for small, noncoding RNAs in bacteria. *Nucleic Acids Res.* 34, 2791-2802.

Tomar, S.K., Dhimole, N., Chatterjee, M., Prakash, B., (2009). Distinct GDP/GTP bound states of the tandem G-domains of EngA regulate ribosome binding. *Nucleic Acids Res.* 37, 2359-2370.

Tsui, H.C., Feng, G., Winkler, M.E., (1996). Transcription of the *mutL* repair, *miaA* tRNA modification, *hfq* pleiotropic regulator, and *hflA* region protease genes of *Escherichia coli* K-12 from clustered Esigma32-specific promoters during heat shock. *J. Bacteriol.* 178, 5719-5731.

Tsui, H.C., Feng, G., Winkler, M.E., (1997). Negative regulation of *mutS* and *mutH* repair gene expression by the Hfq and RpoS global regulators of *Escherichia coli* K-12. *J. Bacteriol.* 179, 7476-7487.

Tsui, H.C.T., Leung, H.C.E., Winkler, M.E., (1994). Characterization of broadly pleiotropic phenotypes caused by an *hfq* insertion mutation in *Escherichia coli* K 12. Mol. Microbiol. 13, 35-49.

Tsui, H.C.T., Winkler, M.E., (1994).Transcriptional patterns of the *mutL-miaA* superoperon of *Escherichia coli* K-12 suggest a model for posttranscriptional regulation. *Biochimie* 76, 1168-1177.

Udekwu, K.I., Darfeuille, F., Vogel, J., Reimegard, J., Holmqvist, E., Wagner, E.G.H., (2005). Hfq-dependent regulation of OmpA synthesis is mediated by an antisense RNA. *Genes Dev.* 19, 2355-2366.

Udekwu, K.I., Wagner, E.G.H., (2007). Sigma E controls biogenesis of the antisense RNA MicA. *Nucleic Acids Res.* 35, 1279-1288.

Valentin-Hansen, P., Eriksen, M., Udesen, C., (2004). The bacterial Sm-like protein Hfq: a key player in RNA transactions. *Mol. Microbiol.* 5, 1525-1533.

Vanderpool, C.K., Gottesman, S., (2004). Involvement of a novel transcriptional activator and small RNA in post-transcriptional regulation of the glucose phosphoenolpyruvate phosphotransferase system. *Mol. Microbiol.* 54, 1076-1089.

Vazquez-Boland, J.A., Kuhn, M., Berche, P., Chakraborty, T., Domínguez-Bernal, G., Goebel, W., González-Zorn, B., Wehland, J., Kreft, J., (2001). *Listeria* pathogenesis and molecular virulence determinants. *Clin. Microbiol. Rev.* 14, 584-640.

Vecerek, B., Moll, I., Blasi, U., (2005). Translational autocontrol of the *Escherichia coli hfq* RNA chaperone gene. *RNA* 11, 976-984.

Vogel, J., Bartels, V., Tang, T.H., Churakov, G., Slagter-Jager, J.G., Huttenhofer, A., Wagner, E.G.H., (2003). RNomics in *Escherichia coli* detects new sRNA species and indicates parallel transcriptional output in bacteria. *Nucleic Acids Res.* 31, 6435-6443.

Vogel, J., Luisi, B.F., (2011). Hfq and its constellation of RNA. *Nat. Rev. Microbiol.* 9, 578-589.

Wassarman, K.M., Repoila, F., Rosenow, C., Storz, G., Gottesman, S., (2001). Identification of novel small RNAs using comparative genomics and microarrays. *Genes Dev.* 15, 1637-1651.

Waters, L.S., Storz, G., (2009). Regulatory RNAs in Bacteria. *Cell* 136, 615-628.

Wilusz, C.J., Wilusz, J., (2005). Eukaryotic Lsm proteins: lessons from bacteria. *Nat. Struct. Mol. Biol.* 12, 1031-1036.

In: Prokaryotes
Editor: Marina Nisnevitch
ISBN: 978-1-63321-592-4

Chapter 6

Effects of the Cytolethal Distending Toxin from Gram-Negative Bacteria on Mammalian Cells

***Elisabeth Bezine*[1,2], *Julien Vignard*[1] and *Gladys Mirey*[1,3,*]**
[1]INRA, UMR1331, Toxalim, Research Centre in Food Toxicology, Toulouse, France
[2]Université de Toulouse, INP, UMR1331, Toxalim, Toulouse, France
[3]Université de Toulouse, UPS, UMR1331, Toxalim, Toulouse, France

Abstract

The cytolethal distending toxin (CDT) is produced by many pathogenic Gram-negative bacteria (*Escherichia coli, Helicobacter hepaticus, Haemophilus ducreyi* and others) colonizing the oral cavity, the intestinal tract, the liver, etc. Depending on the affected organs, these bacteria are responsible for different pathologies. However, as a virulence factor, CDT has been shown to allow bacteria to persistently colonize the body, evade the immune system, induce inflammation and trigger genetic instability. Moreover, *in vivo*, the production of CDT by *Helicobacter hepaticus* induces the development of dysplastic liver nodules, thus defining CDT as a potential carcinogen. Given the important role of the toxin in the physiopathology of these bacteria, it is essential to accurately identify its mode of action.

In human cells, CDT exposure leads to a singular cytotoxicity associated with characteristic cell distension, and induces cell cycle arrest at the G2/M transition that is dependent on the DNA damage response. Thus, CDT has been classified as a cyclomodulin and a genotoxin, as this toxin induces DNA double-strand breaks. It is therefore important to characterize the outcome of CDT-induced DNA damage and the consequences for the intoxicated cells and organisms. Indeed, the presence of unrepaired damage can lead to cell death, whereas effective repair will allow cells to resume cell cycle. However, improper repair of DNA damage can induce genetic instability and lead to cancer.

[*] E-mail: gladys.mirey@toulouse.inra.fr.

After years of study, the CDT mode of action begins to reveal itself, however, many aspects remain to be identified. In this chapter, we briefly present the bacteria producing CDT as well as their cellular niche. We will discuss CDT trafficking within eukaryotic cells, in order to understand cell entry and nuclear targeting of the toxin. We will then consider host cell behavior in response to CDT intoxication, by focusing on morphological changes, cell cycle modulation and cell death response. Finally, we will examine the biochemistry of CDT, specifically regarding its catalytic activities and relate these aspects to the overall physiopathology of the toxin.

Keywords: Gram-negative bacteria; cytolethal distending toxin; virulence factor; pathogenicity; retrograde traffic; DNA damage response; double-strand breaks; cell cycle checkpoints; apoptosis

Introduction

Human beings are constantly in contact with a huge amount of bacterial agents. Some, non-pathogenic commensal bacteria, are essential for our body, whilst pathogenic bacteria are the source of various diseases. For example, pathogenic bacteria are responsible for a lot of food-borne contaminations. Among the most common bacteria, *Salmonella enterica* induces gastroenteritis, whilst *Salmonella enterica* serovar Typhi is the agent responsible for typhoid fever. *Campylobacter jejuni* causes diarrhea and acute intestinal diseases, and some *Escherichia coli* strains are responsible for gastrointestinal infections, etc. The type of induced disease depends on the bacterial strains capacity to induce a disease (pathogenicity of the strain), and on the location of the bacterial niche. However, infection by a pathogenic bacteria strain does not necessarily imply the development of a disease in infected individuals. In fact, the bacteria pathogenicity depends on several factors: the genetic background of the affected individual, the presence of other bacterial strains in the niche that may act synergistically or antagonistically, and finally the virulence of the bacterial strain itself. Indeed, the virulence of bacteria is the ability to induce damage, directly or indirectly, at the cellular level (DNA damage and cell death) and at the tissular or organ level (inflammation, fibrosis, necrosis...), particularly through the production of virulence factors (Casadevall and Pirofski, 1999).

The production of virulence factors allows bacteria to evade or compromise host defenses, in order to survive and/or exit the host and be transmitted to another. Among them can be distinguished exotoxins and endotoxins. Endotoxins are found within the outer membrane of Gram-negative bacteria and enter the systemic circulation of infected host once the bacteria are killed, whereas exotoxins are usually secreted into the surrounding environment of bacteria. This class of toxins varies considerably in specificity, as some act only on certain cell types while others affect a wide range of cells or tissues. Pathogens usually produce these toxins in an inactive form, activated only after having left the bacteria. Roughly, three types of exotoxins can be considered: cytotoxins that kill cells, neurotoxins interfering with the transmission of the neurological signal and enterotoxins, affecting the digestive epithelium. Many exotoxins consist of A and B moieties. The B moiety is involved in the binding of a receptor found on specific target host cells whereas the A moiety is associated with the interference of host cell metabolism. Moreover, a specific bacterial pathogen may produce one or more exotoxins. *Helicobacter pylori*, for example, produces different factors, including CagA, VacA, DupA and OipA, and the production of these toxins has an important impact on the related physiopathology. In fact, some strains expressing CagA and VacA are involved in the development of gastric diseases, whilst production of DupA induces duodenal ulcer cancers (Yamaoka, 2011). This demonstrates the key role of virulence factors produced by bacteria and the importance of characterizing and understanding their mode of action.

Cytoletal Distending Toxin Producing Bacteria

Reciprocally, if bacteria can produce several virulence factors, different bacteria can produce a particular virulence factor. One of these virulence factors, inducing the distension and the death of eukaryotic cells, was discovered by Johnson and Lior in 1987. Because of its action, this factor was named cytolethal distending toxin (CDT). In the two years following its discovery, several bacterial strains, obtained from clinical isolates, were characterized as producing CDT, including *Escherichia coli* (*Ec*CDT) (Johnson and Lior, 1987) and *Campylobacter jejuni* (*Cj*CDT) (Johnson and Lior, 1988). In 1994, the genes encoding CDT, produced by *Escherichia coli*, were cloned (Scott and Kaper, 1994; Pickett et al., 1994), which led to the discovery of novel Gram-negative pathogenic bacterial strains producing CDT, through systematic gene detection. Among most characterized are *Helicobacter hepaticus (HhCDT)* (Chien et al., 2000), *Haemophilus ducreyi (Hd*CDT*)* (Cope et al., 1997), *Aggregatibacter actinomycetemcomitans* (*Aa*CDT) (Sugai et al., 1998) and *Salmonella enterica* serovar Typhi (*St*CDT) (Haghjoo and Galán, 2004), although CDT toxin production appears atypical. To date, no Gram-positive bacteria have been characterized to produce CDT.

Implication of the Cytoletal Distending Toxin in the Pathogenicity

Our body constantly coexists with microorganisms. These relationships can be mutualistic or commensal. There is therefore a balance established between the microbiota and the host, particularly at the involved epithelium and regarding the immune system. A change in the presence or amount of a bacterial strain in the microbiota can modify a mutualistic or commensal situation to a host-pathogen relationship (Sansonetti, 2011). To persist and colonize host organisms, pathogenic bacteria must adapt to an ever-changing microenvironment, renewal of the epithelium, competition with other bacterial strains, etc. To do so, bacteria have developed adherence, invasion and survival strategies in response to environmental signals of the host. In addition, pathogens not only have to colonize the host body to induce disease, but also have to evade the immune system, and finally, damage the host, either at the cellular, tissue or organ level. Whatever the case, the pathogenicity of bacteria is dependent on the production of exotoxins, i.e. virulence factors (Finlay and Falkow, 1997).

Virulence factors may be involved in the adhesion of bacteria to eukaryotic cells, invasion, damage of host cells or in the exhaustion of the immune system (Finlay and Falkow, 1997). Indeed, the interaction of a pathogen with eukaryotic cells almost always leads to the activation of host signaling pathways and to an immune response (Sansonetti, 2011). Different lines of defense are implemented by host organisms to deal with pathogens. First, a physical barrier formed by the epithelial cells, and sometimes mucous, prevents the penetration and invasion of bacteria into host tissues. In case of damages and passage of bacteria through the epithelial barrier, actors of innate immunity (macrophages, neutrophils, mast cells) recognize the bacterial antigen (lipopolysaccharide (LPS) for example) and limit the infection through the production of pro-inflammatory cytokines, chemokines, which will attract other immune effectors. Amongst the immune effectors recruited, T- and B-lymphocytes are key actors of the adaptive immune response. This response will among other things, produce specific antibodies against the bacteria or against the virulence factors produced (McGhee and Fujihashi, 2012; Hayashi et al., 2011). In most cases, activation of the immune system will control and stop the infection. Therefore, pathogens must develop mechanisms, and virulence factors, to escape the immune system in order to persist in the body and induce diseases. Among them, the CDT virulence factor has been shown to be expressed in many pathogenic Gram-negative bacteria. Several examples of CDT involvement in the pathogenicity have been demonstrated. Importantly, depending on the producing bacteria and the induced disease, the involvement of CDT in the pathogenicity may be more or less important. However, for all CDT producing strains, CDT production leads to damage and necrosis of the mucosa through destabilization of the epithelial barrier (Damek-Poprawa et al., 2011; Okuda et al., 1997; Pokkunuri et al., 2012; Wising et al., 2005), allowing bacteria to infect the tissue and bypass the first defense system of the host organism. In addition, numerous studies have shown that CDT is highly cytotoxic to cells of the immune system (Gelfanova et al., 1999; Shenker et al., 2006; Shenker et al., 1999). These results suggest that CDT could induce localized immunodepletion during the bacterial infection. Herein are presented examples of bacteria for which the CDT toxin has been found to be involved in disease development.

Escherichia coli

Escherichia coli (*E. coli*) bacterial strains form the major components of the intestinal flora (Tenaillon et al., 2010). These bacteria, commensal most of the time, in some cases may be involved in the induction of diseases (gastroenteritis, urinary tract infections, meningitis). Some studies have investigated the involvement of *Ec*CDT production in these diseases (Van Bost et al., 2003). The involvement and the importance of *Ec*CDT in the pathogenicity of *E. coli* have not been clearly demonstrated. Moreover, CDT is not expressed in all strains of pathogenic *E. coli* (Johnson et al., 2002; Kim et al., 2009; Pandey et al., 2003). Indeed, many studies have investigated the prevalence of CDT encoding genes in *E. coli* strains involved in diarrhea (1.6 2.7% prevalence) (Kim et al., 2009), meningitis (46%) (Johnson et al., 2002), etc.

These studies suggest that although the production of CDT by *E. coli* is involved in its pathogenicity, CDT is not the only virulence factor produced by *E. coli* responsible for disease induction.

To date, five different CDTs have been characterized in *E. coli* strains (Smith and Bayles, 2006). *Ec*CDT-I and *Ec*CDT-II were characterized in 1994 in strains of serotype 086:H34 (Scott and Kaper, 1994) and 0128:H- (Pickett et al., 1994), respectively. *Ec*CDT-III was found in strains of *E. coli* producing the cytotoxic necrotizing factor-2 (CNF2) (Pérès et al., 1997). *Ec*CDT-IV was found in many strains colonizing humans and some animals (Toth et al., 2003) and, finally, *Ec*CDT-V was discovered in shiga toxin producing *E. coli* (STEC) O157:H7 (Janka et al., 2003). Genes encoding these five CDTs exhibit 54 to 89% sequence similarity (Toth et al., 2003) and are not found in the same DNA structures, suggesting that they were acquired from different means. The genes encoding *Ec*CDT-I are localized into a lambdoid prophage (Asakura et al., 2007), suggesting that they have been acquired following a lysogenic conversion, after a prophage infection. In contrast, genes encoding *Ec*CDT-II and *Ec*CDT-V are found in the bacterial chromosomal DNA. As these later ones are flanked by P2-phage and λ homologous sequences (Janka et al., 2003), this suggests that they were acquired after phage infection and integration of viral material into the bacterial genome. Finally, the genes encoding *Ec*CDT-III are located in the pVir plasmid, carrying the CNF2 virulence genes (Pérès et al., 1997).

Shigella dysenteriae

Among the bacteria of the *Shigella* genus, *Shigella dysenteriae* (*S. dysenteriae*) is the most virulent specie. Interestingly, *S. dysenteriae* is also the only specie where CDT encoding genes were found, in 22% of cases (Okuda et al., 1995). This characterization was made by PCR, and as these genes may present some sequence variability, it is possible that this quantification has been underestimated. *S. dysenteriae* colonizes the colon and is responsible for shigellosis, in some cases characterized by diarrhea and/or dysentery. Okuda et al. have studied the involvement and the importance of *Sd*CDT production in the pathogenicity of *S. dysenteriae*. In a mouse model, the administration of purified *Sd*CDT into the stomach induced diarrhea in the majority of individuals, as well as necrosis of the colon mucosa (Okuda et al., 1997). These results suggest the involvement of CDT in the induction or in the severity of symptoms developed after infection with *S. dysenteriae*.

Campylobacter jejuni

Bacteria of the *Campylobacter* genus colonize the intestine and are the cause of many food contaminations. The specie *Campylobacter jejuni* (*C. jejuni*) is responsible for gastroenteritis characterized by fever and diarrhea of variable severity. Almost all *C. jejuni* strains possess the genes encoding CDT (Bang et al., 2001; Jain et al., 2008). Therefore, PCR tests directed against *Cj*CDT are available to diagnose gastro-enteritis caused by *C. jejuni* (Kabir et al., 2011).

Experiments on human Hela cells showed that *Cj*CDT increases the adhesion and invasion capacity of *C. jejuni* (Jain et al., 2008). In a rat model of post-infectious irritable bowel syndrome (PI-IBS), *Cj*CDT does not increase the percentage of colonization by *C. jejuni* (Pokkunuri et al., 2012). However, CDT seems to increase the duration of *C. jejuni* colonization, suggesting that CDT is involved in the persistence of *C. jejuni* at the intestinal epithelium, and not in the initial colonization. The *in vivo* study of CDT involvement in the pathogenicity of *C. jejuni* was performed in a mouse model (Jain et al., 2008). Intra-gastric treatment of mice with the supernatant of a CDT producing *C. jejuni* culture, showed diarrhea and severe inflammation of the entire gastro-intestinal tract (for 6 out of 10 mice), whereas treatment with the catalytically inactive *Cj*CDT supernatant induces no diarrhea and only a slight colon inflammation (Jain et al., 2008). Moreover, Pokkunuri et al. demonstrated in a PI-IBS rat model that the *Cj*CDT production alters stool forms, and, at a histological level, damages the epithelial barrier and induces an inflammation (Pokkunuri et al., 2012). Altogether, these results demonstrate the importance of *Cj*CDT in the pathogenicity of the bacteria *C. jejuni*.

Haemophilus ducreyi

Haemophilus ducreyi (*H. ducreyi*) is the bacteria responsible for a sexually transmissible disease named chancroid. The niche of *H. ducreyi* is in the genital mucosa (Spinola et al., 2002; Janowicz et al., 2011). This disease develops in different phases: first, the development of erythema within a few hours, before appearance of pustules within 2-3 days. Several weeks after infection, and in the absence of treatment, ulcers and in some cases suppurative lymphadenopathy appear. A temperature-dependent rabbit model for chancroid has been developed and used in many *in vivo* studies (Spinola et al., 2002; Stevens et al., 1999; Lewis et al., 2001; Wising et al., 2005). In this model, *Hd*CDT is not required for the initiation of the disease and the development of early symptoms, such as erythema and pustules (Lewis et al., 2001; Stevens et al., 1999). However, these studies examine the symptoms only after a few days of treatment, therefore CDT involvement in disease progression (such as ulcer development), or remission, has not been studied.

Aggregatibacter actinomycetemcomitans

Aggregatibacter actinomycetemcomitans (*A. actinomycetemcomitans*) is part of the human oral flora and is involved in periodontal disease, but also in cardiac and neurological diseases. The prevalence of *Aa*CDT production in individuals with periodontitis is between 27% and 79% (Bandhaya et al., 2012; Jentsch et al., 2012; Wahasugui et al., 2013) and the presence of the toxin is associated with an increase in gingival bone loss. *Ex vivo* studies on gingival explant models showed that the purified *Aa*CDT induces damages and destabilization of the epithelial barrier (Damek-Poprawa et al., 2011). However, *in vivo*, in a rat model, CDT production by *A. actinomycetemcomitans* does not seem to cause any periodontal disease, nor bone loss (Schreiner et al., 2013).

Salmonella Typhi

Salmonella enterica serovar Typhi (*S.* Typhi) is the pathogen responsible for typhoid fever, which causes more than 200,000 deaths per year worldwide. For a long time, factors involved in the induction of typhoid fever were unknown. Recently, the virulence factor produced by *S.* Typhi has been characterized (Song et al., 2013; Spanò et al., 2008). This factor, named the typhoid toxin, is an atypical A_2B_5 toxin consisting of three different subunits, one of the catalytic A moieties being identical to a CDT constituent (see after) (Song et al., 2013).

A recent *in vivo* study, in a mouse model, showed that systemic administration of purified typhoid toxin induces huge weight loss, a decrease in the rate of circulating immune cells and, after few days, the death of individuals. In contrast, injection of a catalytic mutant of the toxin (mutant of the CDT catalytic moiety) induces none of these typhoid fever symptoms (Song et al., 2013). In conclusion, in the case of *S.* Typhi, the catalytic CDT subunit is clearly involved, and even essential, to the pathogenicity.

Helicobacter hepaticus

Helicobacter hepaticus (*H. hepaticus*) is found in the gastrointestinal mucosa and, in some cases, can colonize the hepato-biliary channels and be found in the liver. For a long time, *H. hepaticus* has been characterized as a mouse pathogen, although this bacteria has also been found in the human body (Murakami et al., 2011). *H. hepaticus* induces an acute hepatitis, inflammation and even liver cancers in a mouse model (Ge et al., 2007). In this immunocompetent mouse model, *Hh*CDT is not involved in the induction of chronic hepatitis or in the colonization of the mucosa by *H. hepaticus.* However, only mice infected with a strain of *H. hepaticus* producing a functional CDT exhibited increased levels of pro-inflammatory mediators and developed hepatic dysplastic nodules (Ge et al., 2007). This study clearly shows CDT involvement in the *H. hepaticus* pathogenicity. In addition, it is also the first study to demonstrate the carcinogenic potential of CDT.

CDT Biogenesis and Secretion by Bacteria

CDT Is a Tripartite AB Toxin

The three genes encoding CDT were characterized in 1994 (Pickett et al., 1994; Scott and Kaper, 1994). The *CdtA*, *CdtB* and *CdtC* genes, organized in one operon, encode the three subunits constituting the CDT holotoxin (Figure 1). Among all bacterial strains producing CDT, the CdtB protein is the most conserved, with at least 45% sequence identity (Hu et al., 2006). On the other hand, CdtA and CdtC exhibit up to 19% sequence identity. All three subunits are required to trigger maximal cytotoxicity, albeit in some cases CDT lacking CdtA retains some toxic activity (Gargi et al., 2012).

The crystal structures of the *Aa*CDT natural form and of a recombinant *Hd*CDT, reconstituted *in vitro*, have been determined (Nesić et al., 2004; Yamada et al., 2006). As with many other exotoxins, CDT adopts an AB functional structure. Depending on the structural properties of the A and B subunits (i.e. mode of assembly, monomer subunits, homo- or hetero-polymer subunits, etc.), many classes of these toxins can be differenciated (Henkel et al., 2010). CDT assembles into a heterotrimeric holotoxin, where CdtB is the catalytic A-subunit.

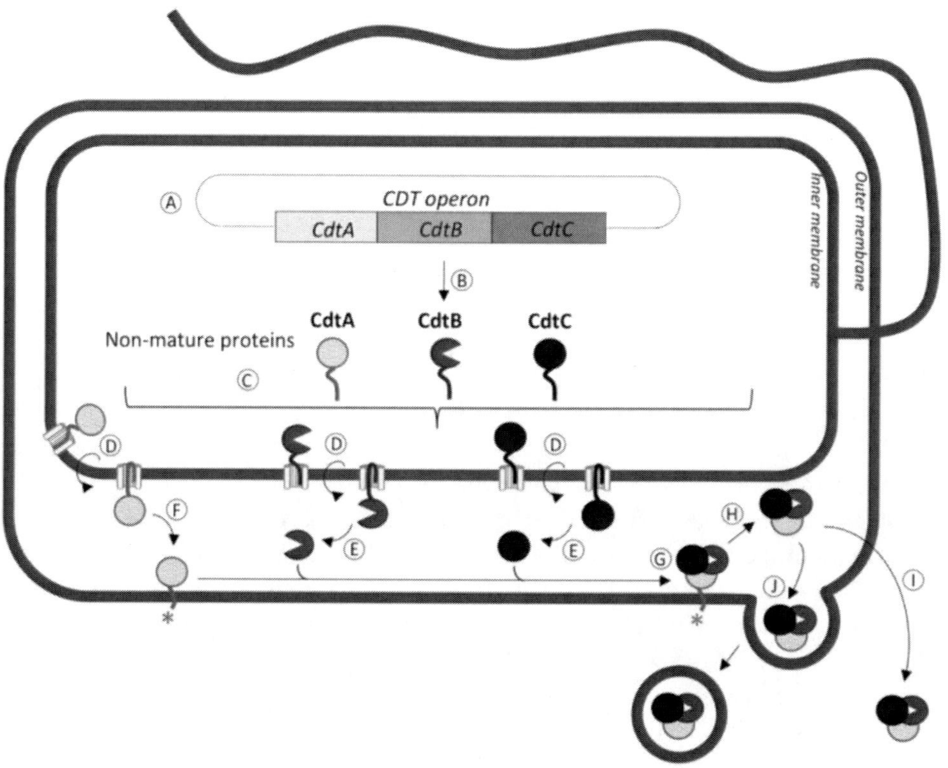

Figure 1. *CDT biogenesis and secretion.* A. CDT operon structure. B. Expression of the three sub-units. C. Localisation of the three sub-units to the inner membrane. D. Periplasm localisation may be via the general secretory pathway. E. Cleavage of the signal peptide leading to the release of CdtB and CdtC into the periplasm. F. Cleavage of the signal peptide and lipid modification of CdtA, leading to his relocalization at the outer membrane. G. Folding of the holotoxin. H. Processing of CdtA, which is N-terminally truncated. Depending on the bacterial strain: I. secretion of the holotoxin in the supernatant or J. secretion of the holotoxin into outer membrane vesicles (OMVs).

The B-moiety, essential for the holotoxin binding to the host cell membrane, is composed of the CdtA and CdtC subunits. Thus, CDT can be classified as an AB_2 exotoxin.

One exception is the CDT-related toxin produced by *S.* Typhi, called the typhoid toxin. Although *S.* Typhi has the *CdtB* gene, it does not possess the *CdtA* and *CdtC* genes. Interestingly, *CdtB* is expressed within the same pathogenicity islet as *PltA* and *PltB* (Spanò et al., 2008), encoding the pertussis-like toxin A (homologous to the pertussis toxin ADP-ribosyltransferase subunit) and the pertussis-like toxin B (homologous to one of the pertussis B subunits) (Kaslow and Burns, 1992) respectively. The structure of this toxin has been recently described (Song et al., 2013), demonstrating that the typhoid toxin is an A_2B_5 toxin. The regulatory subunit (B_5) is composed of a pentameric PltB structure, whereas the catalytic subunit (A_2) is composed of the *St*CdtB/PltA proteins, covalently linked by a disulfide bond.

CDT Biogenesis and Secretion

In an initial report studying CDT biogenesis, using an *E. coli* strain carrying the *AaCDT* genes, it was shown that CdtA is a lipoprotein associated to the outer membrane (Ueno et al., 2006). However, in the periplasm or in the culture supernatant, the CDT holotoxin is found as a complex where CdtA is N-terminally truncated and associated with CdtB and CdtC (Ueno et al., 2006). This suggests that CdtA undergoes lipid modification during the export process from the inner membrane to the outer membrane and, for a second time, a N-terminal processing (Figure 1).

Finally, as CdtA, CdtB and CdtC present a signal sequence, they can be directed to the general secretory pathway leading to CDT genotoxin secretion into the culture supernatant (Zijnge et al., 2012) (Figure 1). However, different studies have showed that certain bacteria release CDT *via* outer membrane vesicles (OMVs). OMVs are constitutively released from some Gram-negative bacteria and have been shown to contain a great variety of virulence factors (Amano et al., 2010; Deatherage and Cookson, 2012). In extra-intestinal pathogenic *E. coli*, the vast majority of secreted CDT was associated to OMVs (Berlanda Scorza et al., 2008). Lindmark et al. studied *C. jejuni* CDT production and release, by OMVs electronic microscopy and bacteria subcellular fractionation: the three CDT subunits have been detected in the periplasmic compartment, and, in the bacterial culture supernatant, extracellular CDT was found tightly associated to OMVs (Lindmark et al., 2009). Isolated OMVs induced the cell distending effects typical of CDT on human intestinal cells, indicating that CDT is present at OMVs under a biologically active form and suggesting that the release of OMVs is a delivery route for *Cj*CDT. In addition, *A. actinomycetemcomitans* OMVs have also been identified as a vehicle for CDT delivery into human cells (Rompikuntal et al., 2012). Concerning the OMVs interaction with the host plasma membrane, the OMVs proteins were found to internalize, in both HeLa cells and human gingival fibroblasts (HGF), *via* the OMVs fusion with lipid rafts (specialized membrane microdomains, enriched in specific lipids, such as cholesterol). In contrast to OMVs isolated from a CDT defective strain, OMVs isolated from a wild type CDT strain induced a characteristic cytolethal distending effect on HeLa and HGF cells, indicating that the OMV-associated CDT was biologically active (Rompikuntal et al., 2012). Association of CDT with OMVs was observed in *A. actinomycetemcomitans* isolates belonging to a, b and c serotypes, indicating that OMV-mediated release of CDT may be conserved in all *A. actinomycetemcomitans* genus. To conclude, whereas CDT may be freely secreted into the extracellular environment (Figure 2A), it is clearly apparent that CDT is also associated to OMVs (Figure 2B). The detailed mechanisms of CDT secretion still remain to be elucidated.

In contrast to these secretory pathways, the *S.* Typhi toxin is once again an exception (Spanò et al., 2008). Infection studies revealed that bacterial uptake into host cells triggers CdtB/PltA/PltB expression, leading to the formation of an intracellular multipartite toxin. Interestingly, following its production, the typhoid toxin is secreted into the extracellular medium. Only after this step, the toxin can interact in an autocrine and paracrine way with the eukaryotic plasma membrane and be internalized to exert its cytotoxic activity (Figure 2C). Inhibition of typhoid toxin export into the extracellular medium inhibits its effects, demonstrating that this step is limiting for toxin activity (Spanò et al., 2008).

All the Same but Different: CDT Host-Cell Binding and Trafficking

Even if all CDT toxins induce the same characteristic effects on eukaryotic cells, they are produced by bacterial strains located within different niches.

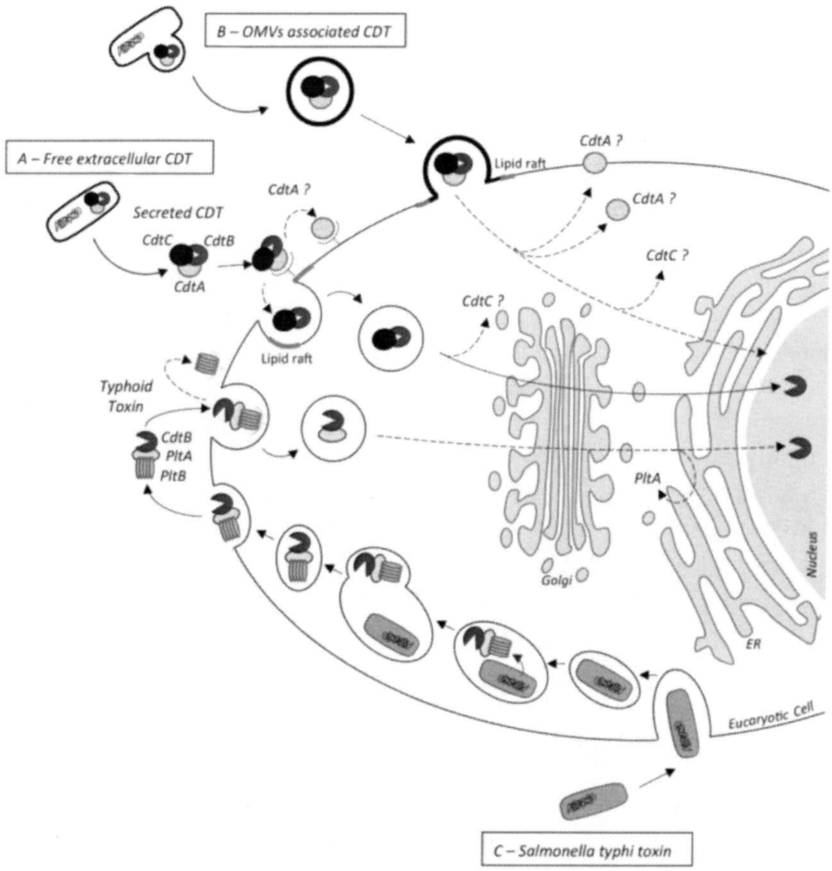

Figure 2. *CDT binding and trafficking in host cell.* Global representation of the CDT binding and trafficking steps. A. Free extracellular CDT. Mainly CdtA, and in a lesser extent CdtC, are involved in the CDT binding to the host cell surface, probably through the interaction with a cargo/receptor. Rafts microdomains may also be important for the binding to the membrane. The CdtA subunit may stay, or be rapidly recycled, at the cell surface, whereas CdtC is found both on the cell surface and in the cytosol and may participate in the CdtB trafficking. CdtB follows a retrograde endosome-Golgi trafficking to translocate at the ER and, later, at the nucleus. However, it has to be noted that different CDT toxins may use different cargoes and traffic through different routes (see text for details). B. Outer Membrane Vesicles (OMVs) associated CDT. After production by bacteria, OMVs are fusing with the host cell membrane. Lipid raft microdomains are important for this fusion step. The CDT toxin is released in the host-cell cytoplasm and CdtB is relocalized to the nucleus of intoxicated cells, apparently independently of the retrograde transport. The CdtA and CdtC outcomes are not clearly understood. C. *Salmonella* Typhi toxin. After its intracellular uptake, *S.* Typhi expresses the CdtB, PltA and PltB subunits, forming the multipartite typhoid toxin. Following its production, the typhoid toxin is secreted into the extracellular medium. The typhoid toxin can bind, in an autocrine and paracrine way, host cells membrane, leading to endocytosis and retrograde trafficking of PltA and CdtB to the ER where they split apart.

This implies that all CDT toxins are not secreted in the same microenvironment (mucus or not, different cell types, highly proliferating host cells or not) and raises different questions regarding the specificity of the CDT holotoxins (Gargi et al., 2012). Are all CDTs the same? Are they acting the same way in all these diverse biological niches? As explained above, there is large sequence variability between CdtC and CdtA homologs. Some authors have hypothesized that this variability may allow CDT to interact specifically with the different cell types residing in the different bacterial niches. Although CdtA and CdtC are clearly involved in the binding of CDT to host cells, a number of questions remain unanswered. Firstly, it is still not clear how CdtA and CdtC contribute to the binding of CDT on the host cell membrane, and furthermore the nature of the CDT receptor remains unknown. Secondly, the internalization and the trafficking of the toxin are still obscure. Hereafter, we present an overview surrounding the current knowledge regarding these different points.

CDT Binding at the Host Membrane

CdtA and CdtC Bind to the Host Plasma Membrane

It is known that CdtA, CdtB and CdtC interact with one another to form an active tripartite holotoxin (Lara-Tejero and Galàn, 2001). CdtA and CdtC are essential for CDT binding to the host cell surface, as *Cj*CdtA or *Cj*CdtC subunits can specifically bind to HeLa cell surface whereas *Cj*CdtB cannot (Lee et al., 2003). Interestingly, mutants containing deletions of regions important for the CdtA/CdtC interaction are still able to bind to the HeLa cell surface (Lee et al., 2003).

The study of the CDT 3D structure by X ray crystallography brought new information on the potential role of CdtC and CdtA in cell surface binding. Indeed, these studies have revealed the presence of an aromatic patch in the structure of CdtA and a groove formed at the CdtA and CdtC interface. The aromatic patch and the groove were shown as key elements of the CDT holotoxin, as mutations in these structures impair host cell intoxication, with or without impairing toxin binding to host cells or toxin stability (Nesić et al., 2004; Nesic and Stebbins, 2005). This key role in the toxin function was recently confirmed by mutational analysis of the CdtA W115 residue. Located in the aromatic patch, this residue exhibits a critical role for host cell binding, without interfering with the holotoxin assembly (Li et al., 2013).

Intriguingly, another point mutation (Y181A), outside of the aromatic patch, has been shown to impair the CDT binding (Li et al., 2013), suggesting that other domain(s) may be involved in host cell binding. Finally, *Ec*CdtA-II and *Ec*CdtC-II subunits may form a binding platform for *Ec*CdtB-II, as *Ec*CdtB-II binds to the HeLa cell surface in a *Ec*CdtA-II and *Ec*CdtC-II dependent manner, as shown by cell cytometry (Mcsweeney and Dreyfus, 2005). Altogether, these results lead to the consensus that CdtA and CdtC form a heterodimeric complex allowing for CDT attachment to the host cell membrane (Figure 2A). This is supported by the fact that *Aa*CdtA and *Aa*CdtC are localized to the plasma membrane (Damek-Poprawa et al., 2012).

Binding to Sugar and to Cholesterol

Interestingly, CdtA and CdtC both display a structure resembling that of lectins (Nesić et al., 2004), characterized as carbohydrate-binding proteins displaying specificity to varying sugars. More precisely, CdtA and CdtC structures are homologous to the lectin-like B-chain repeats of the ricin toxin, important for ricin toxin cellular uptake. It has been shown that CdtA and CdtC structures are conserved between all CDT subtypes, suggesting that they are essential and have key functions in the cell surface binding and holotoxin internalization (Nesic and Stebbins, 2005). In addition, CdtA and CdtC seem to compete for a similar receptor (or binding protein) on the HeLa cell surface, as preincubating cells with specific lectins blocks CdtA and CdtC binding (Mcsweeney and Dreyfus, 2005). This is confirmed by the fact that CdtA or CdtC preincubation with specific glycoproteins prevents their binding to host cells. Moreover, removal of N-linked, but not O-linked, sugars at the host cell surface impedes CDT intoxication (Mcsweeney and Dreyfus, 2005). At the same time, another study showed that CDT could bind to specific glycolipids (Mise et al., 2005). Particularly, gangliosides (ceramide-based oligosaccharide with sialic acids linked on the sugar chain) may be involved, as the pre-incubation of CDT with GM3 (monosialodihexosyl ganglioside)-enriched liposomes reduced CDT toxicity on host cells (Mise et al., 2005). Moreover, *Aa*CDT effects also depend on cholesterol, as cholesterol-depletion reduces CDT binding and cholesterol-repletion restores the binding potential of CDT (Boesze-Battaglia et al., 2009). Further analyses indicate that the CdtC subunit contains a cholesterol recognition/interaction amino acid consensus (CRAC) site, mutation of which results in decreased CDT binding (Boesze-Battaglia et al., 2009). Rafts microdomains are enriched in cholesterol and sphingolipids, explaining why rafts microdomains are required for binding and intoxication by CDT (Boesze-Battaglia et al., 2006; Guerra et al., 2005). Finally, cell intoxication by *Cj*CDT is also achieved through CDT association to cholesterol-rich microdomains (Lin et al., 2011), as a mutation in the *Cj*CdtC CRAC-like region impacts the cholesterol-binding and leads to an impaired CDT intoxication (Lai et al., 2013).

CDTs from *H. ducreyi*, *A. actinomycetemcomitans*, *E. coli* and *C. jejuni* may differ in their abilities to intoxicate host cells, leading to a variable cell target tropism. Enriching the cholesterol-content in host cells enhances the effects of all CDTs tested except *Cj*CDT, whilst the absence of cell surface sialic acid sensitized CHO cells to all CDTs with the exception of *Cj*CDT (Eshraghi et al., 2010). In contrast, it has to be noted that fucosylated N-linked glycans and glycolipids were not required for intoxication by any of the CDTs tested in this study. As stated by the authors, *Aa*CDT and *Hd*CDT behave similarly, whereas *Ec*CDT-III and *Cj*CDT display distinct host cell preferences. This is consistent with the homology presented by these four CDTs, and also illustrates that CDTs produced by different pathogens may possess unique features to intoxicate host cells and may do so through different host receptors (Eshraghi et al., 2010). The CDT tropism may thus correlate to the amino-acid sequence of the CDT-binding subunits. Thus, although *A. actinomycetemcomitans* and *H. ducreyi* have different niches, their CdtA and CdtC subunits display high sequence homologies and respond similarly to host factors alterations (such as glycans, cholesterol content, etc.).

In addition, these niches (oral cavity and genital mucosa respectively) are however characterized by a similar epithelial organization (stratified epithelia), underlining again the necessity of a similar mode of action. Reciprocally, *E. coli* and *C. jejuni* have similar niches but their binding subunits are divergent and display different target cell preferences (Eshraghi et al., 2010). The CDT receptor(s) identification for these different CDT toxins may ultimately provide insights regarding cellular tropism and shed light on host CDT-producing bacteria interactions.

A Specific Membrane Receptor for CDT?

As stated before, the divergent results on factors that mediate CDT binding to the host cell surface may depend on the receptor specificity inherent to each CDT. To underline this specificity, it is important to note that CDT toxins may be specific to host species, as *Hd*CDT cannot intoxicate rodent cells (Cortes-Bratti et al., 1999; Guerra et al., 2005). Recently, two genetic screens conducted on the KBM7 chronic myeloid leukemia cell line identified several genes important for CDT intoxication (Carette et al., 2009; Carette et al., 2011). These host genes have not been found with other toxin screens (i.e. diphtheria toxin, ricin, etc.) and may therefore be considered specific to the CDT intoxication. Different CDT toxins share some common host factors, such as the SGMS1 gene, encoding the sphingomyelin synthase 1, implied in the host-cell sensitivity for all CDTs tested (*E. coli*, *C. jejuni*, *A. actinomycetemcomitans* and *H. ducreyi*).

However, CDTs also exploit specific host proteins, which may help in defining a characteristic profile for each CDT. *Ec*CDT-I specifically requires TMEM181, a putative G protein-coupled receptor, expressed at the cell surface, which has been shown to co-immunoprecipitate with CDT (Carette et al., 2009).

GLG1 (Golgi glycoprotein 1) and ATP6V0A2 (the vacuolar ATPase subunit 2) were also found as important genes for host cell sensitivity to *Ec*CDT-I (Carette et al., 2011), although their cellular localization does not make them good receptor candidates. *Aa*CDT shows a gene pattern similar to the *Ec*CDT-I, except that it does not require TMEM181 but, instead, relies on the SYNGR2 (synaptogyrin 2), an ubiquitous integral membrane protein. *Cj*CDT depends on different host factors: besides proteins involved in vesicular transport, three specific plasma membrane proteins were identified as essential: TMEM127, GPR107 and TM9SF4. Because the SGMS1 depletion reduces levels of sphingomyelin, a key component of lipid rafts, this confirms and underlines the role of these microdomains in the CDT intoxication process.

In addition, these data suggest that the TMEM181 protein may be the *bona fide Ec*CDT-I receptor, however it could not be ruled out that TMEM181 plays an indirect role, yet to be identified, in the receptor-toxin interaction and/or trafficking.

CDT Trafficking into the Mammalian Cells and CdtB Nuclear Localization

Recently, through live cell microscopy analysis, *Aa*CdtA has been found at the cell surface, whereas *Aa*CdtC is found both on the cell surface and in the cytosol, and *Aa*CdtB is found at the ER and later at the nucleus (Damek-Poprawa et al., 2012) (Figure 2A).

CdtA and CdtC localization to the host cell surface is in agreement with their role in binding to the plasma membrane (see above). However, the intracellular localization of CdtB and CdtC, and more precisely the observation that the nucleus represents the final destination of CdtB, implicate a programmed CDT holotoxin disassembly to ensure an effective cellular trafficking of CdtB.

We will present in this part the lines of evidence depicting the cellular road taken by CdtB to reach the nucleus.

Importance of Endocytosis

A great variety of bacterial toxins use cellular trafficking to enter host cells (Spooner and Watson, 2010). Apart from phagocytosis, happening in certain cell types (macrophages, monocytes, etc.), different endocytic pathways have been described in eukaryotic cells. Endocytic pathways generally involve several steps, as the membrane invagination and the endocytic vesicle pinch off. The invagination step can define different endocytic processes, mainly clathrin coat dependent endocytosis, clathrin-independent but caveolin-dependent endocytosis and clathrin- and caveolin-independent endocytosis, as in bulk-phase pinocytosis (Doherty and Mcmahon, 2009).

Molecule-specific endocytosis is achieved through receptor-mediated endocytosis. However, one can consider that host receptor binding by the toxin is a hijack from their cellular function. Thus, host cell receptors can be seen as cargos exploited by CDTs to enter host cells (Figure 2A). *Hd*CDT is rapidly internalized from the cell surface and subsequent CDT effects seem dependent on clathrin and dynamin (Cortes-Bratti et al., 2000), respectively involved at the early stage of endocytic vesicle formation and at the pinch-off step; *Hd*CDT may therefore enter host cells through receptor-mediated endocytosis. However, this does not rule out the possibility that CDT can enter through other endocytic pathways, such as caveolae, given the importance of cholesterol-rich domains (Boesze-Battaglia et al., 2006, 2009; Guerra et al., 2005; Lin et al., 2011). Lately, it has been shown that CDT intoxication of CHO cells could occur despite significant depletion of membrane cholesterol, ultimately underlining the potential role of alternative endocytic pathways (Damek-Poprawa et al., 2012).

Retrograde Trafficking

After cell entry by endocytosis, molecules can trafficked through different routes, for example to be degraded in the lysosome, or to be relocated within host cells. Early endosomes can be seen as a traffic hub, where decisions are made: endocytosed molecules can be recycled to the plasma membrane, transferred directly to the trans-Golgi network (TGN) or passed to late endosomes, where traffic pathways lead to TGN or to lysosome. If transferred to TGN, molecules can be retrogradely transported to the endoplasmic reticulum (ER) (Spooner and Lord, 2012).

CDT intoxication is inhibited when fusion of early endosomes with downstream compartments is blocked, or when Golgi is disrupted (Cortes-Bratti et al., 2000). This suggests the involvement of a retrograde endosome-Golgi trafficking route (Figure 2A). As assessed by glycosylation studies, *Hd*CDT is transported from the TGN to the ER, however, this occurs in the absence of the ER-associated degradation (ERAD) pathway (Guerra et al., 2005). Interestingly, if many ER-translocating toxins exploit the ERAD pathway to transit from the ER to the cytosol, CDT translocation is atypical and does not require protein unfolding. To corroborate this, the CdtB subunit could not be seen freely in the intoxicated cells cytosol by confocal microscopy analysis and biochemical assays (Guerra et al., 2009). Thus, *Hd*CDT translocation to the nucleus does not involve the classic ERAD pathways, usually followed by retrogradely transported toxins, and may be directly translocated from the ER to the nucleus.

Recently, the intoxication mechanisms of CDTs from *Ec*CDT-III and *Hd*CDT, sharing limited sequence homology, were compared (Gargi et al., 2013). If both toxins require the Golgi compartment to traffic, they also present distinct features. For example, endosomal disruption blocks *Hd*CDT transport but has no effect on *Ec*CDT-III localization. In addition, cells expressing dominant negative Rab7 (a small GTPase involved in the late endocytic pathway) present no effects after *Hd*CdtB treatment but are affected by *Ec*CDT-III. These results support that *Hd*CDT, but not *Ec*CDT-III, traffic through the late endosomal compartment. Moreover, *Ec*CDT-III and *Hd*CDT may use different cargos, as competitive binding studies suggest that they bind to different cell surface determinants (Gargi et al., 2013). Altogether, these results reinforce that different CDT toxins may use distinct trafficking pathways, possibly reflecting different cargo interactions at the host cell surface.

CdtB Nuclear Localization

Several studies have tried to characterize how CdtB is translocated to the nucleus, mainly through observations of CdtB localization after cellular microinjection (Lara-Tejero and Galàn, 2000; Nishikubo et al., 2003). Four hours after being microinjected into the cytoplasm, *Cj*CdtB and *Aa*CdtB are exclusively located to the nucleus, demonstrating that CdtB trafficking to the nucleus does not require the two other CDT subunits. In order to identify domains important for this nuclear transport, truncated forms of *Aa*CdtB have been fused to the green fluorescent protein (GFP) and expressed in HeLa cells (Nishikubo et al., 2003). An 11 amino-acid sequence, within the N-terminal region of CdtB, has been found essential for nuclear localization and for toxin-induced cellular effects. The replacement of this N-terminal sequence by the Nuclear Localization Signal (NLS) of the SV40 Large T-antigen allows recovering CDT effects (Nishikubo et al., 2003). Thus, CdtB nuclear entry is crucial for CDT cytotoxicity and is governed by a putative NLS sequence. In addition, two potential NLS have been identified in the C-terminal part of *Ec*CdtB-II (McSweeney and Dreyfus, 2004). Interestingly, deletion of each of these NLS sequences produces a differential localization of the active toxin subunit, suggesting a specific function for each NLS sequence.

The Remaining Black-Boxes

The current model of CDT trafficking implies a stepwise holotoxin disassembly with: 1) CdtA remaining bound to the plasma membrane following initial toxin binding to a yet unidentified receptor, 2) endocytosis of the CdtB-CdtC heterodimer and 3) retrograde trafficking of CdtB to the nucleus (Figure 2A). However, a clear function for CdtC remains to be defined. CdtC could for example impede the CdtB function, as structural data may suggest (Nesić et al., 2004). Even though, some confusion was brought from CdtC trafficking data, as the intracellular delivery of recombinant *Aa*CdtB or *Aa*CdtC was shown to induce CDT characteristic effects on CHO cells (Mao and DiRienzo, 2002). These characteristic CDT effects were also found after HEp-2 epithelial cells were exposed to both *Aa*CdtB and *Aa*CdtC (Akifusa et al., 2005). Yet the CdtC contribution to CDT effects is obscure, and it could well be due to CdtC-related activity on cell trafficking. In addition, *Aa*CdtC mutants predicted to impair receptor binding failed to show internalization defects, indicating that CdtC may only have a minor role in receptor binding (Damek-Poprawa et al., 2012) and may in fact facilitate CdtB internalization, potentially by helping in the first steps of the CdtB trafficking.

Although the majority of CDT internalization studies have been performed using recombinant purified holotoxin, it should not be forgotten that both *Ec*CDT-II, *Cj*CDT and *Aa*CDT may be secreted through OMVs (Berlanda Scorza et al., 2008; Lindmark et al., 2009; Rompikuntal et al., 2012). After treatment with *A. actinomycetemcomitans* OMVs, the CdtB subunit has been localized inside the nucleus of intoxicated cells (Rompikuntal et al., 2012). Whether the delivery of OMV-associated CDT follows the same trafficking as the soluble toxin is doubtful, as differences between these pathways are already known (Amano et al., 2010; Ellis et al., 2010).

Particularly, Rompikuntal *et al.* have shown that OMV internalization is dependent on the lipid raft microdomains of the host cell membrane but is independent of retrograde transport (Figure 2B) (Rompikuntal et al., 2012). In conclusion, many questions still need to be answered, particularly regarding CDT binding, entry, trafficking and nuclear translocation in host cells, as well as the existence of a common pathway to all CDTs or if specific trafficking pathways exist for each CDT.

The Host Cell Response to CDT

At the cellular level, intoxication with CDT causes cell cycle arrest at G2/M accompanied by nuclear and cytoplasmic enlargement, a phenomenon called cellular distention, which eventually lead to apoptotic cell death. Such cell cycle alteration and cytotoxicity effects have been documented with various CDT toxins, originated from diverse bacteria including *A. actinomycetemcomitans* (Ohguchi et al., 1998; Shenker et al., 1999; Sugai et al., 1998), *E. coli* (Comayras et al., 1997; Pérès et al., 1997), *C. jejuni* (Whitehouse et al., 1998), *H. ducreyi* (Cortes-Bratti et al., 1999; Cortes-Bratti et al., 2001), *H. hepaticus* (Young et al., 2000) or *S.* Typhi (Haghjoo and Galán, 2004), and in a broad range of human cell lines. Interestingly, some cellular features of CDT intoxication can be recapitulated in *S. cerevisiae* (Hassane et al., 2001), suggesting that the mode of action of CDT cytotoxicity involves a host target shared from yeast to human.

So far, several lines of evidence indicate that CdtB, the catalytic component of the toxin, attacks host genomic DNA by inducing strand breaks, activating the DNA damage response (DDR), leading to cell cycle arrest and cell death. In this part, we will detail the data supporting this model.

Cellular Features of CDT Intoxication

Morphological Changes

Regardless of CDT origin, intoxicated cells undergo sequential morphological alterations before dying (Figure 3). The timing of events depends on the cell line that has been intoxicated, probably due to the variability in the sensitivity to the toxin.

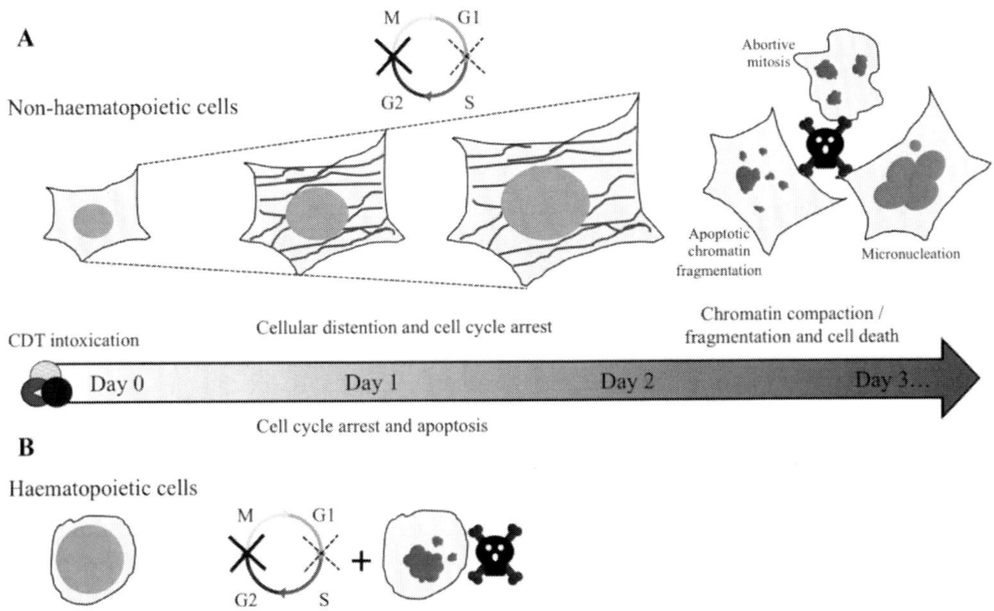

Figure 3. *CDT cellular effects.* This figure depicts the morphological alteration over time of CDT intoxicated cells from non-haematopoietic (A) or haematopoietic (B) lineage. A. Cytoplasmic and nuclear distension occur within the first day of intoxication and extend until three days. At the same time, cells progressively block at G2/M and potentially at G1/S (probably depending on their p53 status). The timing of cell cycle arrest depends on the cell cycle duration and the cellular sensitivity to CDT (i.e. the dose of CDT and the susceptibility of a specific cell line to a defined CDT). After two days, cells dye by either apoptosis or necrosis. A few part of G2/M arrested cells entered an abortive mitosis. Otherwise, cell death is accompanied by chromatin compaction and fragmentation and eventually micronucleation. B. Lymphocytes (and more probably all haematopoietic cells) intoxicated by CDT are much more sensitive than non-haematopoietic cells as they show a rapid apoptotic response even before any obvious cell cycle arrest. As for non-haematopoietic cells, the timing and occurrence of cell cycle arrest and cell death depends on the cellular sensitivity to CDT.

Cellular distension, the first morphological change to be documented, starts to be observable 15 hours after CDT treatment, and increases progressively to reach between four- and sevenfold the normal cell-size (Aragon et al., 1997; De Rycke et al., 2000; Pérès et al., 1997; Sugai et al., 1998; Whitehouse et al., 1998) (Figure 3A). Lymphocytes however represent an exception: albeit these cells are more sensitive to CDT, they do not show any distension phenotype (Shenker et al., 1999) as they die prematurely (Figure 3B). Concomitantly to cytoplasm expanding, the nucleus also enlarges and actin stress fibers are formed (Aragon et al., 1997; Cortes-Bratti et al., 1999; De Rycke et al., 2000; Pérès et al., 1997; Sugai et al., 1998; Young et al., 2000) (Figure 3A). After two days of CDT exposure, several morphological abnormalities are visualized in a subset of cells (De Rycke et al., 2000; Whitehouse et al., 1998). Thus, a small part of intoxicated cells seems to engage in an abortive form of mitosis, characterized by rounding up, chromatin compaction and fragmentation, granular morphology and distorted spindle structure with centrosome amplification. A more important fraction of cells display irregular chromatin condensation, as well as invagination of the nuclear envelope, progressive lobulation and eventually micronucleation. It has to be noticed that such chromatin condensation and degradation are characteristic of apoptotic cells (Ohguchi et al., 1998).

Cell Cycle Arrest

An overall feature of CDT exposure is cell cycle arrest at the end of the G2-phase, in a variety of proliferating cells including lymphocytes (Cortes-Bratti et al., 2001). Progression through eukaryotic cell cycle requires the sequential activation of heterodimeric CDK-cyclin complexes (Diaz-Moralli et al., 2013) (Figure 4). The CDK1-cyclin B complex, formed during G2, controls the initiation of mitosis upon regulated concentration and localization of cyclin B1, and phosphorylation of CDK1 that dictates its activation state. Cyclin B1 progressively accumulates in the cytoplasm between S and the end of G2, before translocating to the nucleus at early mitosis (Pines and Hunter, 1991). Upon CDT intoxication, cyclin B1 levels increase up to a normal G2 level, but remain outside of the nucleus, confirming that cells accumulate at G2 without proceeding to mitotic entry (Comayras et al., 1997; Fedor et al., 2013). The G2/M transition necessitates CDK1 activation by CDC25-mediated dephosphorylation at T14/Y15 (Shen and Huang, 2012). Cells treated with various CDTs exhibit a strong increase in phosphorylated (inactive) CDK1 (Bielaszewska et al., 2005; Comayras et al., 1997; Cortes-Bratti et al., 1999; Shenker et al., 1999; Whitehouse et al., 1998), which can be reactivated *in vitro* with recombinant CDC25 (Sert et al., 1999). This lack of CDK1 activation can be explained at least in part by the CDT-induced sequestration of CDC25C, out of the nucleus (Alby et al., 2001). These results support that CDT treatment inhibits CDC25 nuclear localization, which in turn fails to activate CDK1, preventing mitotic entry and thus blocking cells in G2. Indeed, overexpression of recombinant CDC25B or CDC25C abrogates CDT-induced G2 arrest, enabling intoxicated cells to enter an abortive mitosis (Escalas et al., 2000).

In addition to the well-documented G2/M block induced by CDT, a subset of cell lines also exhibit cell cycle arrest at the G1/S transition after treatment with CDT from various origins (Cortes-Bratti et al. 2001; Hassane et al. 2003; Belibasakis et al. 2004; Bielaszewska et al. 2005) (Figure 3). This particular response to CDT seems to rely on the G1/S checkpoint that depends on the activation of the p53 tumor suppressor and its downstream effector p21 (Diaz-Moralli et al., 2013). Thus, the variation in the p53 status among the different cell types may explain the discrepancy observed in regards to the cell cycle alterations induced by CDT. Besides, we have recently demonstrated that CDT also induces a slight S-phase slow-down that is difficult to observe by classical flow-cytometry analyses (Fedor et al., 2013). Therefore, the CDT-mediated proliferation arrest may neither be cell-cycle phase nor bacterial species specific, but rather be due to overall cellular responses (like checkpoints) that might differ between cell types (Belibasakis et al., 2004).

Cell Death

The final characteristic of CDT intoxication is generally the programmed cell death by an apoptotic pathway (Figure 3). Apoptosis is an essential process regulating development and tissue homeostasis in multicellular organisms by elimination of unnecessary or damaged cells, in an immunologically silent manner (Gao and Kwaik, 2000). In addition, this mechanism modulates the pathogenesis of different infectious diseases by contributing to an antimicrobial immune response. On the other hand, induction of apoptosis by bacterial pathogens may facilitate colonization, persistent infection and chronic disease (Jinadasa et al., 2011).

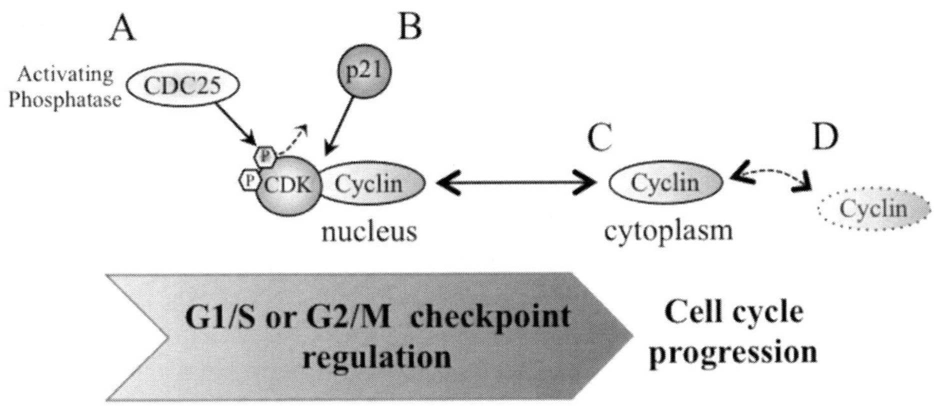

Figure 4. *Model of the DNA damage checkpoints-mediated cyclin-CDK complex regulation*. The cyclin-Cyclin Dependent Kinase (cyclin-CDK) complex is formed by the association of a regulatory cyclin with a kinase, which catalytic activity is highly dependent of the cyclin association. The cyclin-CDK complexes are key components of the cell cycle regulation, particularly at the G1/S and G2/M checkpoints. Specific cyclins and CDK exist throughout the cell cycle phases. A global schema for the regulation of these cyclin-CDK complexes through checkpoint activation can be drawn. Globally, 4 different levels of regulation can be distinguished. A. CDK Phosphorylation status. The activity of the cyclin-CDK complex is controlled by a phosphorylation-dephosphorylation equilibrium. CDC25 proteins are well-documented phosphatases involved in the activation of the cyclin-CDK complexes. CDC25 activation can be regulated by the DNA damage checkpoints. B. p21 CDK inhibitor expression. p21 is able to interact with the CDKs in order to block their activity. p21 expression is induced after the transcription factor p53 is activated, for example after DNA has been damaged. C. The cyclin localization is also a critical step in the cyclin-CDK complex regulation. If bloked in the cytoplasm, cyclins are unable to associate and activate CDKs. D. Cyclin synthesis or degradation. Finally, the synthesis, as well as the degradation of cyclin subunits has been proven key in the regulation of the cyclin-CDK complex.

Apoptosis can be separated into two major pathways (Taylor et al., 2008). The extrinsic pathway is initiated by interaction between extracellular ligands and transmembrane death receptors, resulting in caspase-8 and subsequent caspase-3 and -7 activation, and eventually, cell death. In the intrinsic pathway, activation of the BH3-only pro-apoptotic proteins leads to the alteration of the outer mitochondrial membrane permeability, resulting in the activation of caspase-9 and downstream elements, and subsequent apoptosis. Both pathways imply the activation of caspases that in turn activate specific nucleases (caspase-activated DNases) involved in DNA fragmentation and condensation within the dying cells (Enari et al., 1998; Liu et al., 1997). Cells displaying a wild-type p53 genotype display greater sensitivity to apoptosis, as p53 transcriptionally upregulates many BH3-only proteins (Shibue and

Taniguchi, 2006). However, BH3-only proteins can also be activated by alternative, albeit less efficient, mechanisms in p53-null cells.

CDT-mediated apoptosis has been shown to follow the intrinsic mitochondrial pathway for toxins produced by *H. hepaticus* (Liyanage et al., 2010), *A. actinomycetemcomitans* (Ohara et al., 2004; Shenker et al., 2001), *C. jejuni* (Hickey et al., 2005) and *H. ducreyi* (Wising et al., 2005). Moreover, overexpression of BCL-2, an anti-apoptotic protein involved in the intrinsic pathway, inhibits apoptosis induced by *A. actinomycetemcomitans* CDT (Ohguchi et al., 1998; Shenker et al., 2001). Activation of the mitochondrial pathway-related caspases has been reported in several studies (Shenker et al., 2001; Ohara et al., 2004; Wising, et al., 2005; Liyanage et al., 2010), but among the analyzed cell lines, a few also employ the extrinsic pathway through the activation of caspase-8, which might correlate to their p53 negative status (Hickey et al., 2005; Ohara et al., 2004; Shenker et al., 2001). Finally, a caspase-independent apoptotic pathway has been observed as a late cell-death response to CDT that involves the accumulation of reactive oxygen species and mitochondrial membrane alterations (Ohara et al., 2008).

CDT can induce apoptosis in all cell types tested so far including fibroblasts, epithelial, endothelial or haematopoietic cells, proliferating or non-proliferating primary cells, and established cell lines (Jinadasa et al., 2011). However, haematopoietic cell lines seems more prone to induce a rapid apoptotic response within the first day of CDT exposure, and in some cases before any obvious cell cycle arrest (Ohguchi et al., 1998; Gelfanova et al., 1999; Cortes-Bratti et al., 2001; Ohara et al., 2004). Hence, when tested in identical conditions, epithelial cells, keratinocytes or fibroblasts exhibit a mild apoptotic response compared to monocytes or T cells (Wising et al., 2005). Instead, cells from non-haematopoietic lineages die either by apoptosis or necrosis, the latter being perhaps a consequence of abortive mitosis (see above). The ratio between apoptotic and necrotic cell death is dose-dependent and fluctuate with the timing of treatment. Moreover, apoptotic/necrotic responses of intoxicated cells from similar lineages might depend on the cell types as shown for endothelial cells (Bielaszewska et al., 2005).

To summarize, haematopoietic cells show the most dramatic apoptotic response to CDT, and p53-deficient cells are commonly more resistant to CDT-induced apoptosis, as well as cells exhibiting a high BCL-2 expression level. However, other parameters have to be considered, like cellular susceptibility to the CDT bacterial source, as well as exposure time or toxin concentration. As a matter of fact, all these parameters modulate the balance between the different cellular outcomes after CDT exposure – i.e. prolonged cell cycle arrest, apoptosis, necrosis or even senescence (Blazkova et al., 2010).

CDT Triggers a DNA Damage-Related Response

Interestingly, many studies pointed out that morphological and cell cycle defects induced by CDT are similar to those observed with some DNA damaging agents. As soon as CDT-mediated cell cycle defects were reported, similarities between the checkpoints activated during the DDR have been noticed (Comayras et al., 1997; Whitehouse et al., 1998). The DDR is a multistep process that senses DNA damage and triggers a signaling cascade that activates cell cycle checkpoint arrest at the G1/S or G2/M boundary, to allow sufficient time to repair DNA lesions or to promote programmed cell death if too much damage has been sustained (Deckbar et al., 2011). In other words, DNA damage-mediated G1/S and G2/M checkpoints interrupt cell cycle progression, in order to prevent the succession of genomic instability at crucial steps including DNA replication during S phase and chromosome segregation during mitosis. In addition, when arising during S phase, DNA lesions stall the replication forks and trigger an intra-S checkpoint that slows down (but does not block) the replication progression (Ghosal and Chen, 2013; Jones and Petermann, 2012). When treated with a lethal dose of etoposide, a topoisomerase II poison that induces DNA double strand breaks (DSBs) by inhibiting the ligation of DNA ends produced after the cleavage intermediate step (Liu, 1989), HeLa cells present patterns similar to those observed after *Ec*CDT-I exposure, including G2/M block, loss of cell viability and morphological alterations (Sert et al., 1999). Furthermore, *Hd*CDT and ionizing radiations induce comparable cell cycle arrests on HL fibroblasts or on HEp-2 cells (Cortes-Bratti et al., 2001). Yet, ionizing radiation-mediated DDR activation is a well-known consequence of DSB induction (Deckbar et al., 2011; Vignard et al., 2013). Finally, De Rycke et al. pointed out that the sequential morphological changes promoted by *Ec*CDT are reminiscent of those due to etoposide or irradiation (De Rycke et al., 2000). Taken together, these reports strongly suggest that CDT intoxication elicits a cellular response related to DSB signaling.

Phosphatidylinositol 3-kinase related protein kinases (PIKK) are central players in the DDR. ATM (ataxia-telangiectasia-mutated) drives the G1/S checkpoint, which also requires active p53, whereas the intra-S checkpoint depends mostly on ATR (ATM- and Rad3-related). Either ATM or ATR can activate the G2/M checkpoint. Whilst ATR responds to single-stranded DNA (ssDNA) that are typically formed at replication forks blocked by a broad range of lesions (Zou and Elledge, 2003), ATM is specifically activated by DSBs through the MRE11-RAD50-NBS1 (MRN) complex (Lee and Paull, 2005) (Figure 5A and B). MRN is a damage sensor that is rapidly localized to DSB sites, where NBS1 recruits inactive ATM dimers (Falck et al., 2005; Lee and Paull, 2005). Autophosphorylation of ATM at serine 1981 induces dimer dissociation, releasing free active monomers that can then phosphorylate downstream substrates including H2AX, CHK2 or p53 (Bakkenist and Kastan, 2003; Matsuoka et al., 2007) (Figure 5C). H2AX is an H2A histone variant that is rapidly phosphorylated on serine 139 (referred to as γH2AX) on several Megabases surrounding DSB sites (Rogakou et al., 1999; Rogakou et al., 1998). The Checkpoint kinase 2 (CHK2) plays a central role in the G2/M checkpoint activation by phosphorylating p53, CDC25A and CDC25C (Ahn et al., 2004), as phosphorylated CDC25 proteins are inactive and cannot activate CDK1, thereby preventing mitotic entry (see above and Figure 4).

CDT Activates the ATM-Dependent DSB Signaling Pathway

The accumulation of DDR proteins to DSB sites has been widely analyzed by immunocytological fluorescent observations of docking structures called foci (Vignard et al., 2013). Strikingly, CDT exposure recapitulates the different steps of DSB signaling (Figure 5). First, the toxin induces foci formation of the three MRN complex members, as shown with CDT from *H. ducreyi* for MRE11 (Li et al., 2002), from *C. jejuni* for RAD50 (Hassane et al., 2003) or from *H. hepaticus* for NBS1 (Guerra et al., 2010) (Figure 5B). Infection of gingivial epithelial cells with *A. actinomycetemcomitans* producing CDT seems to increase ATM protein levels, suggesting that the host response to CDT intoxication involves the ATM-related pathway (Alaoui-El-Azher et al., 2010). The accumulation of γH2AX is currently the most assessed DSB biomarker, as it signals DSBs very quickly after their formation and disappears once they are repaired. In response to CDT produced by *H. hepaticus*, *C. jejuni*, *H. ducreyi* and *E. coli,* many studies have shown that H2AX is phosphorylated and γH2AX foci are formed (Bielaszewska et al., 2005; Blazkova et al., 2010; Fedor et al., 2013; Hassane et al., 2003; Li and Sharipo, 2002; Liyanage et al., 2010) (Figure 5C). Moreover, 53BP1, another well-used DSB marker, also forms foci after CDT exposure (Figure 5D), which colocalize with γH2AX foci, strongly supporting the presence of DSBs (Blazkova et al., 2010; Fedor et al., 2013).

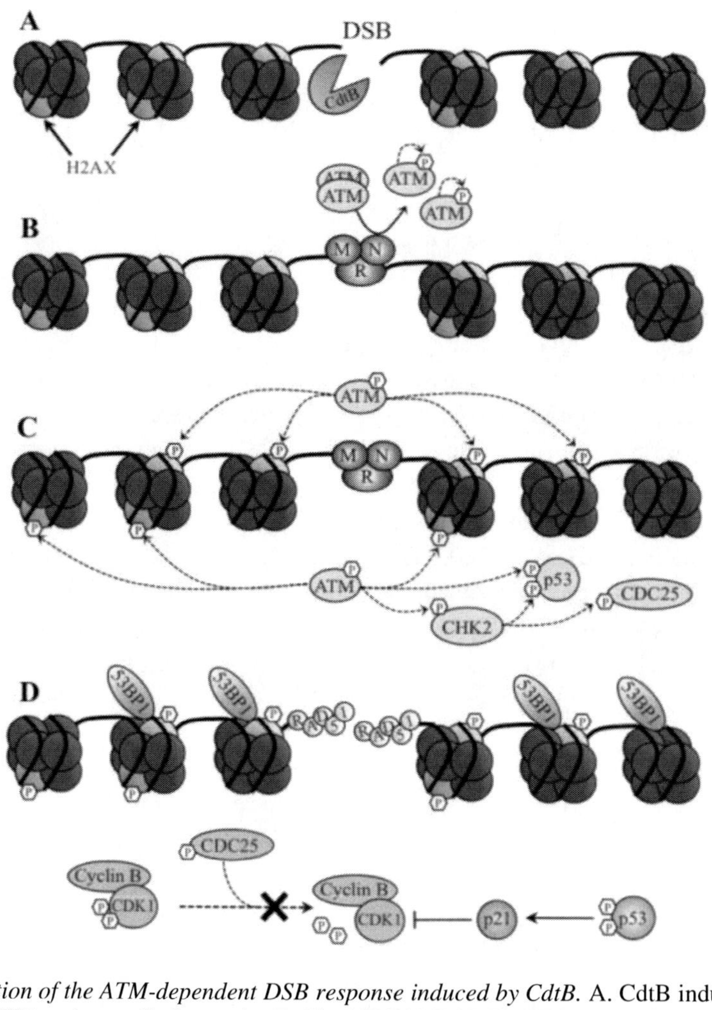

Figure 5. *Activation of the ATM-dependent DSB response induced by CdtB.* A. CdtB induces double-strand breaks (DSB) on host cell chromatin. B. The MRE11-RAD50-NBS1 complex (MRN) binds DSB ends and recruits inactive ATM dimmer through NBS1. ATM autophosphorylation induces release of free active monomers. C. ATM phosphorylates downstream effectors including H2AX (reappointed γH2AX), p53 (stabilization) and CHK2 (activation). Once activated, CHK2 phosphorylates p53 (stabilization) and CDC25 phosphatases (inactivation). D. 53BP1 and RAD51 are recruited to DSB site. Maintaining CDC25 in the inactive form induces the accumulation of phosphorylated (inactive) CDK1. In parallel, p53 transcriptionally upregulates the CDK1 inhibitor p21. CDK1 inhibition prevents mitotic entry.

Finally, CDT intoxication of fibroblasts promotes ATM-dependent CHK2 activation (Cortes-Bratti et al., 2001; Alby et al., 2001; Blazkova et al., 2010; Alaoui-El-Azher et al., 2010). As ATM and CHK2 kinase activities are required for p53 stabilization and DNA damage-related checkpoint activation (Ahn et al., 2004), several studies have shown that CDT induces p53 phosphorylation and stabilization, as well as accumulation of its downstream effector p21, a cyclin-dependent kinase inhibitor (Cortes-Bratti et al., 2001; Sato et al., 2002; Li and Sharipo 2002; Yamamoto et al., 2004; Blazkova et al., 2010) (Figure 5D).

The Cellular Outcomes Induced by CDT Are Associated with the DDR

Thus, the cell cycle defects induced by CDT seem to rely on the ATM-CHK2 pathway that activate p53 and p21, and inactivate CDC25 proteins, both participating to inhibit cyclin-dependent kinase activities and eventually lead to cell cycle arrest (Deckbar et al., 2011). Pharmacological inhibition of PIKK with caffeine or LY294002 prevents CHK2 and CDK1 phosphorylation and partially reverses cell cycle arrest of fibroblasts exposed to CDT produced by *E. coli, A. actinomycetemcomitans* or *H. ducreyi* (Cortes-Bratti et al., 2001; Alby et al., 2001; Alaoui-El-Azher et al., 2010). However, theses chemical inhibitors are not specific to ATM, and involvement of other PIKK members in response to CDT cannot be ruled out. Interestingly, use of wortmannin, another PIKK inhibitor, at a concentration that blocks ATM but not ATR kinase activity, does not affect CDT-induced CHK2 phosphorylation (Alby et al., 2001). Whether this response depends on ATR still needs to be established, but recent work from our lab demonstrated that CDT treatment provokes replicative stress prior to blocking cells at G2/M, also supporting an ATR-mediated signaling pathway (Fedor et al., 2013). However, ATM-deficient lymphoblastoid cell lines show a delayed CDT response, characterized by the progression through the G2/M block and a diminution of γH2AX levels, p53 stabilization and RhoA-dependent actin stress fiber formation (Cortes-Bratti et al., 2001; Li and Sharipo, 2002; Frisan et al., 2003). Besides, the apoptotic response to CDT is attenuated when ATM or CHK2 protein levels are depleted (Alaoui-El-Azher et al., 2010). Altogether, these studies clearly link DNA damage signaling to the cellular outcome following CDT intoxication. In agreement with this model, different genetic screens in yeast identified numerous DNA repair genes to be essential for surviving upon CDT intoxication (Guerra et al., 2011; Kitagawa et al., 2007; Matangkasombut et al., 2010). Lastly, HeLa cells deficient for RAD51, a key component of the homologous recombination repair machinery, are hypersensitive to *E. coli* CDT treatment (Fedor et al., 2013), corroborating that DNA damage repair activity is crucial for cells to resist CDT mediated cytotoxicity.

CdtB Supports the CDT Genotoxic Potential

CdtB Delivery in Host Cell Can Mimic CDT Intoxication

As stated above, CDT is an AB toxin, in which CdtB represents the catalytic A-subunit (Nesić et al., 2004). CdtA and CdtC form a heterodimeric B-subunit that mediates toxin binding to host cells and allows cellular uptake of CdtB. Interestingly, *S.* Typhi, devoided of CdtA and CdtC genes, forms an effective typhoid toxin composed by CdtB, PltA and PltB (Song et al., 2013). Intriguingly, a recombinant *C. jejuni* holotoxin reconstituted *in vitro*, in which *Cj*CdtB has been substituted by *S.* Typhi CdtB, still retains its cyclomodulin and cytotoxic activities (Haghjoo and Galán, 2004). As CdtB from *S.* Typhi and *C. jejuni* are among the most divergent in the CdtB family (Gargi et al., 2012), this result suggests that, although divergent, these CdtB proteins share similar structural properties and mode of action. Moreover, CdtB delivery into host cells is sufficient to elicit the CDT-related cellular defects, as demonstrated by transient expression or direct introduction into host cells of CdtB from *H. ducreyi* (Li and Sharipo, 2002; Wising et al., 2010), *C. jejuni* (Lara-Tejero and Galàn, 2000), *E. coli* (Elwell et al., 2001; McSweeney and Dreyfus, 2004), *A. actinomycetemcomitans* (Mao and DiRienzo, 2002; Nishikubo et al., 2003), *H. hepaticus* (Liyanage et al., 2013) or *S.* Typhi (Haghjoo and Galán, 2004).

Structural and Biochemical Features of CdtB

In silico analyses of the predicted *E. coli* and *C. jejuni* CdtB amino-acid sequences revealed a limited, but relevant, sequence homology with the well documented mammalian deoxyribonuclease (DNase) I (Elwell and Dreyfus, 2000; Lara-Tejero and Galàn, 2000). More precisely, residues from DNase I shown to be essential for enzymatic activity are strictly conserved in all CdtBs homologs - these include residues involved in catalytic activity, divalent cation binding and DNA binding (Pan et al., 1998). Furthermore, the crystal structures of CDT produced by *H. ducreyi* (Nesić et al., 2004), *A. actinomycetemcomitans* (Yamada et al., 2006) and *E. coli* CdtB alone (Hontz et al., 2006) highlighted the structural similarities between CdtB and DNase I, especially concerning their active site and their DNA binding motif. Based on these studies, comparative modeling and structural analyses of CdtB from various origins reinforced the assumption of a conserved three-dimensional feature among all CDTs (Hu et al., 2006) and, by extension with DNase I similarities, of a nuclease-mediated cellular toxicity for CDT.

Therefore, CdtB nuclease activity has been assessed, after incubating plasmid DNA with the CDT holotoxin or with CdtB alone, in order to evaluate the conversion of the supercoiled plasmid to the relaxed (nicking activity) or the linear (DSB activity) form. The nuclease activity of CdtB was first observed with supernatant or periplasmic extracts of bacteria expressing *E. coli* CdtB alone or the CDT holotoxin (Elwell and Dreyfus, 2000). However, a substantial nicking activity, independently of CdtB, demonstrated that a preliminary purification step is necessary to prevent extract-mediated plasmid degradation. Thus, a recombinant *H. hepaticus* CdtB has been purified to homogeneity and biochemical assays emphasize the similarities with DNase I activity, i.e. magnesium and calcium dependencies, and inhibition by zinc and G-actin (Dassanayake et al., 2005). Nevertheless, this study also pointed out that CdtB activity is at least 100 fold weaker than the one observed for the DNase I, a difference corroborated with *H. ducreyi* CDT (Li and Sharipo, 2002). Finally, chimeric constructs between half parts of *Aa*CdtB and human DNase I retain nuclease activity similar to CdtB, strongly supporting a close structural and functional relationship between these two proteins (DiRienzo et al., 2009). Interestingly, it has been shown that *Hd*CdtB activity is inhibited in the context of the holoenzyme, and structural analysis revealed that the N-terminus tail of CdtC occludes the active site of CdtB, by interacting with several key residues (Nesić et al., 2004). In agreement with that, CdtB nuclease activity is restored when the tripartite complex is formed with a truncated form of CdtC missing the N-terminus tail. Modeling data suggest that this CdtC inhibitory role can be extrapolated to other CDTs (Hu et al., 2006). Thereby, *Aa*CdtB completely lost its nuclease activity when properly complexed with CdtA and CdtC (Cao et al. 2005), suggesting that once within the host cell, and during intracellular trafficking, CdtB is sequestered and has to be freed in order to be activated. Finally, molecular dynamics simulations indicate that CdtB adopts a conformational change when released from the complex, stabilizing the active site by maintaining the DNA binding groove in a closed form (Hu and Stebbins, 2006).

CDT Cytotoxicity Relies on the Nuclease Activity of CdtB

Sequence and structural comparison between CdtB and DNase I enabled the identification of residues important for CdtB nuclease activity. Among them, three catalytic residues of DNase I, namely H134, D212 and H252, are conserved within CdtB bacterial homologs. Mutations of these residues completely abolish CdtB nuclease activity (Elwell and Dreyfus, 2000; Nesić et al., 2004). Moreover, CDT holoenzymes from various origins that contain these CdtB catalytic mutants do not elicit the DDR cellular response nor possess any cytotoxic activity (Elwell and Dreyfus, 2000; Fedor et al., 2013; Haghjoo and Galán, 2004; Hassane et al., 2001, 2003; M. Lara-Tejero and Galàn, 2000; Nesić et al., 2004; Song et al., 2013). In the same way, metal ion binding residues seems essential for nuclease and cytotoxic activities (Elwell and Dreyfus, 2000; Lara-Tejero and Galàn, 2000). The CDT nuclease activity has been investigated *in vivo* by direct visualization of genomic DNA fragmentation after pulse field gel electrophoresis or after single cell gel electrophoresis assays. It appears that the chromosomal fragmentation observed in *Cj*CdtB expressing yeasts depends on an intact catalytic site (Hassane et al., 2001) and the cellular DNA strand breaks rapidly observed after *Ec*CDT exposure are abrogated by the mutation of a key catalytic residue (Fedor et al., 2013). A similar observation has been done with a *Hd*CDT divalent cation-binding mutant, unable to induce DSBs on genomic DNA (Guerra et al., 2005).

Mutations of the putative DNA binding residues lead to a diminution of the nuclease activity, instead of abolition (Nesić et al., 2004), without inducing a typical G2/M arrest, suggesting a dose-dependent cellular defect associated with the CdtB activity. This is supported by the fact that the pathogenic potential of *A. actinomycetemcomitans* isolates correlate to a single nucleotide polymorphism in the *cdtB* gene, modulating the nuclease activity of CdtB (Nishikubo et al., 2006).

Finally, chronic exposure to sublethal doses of *H. hepaticus* or *H. ducreyi* CDTs, that induces minor genotoxic stress without altering cell cycle nor viability, causes mutations and chromosomal aberrations, both being hallmarks of misrepaired DNA lesions (Guidi et al., 2013).

Moreover, these cells present other characteristics of transformed cells: proliferation independent of anchorage signals, inhibition of the DDR leading to uncontrolled proliferation and cell death escape. Altogether, these analyses demonstrate that CdtB-induced DNA damage are strongly associated with cellular toxicity and explain why CDT is classified as a genotoxin and even a potential carcinogen.

The Phosphatase Paradigm

Intriguingly, computational analyses based on sequence homology also predict that CdtB share significant sequence conservation with signaling proteins, including inositol polyphosphate 5-phosphatases (IP5P) (Dlakic, 2001). More recently, structural similarities have been detected between CdtB and the IP5P family, with a strong conservation of key residues in the active site (Shenker et al., 2007). This led the authors to test whether CDT cytotoxicity could be explained by a potential phosphatase activity. Biochemical analyses showed that CdtB hydrolyses phosphatidylinositol-3,4,5-triphosphate (PIP_3) to phosphatidylinositol-3,4-diphosphate. PIP_3 is also the preferential substrate of the PTEN and SHIP1 phosphatases, both playing a regulatory role in the Akt signaling pathway. Interestingly, lymphoid cells that are defective for these phosphatases, and therefore maintain a high level of PIP_3, are highly sensitive to *Aa*CDT. Moreover, CDT intoxication of these cells induces a strong decrease in the basal PIP_3 level. It is therefore tempting to speculate that CdtB-related phosphatase activity may represent an alternative cytotoxic pathway, particularly efficient in lymphocytes as the viability of many leukemic cell lines depends on an elevated level of PIP_3 (Shenker et al., 2007). The higher CDT susceptibility of lymphocytes over fibroblasts, which may be attributable to their specific rapid apoptotic response before any distention phenotype and even before any clear G2/M arrest (Shenker et al., 1999; Cortes-Bratti et al., 2001), reinforces this hypothesis. Mutational analyses indicate that CDT-mediated cell cycle arrest of lymphoid cells specifically involves the CdtB phosphatase activity (Shenker et al., 2007). However, contradictory results came from another study demonstrating that a non-proliferative monocyte cell line is sensitive to an *Aa*CdtB phosphatase mutant, but resistant to the wild type CDT when co-treated with specific PIP_3 phosphatase inhibitors (Rabin et al., 2009). In yeast, a model system devoid of PIP_3, the response to CdtB relies on the DDR machinery and nuclease-dead CdtB mutants show no effect (Hassane et al., 2001; Matangkasombut et al., 2010). More investigations will be necessary to account for the precise function of nuclease and phosphatase CdtB catalytic activities, particularly on different host cell types.

Conclusion

CDT has been the first bacterial toxin known to act as a genotoxin in the nucleus of host target cells. Thus, CDT toxins have been high research subjects for scientists within the last decade. To date, only a few bacterial genotoxins have been described: the CDTs presented in this chapter, the colibactin (Nougayrède et al., 2006) and the *E. coli* Usp (uropathogenic-specific protein) (Nipic et al., 2013). As reported here, CDT intoxicated host cells activate the DDR, in order to counteract the CDT genotoxic activity. This is also the case for colibactin (Cuevas-Ramos et al., 2010). Chronic exposure to genotoxins may lead to genomic instability, as illustrated *in vitro* with CDT (Guidi et al., 2013) or *in vivo* with colibactin (Cuevas-Ramos et al., 2010). Genomic instability has been shown to fuel the transformation process, directly enhancing the risk of inducing tumor development (Janssen and Medema, 2013). However, inflammation is also known to contribute to cancerogenesis (Morrison, 2012), and bacterial pathogens can induce DNA damage in host cells through chronic inflammation (Grivennikov et al., 2010). A new story is emerging, where bacteria are tightly associated with cancer development, through the expression of genotoxins or through diverse genotoxic processes (Aituov et al., 2012; Guerra et al., 2011). In this context, elucidating the precise mechanisms involved in bacterial pathogenicity-associated carcinogenesis will be of particular interest.

Acknowledgment

The authors' recent work has been supported by the Agence Nationale de la Recherche (Grant ANR-10-CESA-011). E. Bezine is a recipient of a Ph.D. fellowship of the French Ministry of Research.

References

Ahn, J., Urist, M., and Prives, C. (2004). The Chk2 protein kinase. *DNA Repair*, 3, 1039–1047. doi:10.1016/j.dnarep.2004.03.033.

Aituov, B., Duisembekova, A., Bulenova, A., and Alibek, K. (2012). Pathogen-driven gastrointestinal cancers: Time for a change in treatment paradigm? *Infectious Agents and Cancer*, 7(1), 18. doi:10.1186/1750-9378-7-18.

Akifusa, S., Heywood, W., Nair, S. P., Stenbeck, G., and Henderson, B. (2005). Mechanism of internalization of the cytolethal distending toxin of Actinobacillus actinomycetemcomitans. *Microbiology*, 151, 1395–1402. doi:10.1099/mic.0.27671-0.

Alaoui-El-Azher, M., Mans, J. J., Baker, H. V., Chen, C., Progulske-Fox, A., Lamont, R. J., and Handfield, M. (2010). Role of the ATM-Checkpoint Kinase 2 Pathway in CDT-Mediated Apoptosis of Gingival Epithelial Cells. *PloS One*, 5(7), 1–10. doi:10.1371/journal.pone.0011714.

Alby, F., Mazars, R., de Rycke, J., Guillou, E., Baldin, V., Darbon, J. M., and Ducommun, B. (2001). Study of the cytolethal distending toxin (CDT)-activated cell cycle checkpoint. Involvement of the CHK2 kinase. *FEBS Letters*, 491(3), 261–265.

Amano, A., Takeuchi, H., and Furuta, N. (2010). Outer membrane vesicles function as offensive weapons in host-parasite interactions. *Microbes and Infection / Institut Pasteur*, 12(11), 791–798. doi:10.1016/j.micinf.2010.05.008.

Aragon, V., Chao, K., and Dreyfus, L. A. (1997). Effect of cytolethal distending toxin on F-actin assembly and cell division in Chinese hamster ovary cells. *Infection and Immunity*, 65(9), 3774–3780.

Asakura, M., Hinenoya, A., Alam, M. S., Shima, K., Zahid, S. H., Shi, L., Sugimoto, Ghosh, A. N., Ramamurthy, T., Faruque, S. F., Balakrish Nair, G., and Yamasaki, S. (2007). An inducible lambdoid prophage encoding cytolethal distending toxin (Cdt-I) and a type III effector protein in enteropathogenic Escherichia coli. *PNAS*, 104(36), 14483–14488.

Bakkenist, C. J., and Kastan, M. B. (2003). DNA damage activates ATM through intermolecular autophosphorylation and dimer dissociation. *Nature*, 421, 499–506. doi:10.1038/nature01368.

Bandhaya, P. P. S., Likittanasombat, K., Hengprasith, B., and Torrungruang, K. (2012). Aggregatibacter actinomycetemcomitans serotypes, the JP2 clone and cytolethal distending toxin genes in a Thai population. *Journal of Clinical Periodontology Periodontology*, 39(6), 519–525.

Bang, D. D., Scheutz, F., Ahrens, P., Pedersen, K., Blom, J., and Madsen, M. (2001). Prevalence of cytolethal distending toxin (cdt) genes and CDT production in Campylobacter spp. isolated from Danish broilers. *Journal of Medical Microbiology*, 50(12), 1087–94.

Belibasakis, G. N., Mattsson, A., Wang, Y., Chen, C., and Johansson, A. (2004). Cell cycle arrest of human gingival fibroblasts and periodontal ligament cells by Actinobacillus actinomycetemcomitans: involvement of the cytolethal distending toxin. *APMIS : Acta Pathologica, Microbiologica, et Immunologica Scandinavica*, 112(10), 674–685. doi:10.1111/j.1600-0463.2004.apm1121006.x

Berlanda Scorza, F., Doro, F., Rodríguez-Ortega, M. J., Stella, M., Liberatori, S., Taddei, A. R., Serino, L., Gomes Moriel, D., Nesta, B., Fontana, M. R., Spagnuolo, A., Pizza, M., Norais, N., and Grandi, G. (2008). Proteomics characterization of outer membrane vesicles from the extraintestinal pathogenic Escherichia coli DeltatolR IHE3034 mutant. *Molecular and Cellular Proteomics : MCP*, 7(3), 473–85. doi:10.1074/mcp.M700295-MCP200.

Bielaszewska, M., Sinha, B., Kuczlus, T., and Karch, H. (2005). Cytolethal Distending Toxin from Shiga Toxin-producing Escherichia coli O157 Causes Irreversible G2/M Arrest , Inhibition of Proliferation , and Death of Human Endothelial Cells. *Infection and Immunity*, 73(1), 552–562. doi:10.1128/IAI.73.1.552.

Blazkova, H., Krejcikova, K., Moudry, P., Frisan, T., Hodny, Z., and Bartek, J. (2010). Bacterial intoxication evokes cellular senescence with persistent DNA damage and cytokine signalling. *Journal of Cellular and Molecular Medicine*, 14(1-2), 357–367. doi:10.1111/j.1582-4934.2009.00862.x

Boesze-Battaglia, K., Besack, D., McKay, T., Zekavat, A., Otis, L., Jordan-Sciutto, K., and Shenker, B. J. (2006). Cholesterol-rich membrane microdomains mediate cell cycle arrest induced by Actinobacillus actinomycetemcomitans cytolethal-distending toxin. *Cellular Microbiology*, 8(5), 823–36. doi:10.1111/j.1462-5822.2005.00669.x

Boesze-Battaglia, K., Brown, A., Walker, L., Besack, D., Zekavat, A., Wrenn, S., Krummenacher, C., and Shenker, B. J. (2009). Cytolethal Distending Toxin-Induced Cell Cycle Arrest of Lymphocytes is Dependent Upon Recognition and Binding to Cholesterol. *The Journal of Biological Chemistry*, 284, 10650–10658. doi:10.1074/jbc.M809094200.

Carette, J. E., Guimaraes, C. P., Varadarajan, M., Park, A. S., Wuethrich, I., Godarova, A., Kotecki, M., Cochran, B. H., Spooner, E., Ploegh, H. L., and Brummelkamp, T. R. (2009). Haploid Genetic Screens in Human Cells Identify Host Factors Used by Pathogens. *Science*, 326, 1231–1235. doi:10.1126/science.1178955.

Carette, J. E., Guimaraes, C. P., Wuethrich, I., Blomen, V. A., Sun, C., Bell, G., Yuan, B., Muellner, M. K., Nijman, S. M., Ploegh, H. L., and Brummelkamp, T. R. (2011). Global gene disruption in human cells to assign genes to phenotypes. *Nat. Biotechnol.*, 29(6), 542–546. doi:10.1038/nbt.1857.Global.

Casadevall, A., and Pirofski, L. (1999). Host-Pathogen Interactions : Redefining the Basic Concepts of Virulence and Pathogenicity. *Infection and Immunity*, 67(8), 3703–3713.

Chien, C.-C., Taylor, N. S., Ge, Z., Schauer, D. B., Young, V. B., and Fox, J. G. (2000). Identification of cdtB homologues and cytolethal distending toxin activity in enterohepatic Helicobacter spp. *Journal of Medical Microbiology*, 49, 525–534.

Comayras, C., Tasca, C., Pe, S. Y., Oswald, E., and Rycke, J. D. E. (1997). Escherichia coli cytolethal distending toxin blocks the HeLa cell cycle at the G2/M transition by preventing cdc2 protein kinase dephosphorylation and activation . Escherichia coli Cytolethal Distending Toxin Blocks the HeLa Cell Cycle at the G 2 / M Tr. *Infection and Immunity*, 65(12), 5088–5095.

Cope, L. D., Lumbley, S., Latimer, J. L., Klesney-Tait, J., Stevens, M. K., Johnson, L. S., Purven, M., Munson, R. S., Lagergard, T., Radolf, J. D., and Hansen, E. J. (1997). A diffusible cytotoxin of Haemophilus ducreyi. *Proceedings of the National Academy of Sciences of the United States of America*, 94(8), 4056–4061.

Cortes-Bratti, X., Chaves-Olarte, E., Lagergård, T., and Thelestam, M. (1999). The cytolethal distending toxin from the chancroid bacterium Haemophilus ducreyi induces cell-cycle arrest in the G2 phase. *The Journal of Clinical Investigation*, 103, 107–115. doi:10.1172/JCI3831

Cortes-Bratti, X., Chaves-Olarte, E., Lagergård, T., and Thelestam, M. (2000). Cellular Internalization of Cytolethal Distending Toxin from Haemophilus ducreyi. *Infection and Immunity*, 68(12), 6903–6911.

Cortes-Bratti, X., Frisan, T., and Thelestam, M. (2001). The cytolethal distending toxins induce DNA damage and cell cycle arrest. *Toxicon : Official Journal of the International Society on Toxinology*, 39, 1729–1736.

Cortes-Bratti, X., Karlsson, C., Lagergård, T., Thelestam, M., and Frisan, T. (2001). The Haemophilus ducreyi Cytolethal Distending Toxin Induces Cell Cycle Arrest and Apoptosis via the DNA Damage Checkpoint Pathways. *The Journal of Biological Chemistry*, 276(7), 5296–5302. doi:10.1074/jbc.M008527200

Cuevas-Ramos, G., Petit, C. R., Marcq, I., Boury, M., Oswald, E., and Nougayrède, J.-P. (2010). Escherichia coli induces DNA damage in vivo and triggers genomic instability in mammalian cells. *Proceedings of the National Academy of Sciences of the United States of America*, 107(25), 11537–11542. doi:10.1073/pnas.1001261107

Damek-Poprawa, M., Haris, M., Volgina, A., Korostoff, J., and DiRienzo, J. M. (2011). Cytolethal distending toxin damages the oral epithelium of gingival explants. *Journal of Dental Research*, 90(7), 874–879. doi:10.1177/0022034511403743

Damek-Poprawa, M., Jang, J. Y., Volgina, A., Korostoff, J., and DiRienzo, J. M. (2012). Localization of Aggregatibacter actinomycetemcomitans Cytolethal Distending Toxin Subunits during Intoxication of Live Cells. *Infection and Immunity*, 80(8), 2761–2770. doi:10.1128/IAI.00385-12

Dassanayake, R. P., Griep, M. a, and Duhamel, G. E. (2005). The cytolethal distending toxin B sub-unit of Helicobacter hepaticus is a Ca^{2+} - and Mg^{2+} -dependent neutral nuclease. *FEMS Microbiology Letters*, 251(2), 219–225. doi:10.1016/j.femsle.2005.08.005

De Rycke, J., Sert, V., Comayras, C., and Tasca, C. (2000). Sequence of lethal events in HeLa cells exposed to the G2 blocking cytolethal distending toxin. *European Journal of Cell Biology*, 79, 192–201.

Deatherage, B. L., and Cookson, B. T. (2012). Membrane Vesicle Release in Bacteria, Eukaryotes, and Archaea: a Conserved Yet Underappreciated Aspect of Microbial Life. *Infection and Immunity*, 80(6), 1948–57. doi:10.1128/IAI.06014-11

Deckbar, D., Jeggo, P. a, and Löbrich, M. (2011). Understanding the limitations of radiation-induced cell cycle checkpoints. *Critical Reviews in Biochemistry and Molecular Biology*, 46(4), 271–283. doi:10.3109/10409238.2011.575764

Diaz-Moralli, S., Tarrado-Castellarnau, M., Miranda, A., and Cascante, M. (2013). Targeting cell cycle regulation in cancer therapy. *Pharmacology and Therapeutics*, 138(2), 255–271. doi:10.1016/j.pharmthera.2013.01.011

DiRienzo, J. M., Cao, L., Volgina, A., Bandelac, G., and Korostoff, J. (2009). Functional and structural characterization of chimeras of a bacterial genotoxin and human type I DNAse. *FEMS Microbiology Letters*, 291, 222–231. doi:10.1111/j.1574-6968.2008.01457.x

Dlakic, M. (2001). Is CdtB a Nuclease or a Phosphatase? *Science*, 291(5504), 547. doi:10.1126/science.291.5504.547a

Doherty, G. J., and Mcmahon, H. T. (2009). Mechanisms of Endocytosis. *Annual Review of Biochemistry*, 78, 857–902. doi:10.1146/annurev.biochem.78.081307.110540

Ellis, T. N., Leiman, S. a, and Kuehn, M. J. (2010). Naturally produced outer membrane vesicles from Pseudomonas aeruginosa elicit a potent innate immune response via combined sensing of both lipopolysaccharide and protein components. *Infection and Immunity*, 78(9), 3822–31. doi:10.1128/IAI.00433-10

Elwell, C. A., and Dreyfus, L. A. (2000). DNase I homologous residues in CdtB are critical for cytolethal distending toxin-mediated cell cycle arrest. *Molecular Microbiology*, 37(4), 952–963.

Elwell, C., Chao, K., Patel, K., and Dreyfus, L. (2001). Escherichia coli CdtB Mediates Cytolethal Distending Toxin Cell Cycle Arrest. *Infection and Immunity*, 69(5), 3418–3422. doi:10.1128/IAI.69.5.3418

Enari, M., Sakahira, H., Yokoyama, H., Okawa, K., Iwamatsu, A., and Nagata, S. (1998). A caspase-activated DNase that degrades DNA during apoptosis , ans its inhibitor ICAD. *Nature*, 393, 43–50.

Escalas, N., Davezac, N., De Rycke, J., Baldin, V., Mazars, R., and Ducommun, B. (2000). Study of the Cytolethal Distending Toxin-Induced Cell Cycle Arrest in HeLa Cells : Involvement of the CDC25 Phosphatase. *Experimental Cell Research*, 257, 206–212. doi:10.1006/excr

Eshraghi, A., Maldonado-Arocho, F. J., Gargi, A., Cardwell, M. M., Prouty, M. G., Blanke, S. R., and Bradley, K. A. (2010). Cytolethal Distending Toxin Family Members Are Differentially Affected by Alterations in Host Glycans and Membrane Cholesterol. *The Journal of Biological Chemistry*, 285(24), 18199–18207. doi:10.1074/jbc.M110.112912

Falck, J., Coates, J., and Jackson, S. P. (2005). Conserved modes of recruitment of ATM, ATR and DNA-PKcs to sites of DNA damage. *Nature*, 434(7033), 605–611. doi:10.1038/nature03442

Fedor, Y., Vignard, J., Nicolau-Travers, M.-L., Boutet-Robinet, E., Watrin, C., Salles, B., and Mirey, G. (2013). From single-strand breaks to double-strand breaks during S-phase: a new mode of action of the Escherichia coli Cytolethal Distending Toxin. *Cellular Microbiology*, 15(1), 1–15. doi:10.1111/cmi.12028

Finlay, B. B., and Falkow, S. (1997). Common themes in microbial pathogenicity revisited. *Microbiology and Molecular Biology Reviews : MMBR*, 61(2), 136–169.

Frisan, T., Cortes-bratti, X., Chaves-olarte, E., Stenerlöw, B., Thelestam, M. (2003). The Haemophilus ducreyi cytolethal distending toxin induces DNA double-strand breaks and promotes ATM-dependent activation of RhoA, 5, 695–707. doi:10.1046/j.1462-5822.2003.00311.x

Gao, L., and Kwaik, Y. A. (2000). Hijacking of apoptotic pathways by bacterial pathogens. *Microbes and Infection*, 2, 1705–1719.

Gargi, A., Reno, M., and Blanke, S. R. (2012). Bacterial toxin modulation of the eukaryotic cell cycle: are all cytolethal distending toxins created equally? *Frontiers in Cellular and Infection Microbiology*, 2, 124. doi:10.3389/fcimb.2012.00124

Gargi, A., Tamilselvam, B., Powers, B., Prouty, M. G., Lincecum, T., Eshraghi, A., Maldonado-Arocho, F. J., Wilson, B. A., Bradley, K. A., and Blanke, S. R. (2013). Cellular Interactions of the Cytolethal Distending Toxins from Escherichia coli and Haemophilus ducreyi. *The Journal of Biological Chemistry*, 288, 7492–7505. doi:10.1074/jbc.M112.448118

Ge, Z., Rogers, A. B., Feng, Y., Lee, A., Xu, S., Taylor, N. S., and Fox, J. G. (2007). Bacterial cytolethal distending toxin promotes the development of dysplasia in a model of microbially induced hepatocarcinogenesis. *Cellular Microbiology*, 9(8), 2070–2080. doi:10.1111/j.1462-5822.2007.00939.x

Gelfanova, V., Hansen, E. J., and Spinola, S. M. (1999). Cytolethal Distending Toxin of Haemophilus ducreyi induces Apoptotic Death of Jurkat T Cells. *Infection and Immunity*, 67(12), 6394–6402.

Ghosal, G., and Chen, J. (2013). DNA damage tolerance: a double-edged sword guarding the genome. *Translational Cancer Research*, 2(3), 107–129. doi:10.3978/j.issn.2218-676X.2013.04.01

Grivennikov, S. I., Greten, F. R., and Karin, M. (2010). Immunity, inflammation, and cancer. *Cell*, 140(6), 883–899. doi:10.1016/j.cell.2010.01.025

Guerra, L., Albihn, A., Tronnersjö, S., Yan, Q., Guidi, R., Stenerlöw, B., Sterzenbach, T., Josenhans, C., Fox, J. G., Schauer, D. B., Thelestam, M., Larsson, L.-G., Henriksson, M., and Frisan, T. (2010). Myc is Required for Activation of the ATM-Dependent Checkpoints in Response to DNA Damage. *PloS One*, 5(1), 1–9. doi:10.1371/journal.pone.0008924

Guerra, L., Guidi, R., and Frisan, T. (2011). Do bacterial genotoxins contribute to chronic inflammation, genomic instability and tumor progression? *The FEBS Journal*, 278, 4577–4588. doi:10.1111/j.1742-4658.2011.08125.x

Guerra, L., Guidi, R., Slot, I., Callegari, S., Sompallae, R., Pickett, C. L., Åström, S., Eisele, F., Wolf, D., Sjögren, C., Masucci, M. G., and Frisan, T. (2011). Bacterial genotoxin triggers FEN1-dependent RhoA activation, cytoskeleton remodeling and cell survival. *Journal of Cell Science*, 124, 2735–2742. doi:10.1242/jcs.085845

Guerra, L., Nemec, K. N., Massey, S., Tatulian, S. A., Thelestam, M., Frisan, T., and Teter, K. (2009). A novel mode of translocation for cytolethal distending toxin. *Biochimica et Biophysica Acta*, 1793(3), 489–495. doi:10.1016/j.bbamcr.2008.11.017

Guerra, L., Teter, K., Lilley, B. N., Stenerlöw, B., Holmes, R. K., Ploegh, H. L., Sandvig, K., Thelestam, M., and Frisan, T. (2005). Cellular internalization of cytolethal distending toxin: a new end to a known pathway. *Cellular Microbiology*, 7(7), 921–934. doi:10.1111/j.1462-5822.2005.00520.x

Guidi, R., Guerra, L., Levi, L., Stenerlöw, B., Fox, J. G., Josenhans, C., Masucci, M. G., and Frisan, T. (2013). Chronic exposure to the cytolethal distending toxins of Gram-negative bacteria promotes genomic instability and altered DNA damage response. *Cellular Microbiology*, 15(1), 98–113. doi:10.1111/cmi.12034

Haghjoo, E., and Galán, J. E. (2004). Salmonella typhi encodes a functional cytolethal distending toxin that is delivered into host cells by a bacterial-internalization pathway. *Proceedings of the National Academy of Sciences of the United States of America*, 101(13), 4614–4619. doi:10.1073/pnas.0400932101

Hassane, D. C., Lee, R. B., Mendenhall, M. D., and Pickett, C. L. (2001). Cytolethal Distending Toxin Demonstrates Genotoxic Activity in a Yeast Model. *Infection and Immunity*, 69(9), 5752–5759. doi:10.1128/IAI.69.9.5752

Hassane, D. C., Lee, R. B., and Pickett, C. L. (2003). Campylobacter jejuni Cytolethal Distending Toxin Promotes DNA Repair Responses in Normal Human Cells. *Infection and Immunity*, 71(1), 541–545. doi:10.1128/IAI.71.1.541

Hayashi, C., Gudino, C. V., Gibson III, F. C., and Genco, C. A. (2011). Pathogen-induced inflammation at sites distant from oral infection: bacterial persistence and induction of cell-specific innate immune inflammatory pathways. *Molecular Oral Microbiology*, 25(5), 305–316. doi:10.1111/j.2041-1014.2010.00582.x.

Henkel, J. S., Baldwin, M. R., and Barbieri, J. T. (2010). Toxins from bacteria. *Exs.*, 100, 1–29.

Hickey, T. E., Majam, G., and Guerry, P. (2005). Intracellular Survival of Campylobacter jejuni in Human Monocytic Cells and Induction of Apoptotic Death by Cytholethal Distending Toxin. *Infection and Immunity*, 73(8), 5194–5197. doi:10.1128/IAI.73.8.5194

Hontz, J. S., Villar-Lecumberri, M. T., Potter, B. M., Yoder, M. D., Dreyfus, L. A., and Laity, J. H. (2006). Differences in Crystal and Solution Structures of the Cytolethal Distending Toxin B Subunit: Relevance to Nuclear Translocation and Functional Activation. *The Journal of Biological Chemistry*, 281(35), 25365–25372. doi:10.1074/jbc.M603727200

Hu, X., Nesic, D., and Stebbins, C. E. (2006). Comparative Structure-Function Analysis of Cytolethal Distending Toxins. *Proteins*, 62(2), 421–434. doi:10.1002/prot.20767

Hu, X., and Stebbins, C. E. (2006). Dynamics and Assembly of the Cytolethal Distending Toxin. *Proteins: Structure, Function, and Bioinformatics*, 65, 843–855. doi:10.1002/prot

Jain, D., Prasad, K. N., Sinha, S., and Husain, N. (2008). Differences in virulence attributes between cytolethal distending toxin positive and negative Campylobacter jejuni strains. *Journal of Medical Microbiology*, 57(Pt 3), 267–72. doi:10.1099/jmm.0.47317-0

Janka, A., Bielaszewska, M., Dobrindt, U., Greune, L., Schmidt, M. A., and Karch, H. (2003). Cytolethal Distending Toxin Gene Cluster in Enterohemorrhagic Escherichia coli O157:H- and O157:H7 : Characterization and Evolutionary Considerations. *Infection and Immunity*, 71(6), 3634–3638. doi:10.1128/IAI.71.6.3634

Janowicz, D., Li, W., and Bauer, M. E. (2011). Host-Pathogen Interplay of Haemophilus ducreyi. *Current Opinion in Infection Disease*, 23(1), 64–69. doi:10.1097/QCO.0b013e328334c0cb.

Janssen, A., and Medema, R. H. (2013). Genetic instability: tipping the balance. *Oncogene*, 32(38), 4459–4470. doi:10.1038/onc.2012.576

Jentsch, H., Cachovan, G., Guentsch, A., Eickholz, P., Pfister, W., and Eick, S. (2012). Characterization of Aggregatibacter actinomycetemcomitans strains in periodontitis patients in Germany. *Clin. Oral. Investig.*, 16(6), 1589–1597.

Jinadasa, R. N., Bloom, S. E., Weiss, R. S., and Duhamel, G. E. (2011). Cytolethal distending toxin : a conserved bacterial genotoxin that blocks cell cycle progression , leading to apoptosis of a broad range of mammalian cell lineages. *Microbiology*, 157, 1851–1875. doi:10.1099/mic.0.049536-0

Johnson, J. R., Oswald, E., O'Bryan, T. T., Kuskowski, M. A., and Spanjaard, L. (2002). Phylogenetic Distribution of Virulence-Associated Genes among Escherichia coli Isolates Associated with Neonatal Bacterial Meningitis in The Netherlands. *The Journal of Infectious Diseases*, 185, 774–784.

Johnson, W. M., and Lior, H. (1987). Response of Chinese Hamster Ovary Cells to a Cytolethal Distending Toxin (CDT) of Escherichia coli and Possible Misinterpretation as Heat-labile (LT) enterotoxin. *FEMS Microbiology Letters*, 43(1), 19–23.

Johnson, W. M., and Lior, H. (1988). A new heat-labile cytolethal distending toxin (CLDT) produced by Campylobacter spp. *Microbiology Pathology*, 4(2), 115–126.

Jones, R. M., and Petermann, E. (2012). Replication fork dynamics and the DNA damage response. *The Biochemical Journal*, 443, 13–26. doi:10.1042/BJ20112100

Kabir, S. M. L., Kikuchi, K., Asakura, M., Shiramaru, S., Tsuruoka, N., Goto, A., Hinenoya, A., and Yamasaki, S. (2011). Evaluation of a cytolethal distending toxin (cdt) gene-based species-specific multiplex PCR assay for the identification of Campylobacter strains isolated from diarrheal patients in Japan. *Japanese Journal of Infectious Diseases*, 64(1), 19–27.

Kaslow, H. R., and Burns, D. L. (1992). Pertussis toxin and target eukaryotic cells: binding, entry , and activation. *The FASEB Journal*, 6, 2685–2690.

Kim, J.-H., Kim, J.-C., Choo, Y.-A., Jang, H.-C., Choi, Y.-H., Chung, J.-K., Cho, S.-H., Park, M.,-S. and Lee, B.-K. (2009). Detection of Cytolethal Distending Toxin and Other Virulence Characteristics of Enteropathogenic Escherichia coli Isolates from Diarrheal Patients in Republic of Korea. *Journal of Microbiology and Biotechnology*, 19(5), 525–529. doi:10.4014/jmb.0801.033

Kitagawa, T., Hoshida, H., and Akada, R. (2007). Genome-Wide Analysis of Cellular Response to Bacterial Genotoxin CdtB in Yeast. *Infection and Immunity*, 75(3), 1393–1402. doi:10.1128/IAI.01321-06

Cao, L., Volgina, A., Huang, C.-M., Korostoff, J., and DiRienzo, J. M. (2005). Characterization of point mutations in the cdtA gene of the cytolethal distending toxin of actinobacillus actinomycetemcomitans. *Molecular Microbiology*, 58(5), 1303–1321.

Lai, C.-H., Lai, C.-K., Lin, Y.-J., Hung, C.-L., Chu, C.-H., Feng, C.-L., Chang, C.-S., and Su, H.-L. (2013). Characterization of Putative Cholesterol Recognition/Interaction Amino Acid Consensus-Like Motif of Campylobacter jejuni Cytolethal Distending Toxin C. *PloS One*, 8(6), e66202. doi:10.1371/journal.pone.0066202

Lara-Tejero, M., and Galàn, J. E. (2000). A Bacterial Toxin That Controls Cell Cycle Progression as a Deoxyribonuclease I-Like Protein. *Science*, 290, 354–357. doi:10.1126/science.290.5490.354

Lara-Tejero, M., and Galàn, J. E. (2001). CdtA , CdtB , and CdtC Form a Tripartite Complex That Is Required for Cytolethal Distending Toxin Activity. *Infection and Immunity*, 69(7), 4358–4365. doi:10.1128/IAI.69.7.4358

Lee, J.-H., and Paull, T. T. (2005). ATM Activation by DNA Double-Strand Breaks Through the Mre11-Rad50-Nbs1 Complex. *Science*, 308, 551–554. doi:10.1126/science.1108297

Lee, R. B., Hassane, D. C., Cottle, D. L., and Pickett, C. L. (2003). Interactions of Campylobacter jejuni Cytolethal Distending Toxin Subunits CdtA and CdtC with HeLa Cells. *Infection and Immunity*, 71(9), 4883–4890. doi:10.1128/IAI.71.9.4883

Lewis, D. A., Stevens, M. K., Latimer, J. O. L., Ward, C. K., Deng, K., Blick, R., Lumbley, S. R., Ison, C. A., and Hansen, E. J. (2001). Characterization of Haemophilus ducreyi cdtA, cdtB, and cdtC Mutants in In Vitro and In Vivo Systems. *Infection and Immunity*, 69(9), 5626–5634. doi:10.1128/IAI.69.9.5626

Li, L., Ding, C., Duan, J., Yang, M., Sun, Y., Wang, X., and Xu, Y. (2013). A new functional site W115 in CdtA is critical for Aggregatibacter actinomycetemcomitans cytolethal distending toxin. *PloS One*, 8(6), e65729. doi:10.1371/journal.pone.0065729

Li, L., and Sharipo, A. (2002). The Haemophilus ducreyi cytolethal distending toxin activates sensors of DNA damage and repair complexes in proliferating and non-proliferating cells. *Cellular Microbiology*, 4, 87–99.

Li, L., Sharipo, A., Chaves-Olarte, E., Masucci, M. G., Levitsky, V., Thelestam, M., and Frisan, T. (2002). The Haemophilus ducreyi cytolethal distending toxin activates sensors of DNA damage and repair complexes in proliferating and non-proliferating cells. *Cellular Microbiology*, 4(2), 87–99.

Lin, C.-D., Lai, C.-K., Lin, Y.-H., Hsieh, J.-T., Sing, Y.-T., Chang, Y.-C., Chen, K.-C., Wang, W.-C., Su, H.-L., and Lai, C.-H. (2011). Cholesterol Depletion Reduces Entry of Campylobacter jejuni Cytolethal Distending Toxin and Attenuates Intoxication of Host Cells. *Infection and Immunity*, 79(9), 3563–3575. doi:10.1128/IAI.05175-11

Lindmark, B., Rompikuntal, P. K., Vaitkevicius, K., Song, T., Mizunoe, Y., Uhlin, B. E., Guerry, P., Wai, S. N. (2009). Outer membrane vesicle-mediated release of cytolethal distending toxin (CDT) from Campylobacter jejuni. *BMC Microbiology*, 9, 220. doi:10.1186/1471-2180-9-220

Liu, L. F. (1989). DNA topoisomerase poisons as antitumor drugs. *Annual Review Biochemistry*, 58, 351–375.

Liu, X., Zou, H., Slaughter, C., and Wang, X. (1997). DFF, a Heterodimeric Protein that Functions Downstream of Caspase-3 to Trigger DNA Fragmentation during Apoptosis. *Cell*, 89(2), 175–184.

Liyanage, N. P. M., Dassanayake, R. P., Kuszynski, C. A., and Duhamel, G. E. (2013). Contribution of Helicobacter hepaticus Cytolethal Distending Toxin Subunits to Human Epithelial Cell Cycle Arrest and Apoptotic Death in vitro. *Helicobacter*, 18(6), 433–443. doi:10.1111/hel.12084

Liyanage, N. P. M., Manthey, K. C., Dassanayake, R. P., Kuszynski, C. A., Oakley, G. G., and Duhamel, G. E. (2010). Helicobacter hepaticus Cytolethal Distending Toxin Causes Cell Death in Intestinal Epithelial Cells via Mitochondrial Apoptotic Pathway. *Helicobacter*, 15, 98–107.

Mao, X., and DiRienzo, J. M. (2002). Functional studies of the recombinant subunits of a cytolethal distending holotoxin. *Cellular Microbiology*, 4(4), 245–255.

Matangkasombut, O., Wattanawaraporn, R., Tsuruda, K., Ohara, M., Sugai, M., and Mongkolsuk, S. (2010). Cytolethal Distending Toxin from Aggregatibacter actinomycetemcomitans Induces DNA Damage , S/G2 Cell Cycle Arrest , and Caspase-Independent Death in a Saccharomyces cerevisiae Model. *Infection and Immunity*, 78(2), 783–792. doi:10.1128/IAI.00857-09

Matsuoka, S., Ballif, B. A., Smogorzewska, A., McDonald, E. R., Hurov, K. E., Luo, J., Bakalarski, C. E., Zhao, Z., Solimini, N., Lerenthal, Y., Shiloh, Y., Gygi, S. P. and Elledge, S. J. (2007). ATM and ATR Substrate Analysis Reveals Extensive Protein

Networks Responsive to DNA Damage. *Science*, 316, 1160–1166. doi:10.1126/science.1140321

McGhee, J. R., and Fujihashi, K. (2012). Inside the mucosal immune system. *PLoS Biology*, 10(9), e1001397. doi:10.1371/journal.pbio.1001397

McSweeney, L. A., and Dreyfus, L. A. (2004). Nuclear localization of the Escherichia coli cytolethal distending toxin CdtB subunit. *Cellular Microbiology*, 6(5), 447–458. doi:10.1111/j.1462-5822.2004.00373.x

Mcsweeney, L. A., and Dreyfus, L. A. (2005). Carbohydrate-Binding Specificity of the Escherichia coli Cytolethal Distending Toxin CdtA-II and CdtC-II Subunits. *Infection and Immunity*, 73(4), 2051–2060. doi:10.1128/IAI.73.4.2051

Mise, K., Akifusa, S., Watarai, S., Ansai, T., Nishihara, T., and Takehara, T. (2005). Involvement of Ganglioside GM3 in G2 M Cell Cycle Arrest of Human Monocytic Cells Induced by Actinobacillus actinomycetemcomitans Cytolethal Distending Toxin. *Infection and Immunity*, 73(8), 4846–4852. doi:10.1128/IAI.73.8.4846

Morrison, W. B. (2012). Inflammation and cancer: a comparative view. *Journal of Veterinary Internal Medicine / American College of Veterinary Internal Medicine*, 26(1), 18–31. doi:10.1111/j.1939-1676.2011.00836.x

Murakami, K., Takahashi, R., Ono, M., Watanabe, K., Okimoto, T., Kodama, M., Abe, D., Kimura, M., and Fujioka, T. (2011). Serodiagnosis of Helicobacter hepaticus infection in patients with liver and gastrointestinal diseases: western blot analysis and ELISA using a highly specific monoclonal antibody for H. hepaticus antigen. *Journal of Gastroenterology*, 46(9), 1120–1126.

Nesić, D., Hsu, Y., and Stebbins, C. E. (2004). Assembly and function of a bacterial genotoxin. *Nature*, 429, 429–433. doi:10.1038/nature02532

Nesic, D., and Stebbins, C. E. (2005). Mechanisms of Assembly and Cellular Interactions for the Bacterial Genotoxin CDT. *PLoS Pathogens*, 1(3), 214–224. doi:10.1371/journal.ppat.0010028

Nipic, D., Podlesek, Z., Budic, M., Crnigoj, M., and Zgur-Bertok, D. (2013). Escherichia coli Uropathogenic-Specific Protein, Usp, Is a Bacteriocin-Like Genotoxin. *The Journal of Infectious Diseases*, 208(10), 1545–1552. doi:10.1093/infdis/jit480

Nishikubo, S., Ohara, M., Ikura, M., Katayanagi, K., Fujiwara, T., Komatsuzawa, H., Kurihara, H., and Sugai, M. (2006). Single Nucleotide Polymorphism in the Cytolethal Distending Toxin B Gene Confers Heterogeneity in the Cytotoxicity of Actinobacillus actinomycetemcomitans. *Infection and Immunity*, 74(12), 7014–7020. doi:10.1128/IAI.02063-05

Nishikubo, S., Ohara, M., Ueno, Y., Ikura, M., Kurihara, H., Komatsuzawa, H., Oswald, E., and Sugai, M. (2003). An N-terminal Segment of the Active Component of the Bacterial Genotoxin Cytolethal Distending Toxin B (CDTB) Directs CDTB Into the Nucleus. *The Journal of Biological Chemistry*, 278(50), 50671–50681. doi:10.1074/jbc.M305062200

Nougayrède, J., Homburg, S., Taieb, F., Boury, M., Brzuszkiewicz, E., Gottschalk, G., Buchrieser, C., Hacker, J., Dobrindt, U., and Oswald, E. (2006). Escherichia coli Induces DNA Double-Strand Breaks in Eukaryotic Cells. *Science*, 313, 848–851. doi:10.1126/science.1127059

Ohara, M., Hayashi, T., Kusunoki, Y., Miyauchi, M., Takata, T., and Sugai, M. (2004). Caspase-2 and Caspase-7 Are Involved in Cytolethal Distending Toxin-Induced

Apoptosis in Jurkat and MOLT-4 T-Cell Lines. *Infection and Immunity*, 72(2), 871–879. doi:10.1128/IAI.72.2.871

Ohara, M., Hayashi, T., Kusunoki, Y., Nakachi, K., Fujiwara, T., Komatsuzawa, H., and Sugai, M. (2008). Cytolethal Distending Toxin induces Caspase-Dependent and -Independent Cell Death in MOLT-4 cells. *Infection and Immunity*, 76(10), 4783–4791. doi:10.1128/IAI.01612-07

Ohguchi, M., Ishisaki, A., Okahashi, N., Koide, M., Koseki, T., Yamato, K., Noguchi, T., and Nishihara, T. (1998). Actinobacillus actinomycetemcomitans Toxin Induces Both Cell Cycle Arrest in the G2/M Phase and Apoptosis. *Infection and Immunity*, 66(12), 5980–5987.

Okuda, J., Fukumoto, M., Takeda, Y., and Nishibuchi, M. (1997). Examination of Diarrheagenicity of Cytolethal Distending Toxin: Suckling Mouse Response to the Products of the cdtABC Genes of Shigella dysenteriae. *Infection and Immunity*, 65(2), 428–433.

Okuda, J., Kurazono, H., and Takeda, Y. (1995). Distribution of the cytolethal distending toxin A gene (cdtA) among species of Shigella and Vibrio, and cloning and sequencing of the cdt gene from Shigella dysenteriae. *Microbial Pathogenesis*, 18(3), 167–72.

Pan, C. Q., Ulmer, J. S., Herzka, A., and Lazarus, R. a. (1998). Mutational analysis of human DNase I at the DNA binding interface: implications for DNA recognition, catalysis, and metal ion dependence. *Protein Science : A Publication of the Protein Society*, 7, 628–636. doi:10.1002/pro.5560070312

Pandey, M., Khan, A., Das, S. C., Sarkar, B., Kahali, S., Chakraborty, S., Chattopadhyay, S., Yamasaki, S., Takeda, Y., Nair, G. B., and Ramamurthy, T. (2003). Association of Cytolethal Distending Toxin Locus cdtB with Enteropathogenic Escherichia coli Isolated from Patients with Acute Diarrhea in Calcutta , India. *Journal of Clinical Microbiology*, 41(11), 5277–5281. doi:10.1128/JCM.41.11.5277

Pérès, S. Y., Marchès, O., Daigle, F., Nougayrède, J. P., Herault, F., Tasca, C., De Rycke, J., and Oswald, E. (1997). A new cytolethal distending toxin (CDT) from Escherichia coli producing CNF2 blocks HeLa cell division in G2/M phase. *Molecular Microbiology*, 24(5), 1095–1107.

Pickett, C. L., Cottle, D. L., Pesci, E. C., and Bikah, G. (1994). Cloning, Sequencing, and Expression of the Escherichia coli Cytolethal Distending Toxin Genes. *Infection and Immunity*, 62(3), 1046–1051.

Pines, J., and Hunter, T. (1991). Human Cyclin A and B1 Are Differentially Located in the Cell and Undergo Cell Cycle-Dependent Nuclear Transport. *The Journal of Cell Biology*, 115(1), 1–17.

Pokkunuri, V., Pimentel, M., Morales, W., Jee, S.-R., Alpern, J., Weitsman, S., Marsh, Z., Low, K., Hwang, L., Khoshini, R., Barlow, G. M., Wang, H., and Chang, C. (2012). Role of Cytolethal Distending Toxin in Altered Stool Form and Bowel Phenotypes in a Rat Model of Post-infectious Irritable Bowel Syndrome. *Journal of Neurogastroenterology and Motility*, 18(4), 434–42. doi:10.5056/jnm.2012.18.4.434

Rabin, S. D. P., Flitton, J. G., and Demuth, D. R. (2009). Aggregatibacter actinomycetemcomitans Cytolethal Distending Toxin Induces Apoptosis in Nonproliferating Macrophages by a Phosphatase-Independent Mechanism. *Infection and Immunity*, 77(8), 3161–3169. doi:10.1128/IAI.01227-08

Rogakou, E. P., Boon, C., Redon, C., and Bonner, W. M. (1999). Megabase Chromatin Domains Involved in DNA Double-Strand Breaks In Vivo. *The Journal of Cell Biology*, 146(5), 905–916.

Rogakou, E. P., Pilch, D. R., Orr, A. H., Ivanova, V. S., and Bonner, W. M. (1998). DNA Double-stranded Breaks Induce Histone H2AX Phosphorylation on Serine 139. *The Journal of Biological Chemistry*, 273(10), 5858–5868.

Rompikuntal, P. K., Thay, B., Khan, M. K., Alanko, J., Penttinen, A.-M., Asikainen, S., Wai, S. N., and Oscarsson, J. (2012). Perinuclear Localization of Internalized Outer Membrane Vesicles Carrying Active Cytolethal Distending Toxin from Aggregatibacter actinomycetemcomitans. *Infection and Immunity*, 80(1), 31–42. doi:10.1128/IAI.06069-11

Sansonetti, P. J. (2011). To be or not to be a pathogen: that is the mucosally relevant question. *Mucosal Immunology*, 4(1), 8–14. doi:10.1038/mi.2010.77

Sato, T., Koseki, T., Yamato, K., Saiki, K., Konishi, K., Yoshikawa, M., Ishikawa, I., and Nishihara, T. (2002). p53-Independent Expression of p21 CIP1/WAF1 in Plasmacytic Cells during G2 Cell Cycle Arrest Induced by Actinobacillus actinomycetemcomitans Cytolethal Distending Toxin. *Infection and Immunity*, 70(2), 528–534. doi:10.1128/IAI.70.2.528

Schreiner, H., Li, Y., Cline, J., Tsiagbe, V. K., and Fine, D. H. (2013). A comparison of Aggregatibacter actinomycetemcomitans (Aa) virulence traits in a rat model for periodontal disease. *PloS One*, 8(7), e69382. doi:10.1371/journal.pone.0069382

Scott, D. a, and Kaper, J. B. (1994). Cloning and sequencing of the genes encoding Escherichia coli cytolethal distending toxin. *Infection and Immunity*, 62(1), 244–251.

Sert, V., Cans, C., Tasca, C., Oswald, E., Ducommun, B., and De Rycke, J. (1999). The bacterial cytolethal distending toxin (CDT) triggers a G2 cell cycle checkpoint in mammalian cells without preliminary induction of DNA strand breaks. *Oncogene*, 18, 6296–6304.

Shen, T., and Huang, S. (2012). The role of Cdc25A in the regulation of cell proliferation and apoptosis. *Anticancer Agents Med. Chem.*, 12(6), 631–639.

Shenker, B. J., Demuth, D. R., and Zekavat, A. (2006). Exposure of Lymphocytes to High Doses of Actinobacillus actinomycetemcomitans Cytolethal Distending Toxin Induces Rapid Onset of Apoptosis-Mediated DNA Fragmentation. *Infection and Immunity*, 74(4), 2080–2092.

Shenker, B. J., Dlakic, M., Walker, L. P., Besack, D., Jaffe, E., LaBelle, E., and Boesze-Battaglia, K. (2007). A Novel Mode of Action for a Microbial-Derivated Immunotoxin: The Cytolethal Distending Toxin Subunit B Exhibits Phosphatidylinositol3,4,5-Triphosphate Phosphatase Activity. *The Journal of Immunology*, 178, 5099–5108.

Shenker, B. J., Hoffmaster, R. H., Zekavat, A., Yamaguchi, N., Lally, E. T., and Demuth, D. R. (2001). Induction of Apoptosis in Human T Cells by Actinobacillus actinomycetemcomitans Cytolethal Distending Toxin Is a consequence of G2 Arrest of the Cell Cycle. *The Journal of Immunology*, 167, 435–441.

Shenker, B. J., Mckay, T., and Datar, S. (1999). Actinobacillus actinomycetemcomitans Immunosuppressive Protein Is a member of the Family of Cytolethal Distending Toxins Capable of Causing a G2 Arrest in Human T cells. *The Journal of Immunology*, 162, 4773–4780.

Shibue, T., and Taniguchi, T. (2006). BH3-only proteins: Integrated control point of apoptosis. *International Journal of Cancer. Journal International Du Cancer*, 119(9), 2036–2043. doi:10.1002/ijc.21751

Smith, J. L., and Bayles, D. O. (2006). The Contribution of Cytolethal Distending Toxin to Bacterial Pathogenesis. *Critical Reviews in Microbiology*, 32, 227–248. doi:10.1080/10408410601023557

Song, J., Gao, X., and Galán, J. E. (2013). Structure and function of the Salmonella Typhi chimaeric A2B5 typhoid toxin. *Nature*, 499, 350–354. doi:10.1038/nature12377

Spanò, S., Ugalde, J. E., and Galán, J. E. (2008). Delivery of a Salmonella Typhi exotoxin from a host intracellular compartment. *Cell Host and Microbe*, 3(1), 30–38. doi:10.1016/j.chom.2007.11.001

Spinola, S. M., Bauer, M. E., and Robert, S. (2002). Immunopathogenesis of Haemophilus ducreyi Infection (Chancroid). *Infection and Immunity*, 70(4), 1667–1676. doi:10.1128/IAI.70.4.1667

Spooner, R. A., and Lord, J. M. (2012). How Ricin and Shiga Toxin Reach the Cytosol of Target Cells: Retrotranslocation from the Endoplasmic Reticulum. *Current Topics in Microbiology and Immunology*, 357, 19–40. doi:10.1007/82

Spooner, R. A., and Watson, P. (2010). Drug targeting: Learning from toxin entry and trafficking in mammalian cells. *Current Opinion in Drug Discovery and Development*, 13(1), 86–95.

Stevens, M. K., Latimer, J. L., Lumbley, S. R., Ward, C. K., Cope, L. D., Lagergard, T., and Hansen, E. J. (1999). Characterization of a Haemophilus ducreyi mutant deficient in expression of cytolethal distending toxin. *Infection and Immunity*, 67(8), 3900–3908.

Sugai, M., Kawamoto, T., Pérès, S. Y., Ueno, Y., Komatsuzawa, H., Fujiwara, T., Kurihara, H., Suginaka, H., and Oswald, E. (1998). The Cell Cycle-Specific Growth-Inhibitory Factor Produced by Actinobacillus actinomycetemcomitans is a Cytolethal Distending Toxin. *Infection and Immunity*, 66(10), 5008–5019.

Taylor, R. C., Cullen, S. P., and Martin, S. J. (2008). Apoptosis: controlled demolition at the cellular level. *Nature Reviews. Molecular Cell Biology*, 9, 231–241. doi:10.1038/nrm2312

Tenaillon, O., Skurnik, D., Picard, B., and Denamur, E. (2010). The population genetics of commensal Escherichia coli. *Nature Reviews. Microbiology*, 8(3), 207–217. doi:10.1038/nrmicro2298

Toth, I., Hérault, F., Beutin, L., and Oswald, E. (2003). Production of Cytolethal Distending Toxins by Pathogenic Escherichia coli Strains Isolated from Human and Animal Sources: Establishment of the Existence of a New cdt Variant (Type IV). *Journal of Clinical Microbiology*, 41(9), 4285–4291. doi:10.1128/JCM.41.9.4285

Ueno, Y., Ohara, M., Kawamoto, T., Fujiwara, T., Komatsuzawa, H., Oswald, E., and Sugai, M. (2006). Biogenesis of the Actinobacillus actinomycetemcomitans Cytolethal Distending Toxin Holotoxin. *Infection and Immunity*, 74(6), 3480–3487. doi:10.1128/IAI.00739-05

Van Bost, S., Roels, S., Oswald, E., and Mainil, J. (2003). Putative roles of the CNF2 and CDTIII toxins in experimental infections with necrotoxigenic Escherichia coli type 2 (NTEC2) strains in calves. *Microbes and Infection*, 5, 1189–1193. doi:10.1016/j.micinf.2003.08.005

Vignard, J., Mirey, G., and Salles, B. (2013). Ionizing-radiation induced DNA double-strand breaks: A direct and indirect lighting up. *Radiotherapy and Oncology*, 108(3), 362–369. doi:10.1016/j.radonc.2013.06.013

Wahasugui, T. C., Nakano, V., Piazza, R. M. F., and Avila-Campos, M. J. (2013). Phenotypic and genotypic features of Aggregatibacter actinomycetemcomitans isolated from patients with periodontal disease. *Diagnostic Microbiology and Infectious Disease*, 75(4), 366–72. doi:10.1016/j.diagmicrobio.2012.12.013

Whitehouse, C. A., Balbo, P. B., Pesci, E. C., Daniel, L., Mirabito, P. M., Pickett, C. L., and Cottle, D. L. (1998). Campylobacter jejuni Cytolethal Distending Toxin Causes a G2 - Phase Cell Cycle Block. *Infection and Immunity*, 66(5), 1934–1940.

Wising, C., Azem, J., Zetterberg, M., Svensson, L. A., Ahlman, K., and Lagerga, T. (2005). Induction of apoptosis/necrosis in various human cell lineages by Haemophilus ducreyi cytolethal distending toxin. *Toxicon*, 45, 767–776. doi:10.1016/j.toxicon.2005.01.016

Wising, C., Magnusson, M., Ahlman, K., Lindholm, L., and Lagergård, T. (2010). Toxic activity of the CdtB component of Haemophilus ducreyi cytolethal distending toxin expressed from an adenovirus 5 vector. *APMIS : Acta Pathologica, Microbiologica, et Immunologica Scandinavica*, 118(2), 143–149. doi:10.1111/j.1600-0463.2009.02573.x

Wising, C., Mölne, L., Jonsson, I.-M., Ahlman, K., and Lagergård, T. (2005). The cytolethal distending toxin of Haemophilus ducreyi aggravates dermal lesions in a rabbit model of chancroid. *Microbes and Infection / Institut Pasteur*, 7, 867–874. doi:10.1016/ j.micinf.2005.02.009

Yamada, T., Komoto, J., Saiki, K., Konishi, K., and Takusagawa, F. (2006). Variation of loop sequence alters stability of cytolethal distending toxin (CDT): Crystal structure of CDT from Actinobacillus actinomycetemcomitans. *Protein Science*, 15, 362–372. doi:10.1110/ ps.051790506.

Yamamoto, K., Tominaga, K., Sukedai, M., Okinaga, T., Iwanaga, K., Nishihara, T., and Fukuda, J.-I. (2004). Delivery of cytolethal distending toxin B induces cell cycle arrest and apoptosis in gingival squamous cell carcinoma in vitro. *European Journal of Oral Sciences*, 112(5), 445–451. doi:10.1111/j.1600-0722.2004.00157.x

Yamaoka, Y. (2011). Mechanisms of disease: Helicobacter pylori virulence factors. *Nature Review Gastroenterology Hepatology*, 7(11), 629–641. doi:10.1038/ nrgastro.2010.154.

Young, V. B., Knox, K. a, and Schauer, D. B. (2000). Cytolethal Distending Toxin Sequence and Activity in the Enterohepatic Pathogen Helicobacter hepaticus. *Infection and Immunity*, 68(1), 184–191.

Zijnge, V., Kieselbach, T., and Oscarsson, J. (2012). Proteomics of protein secretion by Aggregatibacter actinomycetemcomitans. *PloS One*, 7(7), e41662. doi:10.1371/ journal.pone.0041662

Zou, L., and Elledge, S. J. (2003). Sensing DNA Damage Through ATRIP Recognition of RPA-ssDNA Complexes. *Science (New York, N.Y.)*, 300, 1542–1548. doi:10.1126/ science.1083430.

In: Prokaryotes
Editor: Marina Nisnevitch

ISBN: 978-1-63321-592-4

Chapter 7

Cytotoxic Properties and Medical Usages of Cyt Toxins from *Bacillus thuringiensis* Subsp. *Israelensis*

Shmuel Cohen*
Spring Bio-Med LTD, Kiryat-Matalon, Petach-Tikva, Israel

Abstract

Bacillus thuringiensis subsp. *israelensis (Bti)* is a Gram-positive spore-forming bacterium and a highly potent mosquitocidal strain. During the sporulation phase it synthetizes several pore-forming pro-δ-endotoxins: four of the Cry (Crystal) family (Cry4Aa, Cry4Ba, Cry10Aa and Cry11Aa) and three of the Cyt (Cytotoxic) family (Cyt1Aa and a tiny amount of Cyt2Ba and Cyt1Ca). These proteins are assembled together into a crystalline-like body which is dissolved in the alkaline environment of the larval midgut, digested and activated by endoproteases. Cry pro-toxins are composed of three domains. One domain is associated with a specific receptor on the midgut epithelial cells and the other domains undergo conformational change and are inserted into the lipid bilayer. Once a number of these toxins are associated together within the membrane, they create pores that interrupt the osmotic balance of the cells and lead to cell destruction. The Cyt pro-toxins have only one domain which non-specifically destroys various insect and mammalian cells by a pore-forming or a detergent-like mechanism of membrane lysis. Several studies have shown that Cyt toxins serve as receptors for the Crys (e.g., Cry11Aa). They act synergistically to increase the toxicity of Cry toxins and also prevent the development of larval resistance to Crys. We have recently crystallized and deciphered the 3D structure of the active form of Cyt1Aa and CytBa toxins and demonstrated their similar spatial structure and homology to other pore-forming toxins such as Cyt2Aa, volvatoxin A2 (VVA2) and the Erwinia virulence factor (Evf). This chapter summarizes up-to-date findings on the Cyts family and their homology toxins and discusses the potential use of these toxins in the therapeutic field, especially by specific targeting of Cyts to cancer cells.

* Corresponding author: Email: shmuel.cohen@spring-biomed.com.

Keywords: Pore Forming Toxin, Cry, Cyt, *Bacillus thuringiensis israelensis*, Cry 4Aa, Cry4Ba, Cry10Aa, Cry11Aa Cyt1Aa, Cyt2Aa, Cyt2Ba, Cyt1Ca, VVA2, Evf.

Introduction

During recent years, numerous works have been written on the role of *Bacillus thuringiensis* subsp. *israelensis* δ-endotoxins in the context of biological control against agricultural pests and vectors of diseases (Agaisse & Lereclus, 1995; Ben-Dov, 2014; Manceva et al., 2005; Margalit & Dean, 1985; Pardo-Lopez et al., 2013; Schnepf et al., 1998). This chapter will attempt to summarize current data on the Cyts' insecticidal PFTs (pore-forming toxins), their structure, mode of action and possible application for therapeutic purposes.

Bti is a spore-forming Gram-positive bacterium which is used worldwide as an effective pest control against mosquito and black fly larvae which are vectors of several human diseases, primarily malaria and West Nile virus infection (Bravo et al., 2011; Reisen, 2010; Sinka et al., 2012). This bacterium is also effective against Diptera, Hemiptera, Homoptera, Coleoptera, Lepidoptera and Trematoda (Ben-Dov, 2014). This type of pest control is a friendly and safe alternative pesticide, which can replace environmentally toxic chemical pesticides that are also prone to failure in the long run upon artificial selection for resistant pests. Since the finding of *Bti* in a pond in the Negev Desert of Israel 35 years ago, no *Bti* resistant mosquito larvae strains have been found, probably due to the unique δ-endotoxins content of its parasporal crystal (Becker & Ludwig, 1993; Kamgang et al., 2011; Loke et al., 2010; Vasquez et al., 2009).

During the sporulation phase, *Bti* produces two families of PFTs δ-endotoxins: four Cry proteins – Cry4Aa, Cry4Ba, Cry10Aa and Cry11Aa (134, 128, 72 and 78 kDa, respectively) and three Cyt proteins – Cyt1Aa, Cyt2B and Cyt1Ca (27, 29, and 57 kDa, respectively). All are encoded as pro-toxins on one 128 Kb *pBtoxis* plasmid (Ben-Dov et al.,1996; Ben-Dov et al., 1999; Berry et al., 2002). The expression of Cry10Aa, Cyt2Ba and Cyt1Ca is minor, but is assumed to contribute to the overall toxicity of the bacterium (Guerchicoff et al., 1997; Lee et al., 1985; Thorne et al., 1986). The two toxin families are not related to each other either by sequence or by structure (Ben-Dov, 2014; Crickmore et al., 1998). Nevertheless, they assemble together into a crystalline-like structure during the sporulation phase. The C-termini of Cry4Aa and Cry4Ba have conserved sequences that participate in the spontaneous crystallization of the protein due to their intra- and inter-molecular disulfide bonds (Bietlot et al., 1990; Coucheet al., 1987). Cyt1Aa and Cry11Aa lack these sequences and their stabilization and crystallization depend on other proteins, such as P19 and P20 chaperons, which are located in the same operon with Cry11Aa (Agaisse & Lereclus, 1995; Visick & Whiteley, 1991; Wu & Federici, 1993, 1995; Xu, et al., 2001).

Following ingestion by mosquito larvae, the *Bti* is disintegrated and its crystal is dissolved in the alkaline environment of the midgut, its solubilized proteins are proteolyzed and activated by gastric enzymes and associated with the midgut epithelial cells. After oligomerization and conformation changes, the activated Cry and Cyt are inserted into the membrane and affect the integrity and osmotic balance of the cells, leading to cell swelling and death (Bravo et al., 2011). These characteristics enable the use of these toxins in controlling agricultural pests and vector-borne diseases. However, there is also a growing interest in the potential exploitation of these toxins in the therapeutic field due to their additional capability of destroying various bacterial and mammalian cells (Cahan et al., 2008; Yokoyama et al., 1991; Yokoyama et al., 1992; Al-yahyaee & Ellar, 1996; Cohen et al., 2007).

The Cry and Cyt Toxins Activity in Mosquito Larvae

The Cry toxins are specifically active against different species of mosquitoes, insects and nematodes (Bravo et al., 2011), whereas Cyt toxins are mainly active against dipteran insects (Soberon et al., 2013). Indeed, Cyts are formed mainly by *Bt* bacteria that produce Crys which are active against mosquitoes and black flies. Despite the widespread usage of *Bti* as an agriculture pest, no resistance of mosquito larvae to the whole complex of *Bti* was observed. On the contrary, it has been shown that *Aedes aegypti* larvae developed resistance to a single protein Cry4Aa in field conditions (Stalinski et al., 2014).

The toxicity of Cyt1Aa (248 amino acids) against mosquito larvae is low (Bourgouin et al., 1986; Thiery et al., 1997). The proteolyzed active form of Cyt1Aa (activated-Cyt1Aa) is a C- and N-termini truncated monomer of 22-25 kDa (Al-yahyaee & Ellar, 1995) which is three times more toxic than the native pro-toxin (Butko et al. 1996, 1997). This toxin exhibits synergistic activity when administrated with Cry4Ba, Cry4Aa or Cry11Aa and also prevents the development of resistant *Culex quinquefasciatus* larvae in sub-lethal concentrations (Wirth et al., 2005; Wirth et al.,2012; Wirth et al., 2007). In addition, Cyt1Aa induced oligomerization of Cry11Aa and Cry11Aa channels (Munoz-Garay et al., 2009). It was suggested that Cyt1Aa acts as a receptor for Crys and associates with domain II of Cry11Aa and Cry4Ba. This mutual association accelerates the oligomerization and the binding of Cry11Aa to the cell membrane, enhances its activity and prevents the formation of resistant mosquito larvae (Perez et al., 2005; Perez et al., 2007). It was also shown that Cyt1Aa synergizes the toxicity of binary toxins (BinA and BinB) of *Bacillus sphaericus* against *A. aegypti* larvae and also prevents and represses the resistance of *Culex quinquefasciatus* to the bacteria (Gomez et al., 2007; Perez et al., 2007). However, the ability of Cyt1Aa to undergo self-oligomerization is not related to its synergistic activity with Cry11Aa: Cyt1Aa α-C loss of function mutant that does not oligomerize still retains its synergistic activity with Cry11Aa, indicating that these activities are distinct functions of Cyt1Aa (Lopez-Diaz et al., 2013).

Cyt2Ba (263 amino acids) is 41% identical to Cyt1Aa and found in minute quantities in the crystal (Guerchicoff et al., 1997; Nisnevitch et al., 2006). The activated Cyt2Ba is also toxic to mosquito larvae and hemolytic *in vitro* but significantly less than the activated-Cyt1Aa (Juarez-Perez, et al., 2002; Nisnevitch et al., 2006). It was also shown that activated-Cyt2Ba acts synergistically with Cry4Aa, Cry4Ba and Cry11Aa (Manasherob et al., 2006) and with *B. sphaericus* binary toxin (Wirthet al., 2001) against mosquito larvae.

Cyt1Ca (525 amino acids) is 52% identical to Cyt1Aa in its N-terminus (Itsko & Zaritsky, 2007; Manasherob et al., 2006), but its C-terminus contains an additional domain of β-trefoil carbohydrate-binding motif similar to the receptor binding domain of ricin-B lectin type (Itsko et al., 2005). Cyt1Ca has neither hemolytic nor larvicidal activity, however, when its N-terminus was expressed in a recombinant *E. coli*, there was a biomass growth arrest and a significant decrease in viability (Itsko et al., 2005).

The Spatial Structure of Cyt Toxins and Other Related Proteins

The first deciphered 3D structure of a Cyt family member was Cyt2Aa (PDB code: 1CBY) of *Bt kyushuenesis* (Li et al., 1996). Its monomer structure consists of four A-D ampiphilic α-helices in which their hydrophobic residues packed against the β-sheets.

Recently, we deciphered the spatial structure of proteolyted Cyt1Aa (PDB code: 3RON, http://proteopedia.org/wiki/index.php/Journal:JMB:1, (Cohen et al., 2011)) and Cyt2Ba (PDB code: 2RCI, http://proteopedia.org/wiki/index.php/Cyt2Ba, (Cohen et al., 2008)) in resolution of 1.8 and 2.2 Å, respectively.

The structure of these proteins was found to be strictly similar to that of fungal volvatoxin VVA2 and to Cyt2Aa: they all adopt a typical cytolysin fold: one α/β domain consisting of 7-8 β-strands flanked by two layers of α-helices (Li et al., 1996; Lin et al., 2004). This structure is different from that of Cry toxins which consist of domain I that induces formation of cation-selective pores, and domains II and III that bind to specific receptors (Pardo-Lopez et al., 2013; Schnepf et al., 1998). Previous experiments with Cyt1Aa fragments showed that the N-terminus of the protein containing the A and C α-helices is essential for oligomerization and that its absence impaired the toxicity of the protein. On the other hand, the C terminus of Cyt1Aa is involved in membrane permeability (Gazit et al., 1997; Rodriguez-Almazan et al., 2011).

It has been suggested that the soluble Cry1Ac of *Bacillus thuringiensis* subsp. *kurstaki* undergoes conformational change upon proteolytic activation (Choma et al., 1991). However, the strict structural identity of the soluble activated form of Cyt1Aa and Cyt2Ba to that of the native Cyt2Aa shows that the spatial structure of these toxins does not spontaneously change in solution following the proteases cleavage and that the toxicity stems from its association with other monomers or with the cell membrane. Indeed, it was actually demonstrated that cell membrane association, oligomerization and conformational change must occur before penetration and perforation of Cyt1Aa (Lopez-Diaz et al., 2013).

When comparing the sequences and analyzing the similarity of Cyts to other related proteins it was found that Cyt1 family is phylogenetically quite distant than Cyt2 proteins (Soberon et al., 2013). However, Cyt2Aa, Cyt1Aa and Cyt2Ba share a significantly high degree of structural similarity between themselves and to volvatoxin2 (VVA2) of the mushroom *Volvariella volvacea* and to Erwinia virulence factor (Evf) (Cohen et al., 2011; Quevillon-Cheruel et al., 2009; Rigden, 2009), even though the VVA2, Evf and Cyts share low sequence homology (up to 10%). Using the FATCAT program, it was shown that Cyt1Aa is more related to VVA2 than to the Evf (91% structural identity between Cyt1Aa to Cyt2Aa or VVA2, 95% between Cyt1Aa to Cyt2Ba and only 72% between Cyt1Aa to Evf) (Soberon et al., 2013). Despite the high structural similarity between Cyt2Aa, Cyt1Aa and Cyt2Ba, the reason for to the relatively lower toxicity of Cyt2Ba is still unknown. It is interesting to note that VAA2 is also a hemolytic toxin and mutations in domains responsible for oligomerization reduce its hemolytic ability (Weng et al., 2004). It is still unclear whether VVA2 and Evf also exhibit synergy with other toxins, whether they reduce or prevent resistance development and by which mechanism.

The Mode of Action of Cyt Toxins

Unlike Crys associated with specific receptors on the cells membrane (Fernandez-Luna et al., 2010; Fernandez et al., 2005; Gomez et al., 2007; Likitvivatanavong et al., 2011; Pardo-Lopez et al., 2013; Perez et al., 2005), Cyts bind non-specifically to unsaturated lipid in the dipteran lipid bilayer (Canton et al., 2014; Gill et al., 1987; Thomas & Ellar, 1983b). However, one cannot rule out the existence of specific membrane receptor/s for Cyts which increase the *Bt* toxicity against certain insect species (Butko et al., 1997; Li et al., 1996).

There is a controversy regarding the destructive mechanism of Cyts. The pore-forming (or 'umbrella like') theory claims that Cyts' monomers bind to the membrane surface and then undergo a conformational change. The cytolysin fold allows the outer α-helices at the N-terminus to swing out and expose the inner β-sheets toward the cell membrane surface. The oligomerization of a number of monomers creates β-barrel pores wherein several transmembrane subunits of β-sheets penetrate into the lipid membrane and form a channel structure (Li et al., 2001; Li et al., 1996; Parker & Feil, 2005; Promdonkoy & Ellar, 2003; Rodriguez-Almazan et al., 2011). These pores assumed to be cation-selective channels that induce ion equilibrium, net-influx of water, cell swelling and eventually colloid-osmotic lysis (Knowles et al., 1989). Another hypothesis claims that the toxin acts by a detergent-like mechanism (Manceva et al., 2005; Thomas & Ellar, 1983b) as exhibited by various anti-microbial peptides (Shai, 1995). According to this hypothesis, the toxin monomers are non-specifically attached to the surface of cell membrane and their aggregates form nonspecific tears in the cell membrane. Then after, there is a leakage of the cell content and rapid cells death characterized by an "all or none" cells destruction pattern (Butko et al., 1996, 1997). However, there is a compromised theory which holds that the umbrella and the detergent-like models can co-exist depending on the toxin concentration and exposure time: under low concentrations channels may be created whereas a detergent-like destruction could occur under higher concentrations and/or long duration of exposure (Ben-Dov, 2014; Butko, 2003).

Several findings reinforce the assumption that the Cyt β-strands are being inserted into the cell membrane and function as membrane pore oligomeric proteins: The A and C α-helices at the end of the N-terminus of Cyts, are related to oligomerization. Point mutations in these domains significantly reduce their aggregation and hemolytic activity (Promdonkoy et al., 2008; Rodriguez-Almazan et al., 2011). In contrast to the Cyt α-helices that are not long enough, the 5-7 β-strands have the required length to span the bilayer cell membrane. It has been shown that Cyt1Aa β-strands that were associated with lipids were protected from proteolytic activity of proteinase K and were also able to release an encapsulated calcein from liposomes (Du et al., 1999; Rodriguez-Almazan et al., 2011). In addition, fragments of β-strands of Cyt2a labeled with acrylodan (an environment pH-sensitive fluorophor) retained their fluorescence in the presence of lipids (Promdonkoy & Ellar, 2000).

Moreover, our results (Cohen et al., 2011), are also support the pore forming hypothesis: (i) the 3D structure of lipid associated pocket of Cyt1Aa is similar to that of an Erwinia Virulence Factor (Evf) produced by the Drosophila pathogen *Erwinia carotovora* (PDB code: 2W3Y). In contrast to Evf pocket that is covalently bound to palmitate in a hydrophobic pocket, none of the Cyts members contain a palmitoylated Cys residue. We have proposed (Cohen et al., 2011) that unlike Evf, where the lipid "locks" the helical hairpin to the β-sheet thus limiting flexibility, absence of the lipid in the Cyt1Aa structure enables its flexibility and allows the conformational changes of the two surrounding α-helical layers necessary for exposing the hydrophobic β-sheet prior to membrane insertion. (ii) The hemolytic patterns of Cyt1Aa were found to be much similar to that of the α-toxin (β-barrel pore forming) than to that of Triton X-100 and SDS which exhibited an 'all or none' hemolytic patterns.

Anyway, according to the two above mentioned models, the ability of Cyt toxins to penetrate cells membrane and destruct cells, depends on the interaction between the hydrophobic segments of Cyt1Aa and the unsaturated phospholipids on the cell membrane and formation of a stable oligomer structure. Indeed, it has been found that in presence of lipids the activated-

Cyt1Aa aggregates into 400 kDa oligomeric structure which remains stable even at high toxin concentrations, suggesting that this toxin preserves its highly stable oligomeric state (Chow et al., 1989; Du et al., 1999; Manceva et al., 2005). Based on density-gradient assay, it has been suggested that this oligomer is composed of 16 units (Chow et al., 1989).

Usage of Cyt Toxins as an Antibacterial Agent

The expression of cyt1Aa in recombinant acrystalliferous *Bt kurstaki* and *E. coli* induces loss of colony forming, biomass growth arrest, partial lysis, and compaction of nucleoid DNA (Douek et al., 1992; Manasherob et al., 2003; Sazhenskiy et al., 2010). It was assumed that simultaneous associations of Cyt1Aa to DNA and to zwitterionic phospholipids lead to disconnection of DNA from the cytoplasm membrane and nucleoid compaction (Manasherob et al., 2003; Sazhenskiy et al., 2010). However, when cyt1Aa was co-expressed with the helper protein P20, the viability was maintained (Manasherob et al., 2001; Wu & Federici, 1993). It was also found that exogenously administrated activated-Cyt1Aa was bactericidal to *E. coli* which plausibly penetrates via the outer membrane into the cytoplasm.

It was also shown that activated-Cyt1Aa is bacteriostatic to Gram-positive bacterium *S. aureus* despite of the massive peptidoglycan layer wrapping the cytoplasmic membrane (Cahan et al., 2008; Revina et al., 2005; Yudina et al., 2007). Therefore, activated-Cyt1Aa could be exploited alone or in combination with other antibiotics as an anti-bacterial agent.

Specific Targeting of Cyt Toxins to Cancer Cells

In addition to its larvicidal and anti-bacterial activity, Cyt1Aa (especially its activated form) also exhibits hemolytic and cytolytic activity against various invertebrates and mammalian cells *in vitro* and *in-vivo* (Thomas & Ellar, 1983a). In this regard, the membrane acting toxicity of Cyt1Aa represents an attractive therapeutic character since it is enable to overcome drug resistance mechanisms of tumor cells. Indeed, it was reported that activated Cyt2Ba exhibits some cytotoxic activity (20%) against MCF-7 breast cancer cells and when administrated along with Cry4Aa and Cry11Aa the toxicity was doubled (Teixeira Correa et al., 2012). Furthermore, it was found that Cyt1Aa potentiated the antitumor activity of bleomycin against various types of solid tumor cells *in vitro* and *in vivo* and that this synergy effect was significantly enhanced under increased temperatures (Yokoyama et al., 1991; Yokoyama et al., 1992). Another interesting strategy is to conjugant the toxin to a specific ligand that is recognized by target cells. It has been shown that the activated-Cyt1Aa-insulin conjugate was significantly specific to cancer cells highly expressing receptors for insulin (Al-yahyaee & Ellar, 1996). Recently, we also have demonstrated that recombinant conjugation of the C termini of activated-Cyt1Aa to MBPp (myelin basic protein peptide) abolished the non-specific activity of Cyt1Aa against mammalian cells while maintained the membrane cytotoxicity of the toxin. This conjugant was specifically recognized by, and destructed cell hybridoma cell line highly expressing anti-MBPp receptors, whereas this conjugant was not toxic to hybridoma control cells harboring other receptor-type (Cohen et al., 2007).

Conclusion

Following association with cell membrane specific receptors, Cry toxins form an oligomeric structure, penetrate the cell membrane and create pores. On the other hand, Cyt toxins are being inserted into the cell membrane via non-specific interactions and destruct various cells types by a still unclear mechanism. Cyt1Aa, the most toxic member of the Cyt family, represents an attractive cytotoxin for exploitation in the therapeutic field since it is a membrane active toxin enabling to overcome the multidrug resistance of bacterial and cancer cells.

The therapeutic approaches utilizing Cyt toxins or other non-specific membrane-active proteins include: (a) using the toxin alone or in combination with chemotherapy, (b) conjugation of the toxin with a specific ligand for targeting it to cancer cells bearing unique or clonotypic surface receptors. These strategies constitute important therapeutic approaches enabling to broaden the spectrum of anti-cancer drugs. However, they are still facing other challenges such as improving the toxicity of these proteins by point mutations, finding the most toxic fragment peptide of the protein, tailoring the specific conjugated peptide for each cancer cells type and for each patient as well as overcoming the immunologic reaction against the toxin or the conjugant.

Possible solutions, for instance, for the later problem are: encapsulation of the toxins/toxins-ligand conjugates within micro- or nano-particles or finding and utilizing the most active toxic fragment/fragmaents of Cyts instead of using the whole immunogenic active form.

References

Agaisse, H., & Lereclus, D. (1995). How does *Bacillus thuringiensis* produce so much insecticidal crystal protein? *J Bacteriol, 177*, 6027-6032.

Al-yahyaee, S. A., & Ellar, D. J. (1996). Cell targeting of a pore-forming toxin, CytA δ-endotoxin from *Bacillus thuringiensis* subspecies *israelensis*, by conjugating CytA with anti-Thy 1 monoclonal antibodies and insulin. *Bioconjug Chem, 7*, 451-460.

Al-yahyaee, S. A. S., & Ellar, D. J. (1995). Maximal toxicity of cloned CytA δ-endotoxin from *Bacillus thuringiensis* subsp. *israelensis* requires proteolytic processing from both the N- and C-termini. *Microbiology, 141*, 3141-3148

Becker, N., & Ludwig, M. (1993). Investigations on possible resistance in *Aedes vexans* field populations after a 10-year application of *Bacillus thuringiensis israelensis*. *J Am Mosq Control Assoc, 9*, 221-224.

Ben-Dov, E. (2014). *Bacillus thuringiensis* subsp. *israelensis* and its dipteran-specific toxins. *Toxins (Basel), 6*, 1222-1243.

Ben-Dov, E., Einav, M., Peleg, N., Boussiba, S., & Zaritsky, A. (1996). Restriction map of the 125-kilobase plasmid of *Bacillus thuringiensis* subsp. *israelensis* carrying the genes that encode δ-endotoxins active against mosquito larvae. *Appl Environ Microbiol, 62*, 3140-3145.

Ben-Dov, E., Nissan, G., Pelleg, N., Manasherob, R., Boussiba, S., & Zaritsky, A. (1999). Refined, circular restriction map of the *Bacillus thuringiensis* subsp. *israelensis* plasmid carrying the mosquito larvicidal genes. *Plasmid, 42*, 186-191.

Berry, C., O'Neil, S., Ben-Dov, E., Jones, A. F., Murphy, L., Quail, M. A., & Parkhill, J. (2002). Complete sequence and organization of pBtoxis, the toxin-coding plasmid of *Bacillus thuringiensis* subsp. *israelensis*. *Appl Environ Microbiol, 68*, 5082-5095.

Bietlot, H. P., Vishnubhatla, I., Carey, P. R., Pozsgay, M., & Kaplan, H. (1990). Characterization of the cysteine residues and disulphide linkages in the protein crystal of *Bacillus thuringiensis*. *Biochem J, 267*, 309-315.

Bourgouin, C., Klier, A., & Rapoport, G. (1986). Characterization of the genes encoding the haemolytic toxin and the mosquitocidal δ-endotoxin of *Bacillus thuringiensis israelensis*. *Mol Gen Genet, 205*, 390-397.

Bravo, A., Likitvivatanavong, S., Gill, S. S., & Soberon, M. (2011). *Bacillus thuringiensis*: A story of a successful bioinsecticide. *Insect Biochem Mol Biol, 41*, 423-431.

Butko, P. (2003). Cytolytic toxin Cyt1A and its mechanism of membrane damage: data and hypotheses. *Appl Environ Microbiol, 69*, 2415-2422.

Butko, P., Huang, F., Pusztai-Carey, M., & Surewicz, W. K. (1996). Membrane permeabilization induced by cytolytic δ-endotoxin CytA from *Bacillus thuringiensis* var. *israelensis*. *Biochemistry, 35*, 11355-11360.

Butko, P., Huang, F., Pusztai-Carey, M., & Surewicz, W. K. (1997). Interaction of the δ-endotoxin CytA from *Bacillus thuringiensis* var. *israelensis* with lipid membranes. *Biochemistry, 36*, 12862-12868.

Cahan, R., Friman, H., & Nitzan, Y. (2008). Antibacterial activity of Cyt1Aa from *Bacillus thuringiensis* subsp. *israelensis*. *Microbiology, 154*, 3529-3536.

Canton, P. E., Lopez-Diaz, J. A., Gill, S. S., Bravo, A., & Soberon, M. (2014). Membrane binding and oligomer membrane insertion are necessary but insufficient for *Bacillus thuringiensis* Cyt1Aa toxicity. *Peptides, 53*, 286-291.

Choma, C. T., Surewicz, W. K., & Kaplan, H. (1991). The toxic moiety of the *Bacillus thuringiensis* protoxin undergoes a conformational change upon activation. *Biochem Biophys Res Commun, 179*, 933-938.

Chow, E., Singh, G. J., & Gill, S. S. (1989). Binding and aggregation of the 25-kilodalton toxin of *Bacillus thuringiensis* subsp. *israelensis* to cell membranes and alteration by monoclonal antibodies and amino acid modifiers. *Appl Environ Microbiol, 55*, 2779-2788.

Cohen, S., Albeck, S., Ben-Dov, E., Cahan, R., Firer, M., Zaritsky, A., & Dym, O. (2011). Cyt1Aa toxin: crystal structure reveals implications for its membrane-perforating function. *J Mol Biol, 413*, 804-814.

Cohen, S., Cahan, R., Ben-Dov, E., Nisnevitch, M., Zaritsky, A., & Firer, M. A. (2007). Specific targeting to murine myeloma cells of Cyt1Aa toxin from *Bacillus thuringiensis* subspecies *israelensis*. *J Biol Chem, 282*.

Cohen, S., Dym, O., Albeck, S., Ben-Dov, E., Cahan, R., Firer, M., & Zaritsky, A. (2008). High-resolution crystal structure of activated Cyt2Ba monomer from *Bacillus thuringiensis* subsp. *israelensis*. *J Mol Biol, 380*, 820-827.

Couche, G. A., Pfannenstiel, M. A., & Nickerson, K. W. (1987). Structural disulfide bonds in the *Bacillus thuringiensis* subsp. *israelensis* protein crystal. *J Bacteriol, 169*, 3281-3288.

Crickmore, N., Zeigler, D. R., Feitelson, J., Schnepf, E., Van Rie, J., Lereclus, D., & Dean, D. H. (1998). Revision of the nomenclature for the *Bacillus thuringiensis* pesticidal crystal proteins. *Microbiol Mol Biol Rev, 62*, 807-813.

Douek, J., Einav, M., & Zaritsky, A. (1992). Sensitivity to plating of *Escherichia coli* cells expressing the *cryA* gene from *Bacillus thuringiensis* var. *israelensis*. *Mol Gen Genet, 232*), 162-165.

Du, J., Knowles, B. H., Li, J., & Ellar, D. J. (1999). Biochemical characterization of *Bacillus thuringiensis* cytolytic toxins in association with a phospholipid bilayer. *Biochem J, 338*, 185-193.

Fernandez-Luna, M. T., Lanz-Mendoza, H., Gill, S. S., Bravo, A., Soberon, M., & Miranda-Rios, J. (2010). An alpha-amylase is a novel receptor for *Bacillus thuringiensis* ssp. *israelensis* Cry4Ba and Cry11Aa toxins in the malaria vector mosquito A*nopheles albimanus* (Diptera: *Culicidae*). *Environ Microbiol, 12*, 746-757.

Fernandez, L. E., Perez, C., Segovia, L., Rodriguez, M. H., Gill, S. S., Bravo, A., & Soberon, M. (2005). Cry11Aa toxin from *Bacillus thuringiensis* binds its receptor in *Aedes aegypti* mosquito larvae through loop alpha-8 of domain II. *FEBS Lett, 579*, 3508-3514.

Gazit, E., Burshtein, N., Ellar, D. J., Sawyer, T., & Shai, Y. (1997). *Bacillus thuringiensis* cytolytic toxin associates specifically with its synthetic helices A and C in the membrane bound state. Implications for the assembly of oligomeric transmembrane pores. *Biochemistry, 36*, 15546-15554.

Gill, S. S., Singh, G. J., & Hornung, J. M. (1987). Cell membrane interaction of *Bacillus thuringiensis* subsp. *israelensis* cytolytic toxins. *Infect Immun, 55*, 1300-1308.

Gomez, I., Pardo-Lopez, L., Munoz-Garay, C., Fernandez, L. E., Perez, C., Sanchez, J., & Bravo, A. (2007). Role of receptor interaction in the mode of action of insecticidal Cry and Cyt toxins produced by *Bacillus thuringiensis*. *Peptides, 28*, 169-173.

Guerchicoff, A., Ugalde, R. A., & Rubinstein, C. P. (1997). Identification and characterization of a previously undescribed *cyt* gene in *Bacillus thuringiensis* subsp. *israelensis*. *Appl Environ Microbiol, 63*, 2716-2721.

Itsko, M., Manasherob, R., & Zaritsky, A. (2005). Partial restoration of antibacterial activity of the protein encoded by a cryptic open reading frame (*cyt1Ca*) from *Bacillus thuringiensis* subsp. *israelensis* by site-directed mutagenesis. *J Bacteriol, 187*, 6379-6385.

Itsko, M., & Zaritsky, A. (2007). Exposing cryptic antibacterial activity in Cyt1Ca from *Bacillus thuringiensis israelensis* by genetic manipulations. *FEBS Lett, 581*, 1775-1782.

Juarez-Perez, V., Guerchicoff, A., Rubinstein, C., & Delecluse, A. (2002). Characterization of Cyt2Bc toxin from *Bacillus thuringiensis* subsp. *medellin*. *Appl Environ Microbiol, 68*, 1228-1231.

Kamgang, B., Marcombe, S., Chandre, F., Nchoutpouen, E., Nwane, P., Etang, J., & Paupy, C. (2011). Insecticide susceptibility of *Aedes aegypti* and *Aedes albopictus* in Central Africa. *Parasit Vectors, 4*, 79.

Knowles, B. H., Blatt, M. R., Tester, M., Horsnell, J. M., Carroll, J., Menestrina, G., & Ellar, D. J. (1989). A cytolytic δ-endotoxin from *Bacillus thuringiensis* var. *israelensis* forms cation-selective channels in planar lipid bilayers. *FEBS Lett, 244*, 259-262.

Lee, S. G., Eckblad, W., & Bulla, L. A., Jr. (1985). Diversity of protein inclusion bodies and identification of mosquitocidal protein in *Bacillus thuringiensis* subsp. *israelensis*. *Biochem Biophys Res Commun, 126*, 953-960.

Li, J., Derbyshire, D. J., Promdonkoy, B., & Ellar, D. J. (2001). Structural implications for the transformation of the *Bacillus thuringiensis* δ-endotoxins from water-soluble to membrane-inserted forms. *Biochem Soc Trans, 29*, 571-577.

Li, J., Koni, P. A., & Ellar, D. J. (1996). Structure of the mosquitocidal δ-endotoxin CytB from *Bacillus thuringiensis* sp. *kyushuensis* and implications for membrane pore formation. *J Mol Biol, 257*, 129-152.

Likitvivatanavong, S., Chen, J., Evans, A. M., Bravo, A., Soberon, M., & Gill, S. S. (2011). Multiple receptors as targets of Cry toxins in mosquitoes. *J Agric Food Chem, 59*, 2829-2838.

Lin, S. C., Lo, Y. C., Lin, J. Y., & Liaw, Y. C. (2004). Crystal structures and electron micrographs of fungal volvatoxin A2. *J Mol Biol, 343*, 477-491.

Loke, S. R., Andy-Tan, W. A., Benjamin, S., Lee, H. L., & Sofian-Azirun, M. (2010). Susceptibility of field-collected *Aedes aegypti* (L.) (Diptera: *Culicidae*) to *Bacillus thuringiensis israelensis* and *temephos*. *Trop Biomed, 27*, 493-503.

Lopez-Diaz, J. A., Canton, P. E., Gill, S. S., Soberon, M., & Bravo, A. (2013). Oligomerization is a key step in Cyt1Aa membrane insertion and toxicity but not necessary to synergize Cry11Aa toxicity in *Aedes aegypti* larvae. *Environ Microbiol*, doi: 10.1111/1462-2920.12263.

Manasherob, R., Itsko, M., Sela-Baranes, N., Ben-Dov, E., Berry, C., Cohen, S., & Zaritsky, A. (2006). Cyt1Ca from *Bacillus thuringiensis* subsp. *israelensis*: production in *Escherichia coli* and comparison of its biological activities with those of other Cyt-like proteins. *Microbiology, 152*, 2651-2659.

Manasherob, R., Zaritsky, A., Ben-Dov, E., Saxena, D., Barak, Z., & Einav, M. (2001). Effect of accessory proteins P19 and P20 on cytolytic activity of Cyt1Aa from *Bacillus thuringiensis* subsp. *israelensis* in *Escherichia coli*. *Curr Microbiol, 43*, 355-364.

Manasherob, R., Zaritsky, A., Metzler, Y., Ben-Dov, E., Itsko, M., & Fishov, I. (2003). Compaction of the *Escherichia coli* nucleoid caused by Cyt1Aa. *Microbiology, 149*, 3553-3564.

Manceva, S. D., Pusztai-Carey, M., Russo, P. S., & Butko, P. (2005). A detergent-like mechanism of action of the cytolytic toxin Cyt1A from *Bacillus thuringiensis* var. *israelensis*. *Biochemistry, 44*, 589-597.

Margalit J, Dean D. (1985). The story of *Bacillus thuringiensis* var. *israelensis* (*B.t.i.*). *J Am Mosq Control Assoc*, 1, 1-7.

Munoz-Garay, C., Rodriguez-Almazan, C., Aguilar, J. N., Portugal, L., Gomez, I., Saab-Rincon, G., & Bravo, A. (2009). Oligomerization of Cry11Aa from *Bacillus thuringiensis* has an important role in toxicity against *Aedes aegypti*. *Appl Environ Microbiol, 75*, 7548-7550.

Nisnevitch, M., Cohen, S., Ben-Dov, E., Zaritsky, A., Sofer, Y., & Cahan, R. (2006). Cyt2Ba of *Bacillus thuringiensis israelensis*: activation by putative endogenous protease. *Biochem Biophys Res Commun, 344*, 99-105.

Pardo-Lopez, L., Soberon, M., & Bravo, A. (2013). *Bacillus thuringiensis* insecticidal three-domain Cry toxins: mode of action, insect resistance and consequences for crop protection. *FEMS Microbiol Rev, 37*, 3-22.

Parker, M. W., & Feil, S. C. (2005). Pore-forming protein toxins: from structure to function. *Prog Biophys Mol Biol, 88*, 91-142.

Perez, C., Fernandez, L. E., Sun, J., Folch, J. L., Gill, S. S., Soberon, M., & Bravo, A. (2005). *Bacillus thuringiensis* subsp. *israelensis* Cyt1Aa synergizes Cry11Aa toxin by functioning as a membrane-bound receptor. *Proc Natl Acad Sci U S A, 102*, 18303-18308.

Perez, C., Munoz-Garay, C., Portugal, L. C., Sanchez, J., Gill, S. S., Soberon, M., & Bravo, A. (2007). *Bacillus thuringiensis* ssp. *israelensis* Cyt1Aa enhances activity of Cry11Aa toxin by facilitating the formation of a pre-pore oligomeric structure. *Cell Microbiol, 9*, 2931-2937.

Promdonkoy, B., & Ellar, D. J. (2000). Membrane pore architecture of a cytolytic toxin from *Bacillus thuringiensis*. *Biochem J, 350*, 275-282.

Promdonkoy, B., & Ellar, D. J. (2003). Investigation of the pore-forming mechanism of a cytolytic *delta-endotoxin* from *Bacillus thuringiensis*. *Biochem J, 374*, 255-259.

Promdonkoy, B., Rungrod, A., Promdonkoy, P., Pathaichindachote, W., Krittanai, C., & Panyim, S. (2008). Amino acid substitutions in alphaA and alphaC of Cyt2Aa2 alter hemolytic activity and mosquito-larvicidal specificity. *J Biotechnol, 133*, 287-293.

Quevillon-Cheruel, S., Leulliot, N., Muniz, C. A., Vincent, M., Gallay, J., Argentini, M., & van Tilbeurgh, H. (2009). Evf, a virulence factor produced by the Drosophila pathogen *Erwinia carotovora*, is an S-palmitoylated protein with a new fold that binds to lipid vesicles. *J Biol Chem, 284*, 3552-3562.

Reisen, W. K. (2010). Landscape epidemiology of vector-borne diseases. *Annu Rev Entomol, 55*, 461-483.

Revina, L. P., Kostina, L. I., Dronina, M. A., Zalunin, I. A., Chestukhina, G. G., Yudina, T. G., & Izumrudova, A. V. (2005). Novel antibacterial proteins from entomocidal crystals of *Bacillus thuringiensis* ssp. *israelensis*. *Can J Microbiol, 51*, 141-148.

Rigden, D. J. (2009). Does distant homology with Evf reveal a lipid binding site in *Bacillus thuringiensis* cytolytic toxins? *FEBS Lett, 583*, 1555-1560.

Rodriguez-Almazan, C., Ruiz de Escudero, I., Canton, P. E., Munoz-Garay, C., Perez, C., Gill, S. S., & Bravo, A. (2011). The amino- and carboxyl-terminal fragments of the *Bacillus thuringensis* Cyt1Aa toxin have differential roles in toxin oligomerization and pore formation. *Biochemistry, 50*, 388-396.

Sazhenskiy, V., Zaritsky, A., & Itsko, M. (2010). Expression in *Escherichia coli* of the native cyt1Aa from *Bacillus thuringiensis* subsp. *israelensis*. *Appl Environ Microbiol, 76*, 3409-3411.

Schnepf, E., Crickmore, N., Van Rie, J., Lereclus, D., Baum, J., Feitelson, J., & Dean, D. H. (1998). *Bacillus thuringiensis* and its pesticidal crystal proteins. *Microbiol Mol Biol Rev, 62*, 775-806.

Shai, Y. (1995). Molecular recognition between membrane-spanning polypeptides. *Trends Biochem Sci, 20*, 460-464.

Sinka, M. E., Bangs, M. J., Manguin, S., Rubio-Palis, Y., Chareonviriyaphap, T., Coetzee, M., & Hay, S. I. (2012). A global map of dominant malaria vectors. *Parasit Vectors, 5*, 69.

Soberon, M., Lopez-Diaz, J. A., & Bravo, A. (2013). Cyt toxins produced by *Bacillus thuringiensis*: a protein fold conserved in several pathogenic microorganisms. *Peptides, 41*, 87-93.

Stalinski, R., Tetreau, G., Gaude, T., & Despres, L. (2014). Pre-selecting resistance against individual Bti Cry toxins facilitates the development of resistance to the Bti toxins cocktail. *J Invertebr Pathol, 119*, 50-53.

Teixeira Correa, R. F., Ardisson-Araujo, D. M., Monnerat, R. G., & Ribeiro, B. M. (2012). Cytotoxicity analysis of three *Bacillus thuringiensis* subsp. *israelensis delta-endotoxins* towards insect and mammalian cells. *PLoS One, 7*, e46121.

Thiery, I., Delecluse, A., Tamayo, M. C., & Orduz, S. (1997). Identification of a gene for Cyt1A-like hemolysin from *Bacillus thuringiensis* subsp. *medellin* and expression in a crystal-negative *B. thuringiensis* strain. *Appl Environ Microbiol, 63*, 468-473.

Thomas, W. E., & Ellar, D. J. (1983a). *Bacillus thuringiensis* var *israelensis* crystal *delta-endotoxin*: effects on insect and mammalian cells in vitro and in vivo. *J Cell Sci, 60*, 181-197

Thomas, W. E., & Ellar, D. J. (1983b). Mechanism of action of *Bacillus thuringiensis* var *israelensis* insecticidal *delta-endotoxin*. *FEBS Lett, 154*, 362-368.

Thorne, L., Garduno, F., Thompson, T., Decker, D., Zounes, M., Wild, M., & Pollock, T. J. (1986). Structural similarity between the lepidoptera- and diptera-specific insecticidal endotoxin genes of *Bacillus thuringiensis* subsp. "*kurstaki*" and "*israelensis*". *J Bacteriol, 166*, 801-811.

Vasquez, M. I., Violaris, M., Hadjivassilis, A., & Wirth, M. C. (2009). Susceptibility of *Culex pipiens* (Diptera: *Culicidae*) field populations in Cyprus to conventional organic insecticides, *Bacillus thuringiensis* subsp. *israelensis*, and methoprene. *J Med Entomol, 46*, 881-887.

Visick, J. E., & Whiteley, H. R. (1991). Effect of a 20-kilodalton protein from *Bacillus thuringiensis* subsp. *israelensis* on production of the CytA protein by Escherichia coli. *J Bacteriol, 173*, 1748-1756.

Weng, Y. P., Lin, Y. P., Hsu, C. I., & Lin, J. Y. (2004). Functional domains of a pore-forming cardiotoxic protein, volvatoxin A2. *J Biol Chem, 279, 6805*-6814.

Wirth, M. C., Delecluse, A., & Walton, W. E. (2001). Cyt1Ab1 and Cyt2Ba1 from *Bacillus thuringiensis* subsp. *medellin* and *B. thuringiensis* subsp. *israelensis* Synergize Bacillus sphaericus against *Aedes aegypti* and resistant *Culex quinquefasciatus* (Diptera: *Culicidae*). *Appl Environ Microbiol, 67*, 3280-3284.

Wirth, M. C., Park, H. W., Walton, W. E., & Federici, B. A. (2005). Cyt1A of *Bacillus thuringiensis* delays evolution of resistance to Cry11A in the mosquito *Culex quinquefasciatus. Appl Environ Microbiol, 71*, 185-189.

Wirth, M. C., Walton, W. E., & Federici, B. A. (2012). Inheritance, stability, and dominance of cry resistance in *Culex quinquefasciatus* (Diptera: *Culicidae*) selected with the three cry toxins of *Bacillus thuringiensis* subsp. *israelensis. J Med Entomol, 49*, 886-894.

Wirth, M. C., Zaritsky, A., Ben-Dov, E., Manasherob, R., Khasdan, V., Boussiba, S., & Walton, W. E. (2007). Cross-resistance spectra of *Culex quinquefasciatus* resistant to mosquitocidal toxins of *Bacillus thuringiensis* towards recombinant *Escherichia coli* expressing genes from *B. thuringiensis* ssp. *israelensis. Environ Microbiol, 9*, 1393-1401.

Wu, D., & Federici, B. A. (1993). A 20-kilodalton protein preserves cell viability and promotes CytA crystal formation during sporulation in *Bacillus thuringiensis. J Bacteriol, 175*, 5276-5280.

Wu, D., & Federici, B. A. (1995). Improved production of the insecticidal CryIVD protein in *Bacillus thuringiensis* using cryIA(c) promoters to express the gene for an associated 20-kDa protein. *Appl Microbiol Biotechnol, 42*, 697-702.

Xu, Y., Nagai, M., Bagdasarian, M., Smith, T. W., & Walker, E. D. (2001). Expression of the *p20* gene from *Bacillus thuringiensis* H-14 increases Cry11A toxin production and enhances mosquito-larvicidal activity in recombinant Gram-negative bacteria. *Appl Environ Microbiol, 67*, 3010-3015.

Yokoyama Y, Ohmori I, Kohda K, Kawazoe Y. (1991). Potentiation of antitumor activity of bleomycin towards solid tumors in mice by *Bacillus thuringiensis* subsp. *israelensis* toxin. *Anticancer Res*, 11, 1625-1628.

Yokoyama Y, Ohmori I, Suzuki H, Kohda K, Kawazoe Y. (1992). Hyperthermic potentiation of bleomycin cytotoxicity in the presence of *Bacillus thuringiensis* subsp. *israelenis* delta-endotoxin. *Anticancer Res*, 12, 1079-1081.

Yudina, T. G., Brioukhanov, A. L., Zalunin, I. A., Revina, L. P., Shestakov, A. I., Voyushina, N. E., & Netrusov, A. I. (2007). Antimicrobial activity of different proteins and their fragments from *Bacillus thuringiensis* parasporal crystals against *clostridia* and *archaea. Anaerobe, 13*, 6-13.

In: Prokaryotes
Editor: Marina Nisnevitch

ISBN: 978-1-63321-592-4

Chapter 8

The Novel Plasmids of Botulinum Neurotoxigenic Clostridia

Giovanna Franciosa**, Angelo Iacobino and Concetta Scalfaro
Department of Food Safety and Veterinary Public Health
Istituto Superiore di Sanità, Rome, Italy

Abstract

Twenty years after the first discovery of a plasmid harbouring the gene coding for the botulinum neurotoxin type G in a very rare botulinum neurotoxigenic Clostridium species, i.e. *Clostridium argentinense,* large novel plasmids carrying one or even two different botulinum neurotoxin genes are currently being identified in an increasing number of *C. botulinum* strains. Antibiotic resistance megaplasmids have concomitantly been revealed in rare botulinum neurotoxigenic *C. butyricum* strains associated with human botulism.

In this chapter we will review recent findings on the novel plasmids being identified in botulinum neurotoxigenic clostridia and will discuss the potential benefits of those plasmids to their clostridial hosts.

Keywords: *Clostridium botulinum*; *Clostridium butyricum*; clostridia Groups; plasmids; botulinum neurotoxin genes; β-lactamase genes; plasmid conjugation; bacteriophages; CRISPR loci

Introduction

Bacterial genomes are highly dynamic in order to efficiently respond to the environmental selective pressure. Rapid changes in the composition of bacterial genomes are

* Corresponding author: Email: giovanna.franciosa@iss.it.

the driving force of innovation and adaptation shaping the evolution of microbes (Frost et al., 2005; Koonin and Wolf, 2008).

Plasmids, namely autonomously replicating extrachromosomal genetic elements, significantly contribute to bacterial genome diversity: in fact, they harbour genes for nonessential functions that confer adaptive traits on their microbial hosts, including production of toxins, virulence factors, bacteriocins and/or resistance to antibiotics and to heavy metals, and they may be transferred between even distantly related organisms. The potential dissemination of toxin/virulence genes or antibiotic resistance genes through plasmid transfer among bacteria is of concern, as it may lead to the emergence of bacteria with new pathogenic or antibiotic resistance potentials (Norman et al., 2009; Carattoli, 2013).

Studies on the genes coding for the botulinum neurotoxins (BoNTs), i.e. the most toxic substances known to humans and animals, are revealing that they are plasmid-borne in *Clostridium botulinum* strains more often than previously known (Marshall et al., 2007; Smith et al., 2007; Franciosa et al., 2009; Hill et al., 2009; Umeda et al., 2009; Dover et al., 2013; Stringer et al., 2013; Zhang et al., 2013; Raphael et al., 2014; Wangroongsarb et al., 2014).

Besides, rare *C. butyricum* strains that produce BoNT have recently been shown to harbor very large plasmids, some of which carrying β-lactamase genes (Franciosa et al., 2011). Although the outstanding peculiarity of the botulinum neurotoxigenic *C. butyricum* strains mainly consists of their atypical ability to produce BoNT, due to the presence of the BoNT-encoding gene in their genome (McCroskey et al., 1986; Franciosa et al., 1994; Wang et al., 2000), the occurrence of β-lactamase megaplasmids among those strains is also of interest, as they might contribute to the spread of resistance to β-lactams among clinically important pathogens.

Prior to those findings, the only identified BoNT-encoding plasmid was restricted to *C. argentinense*, a very rare botulinum neurotoxigenic Clostridium species isolated from the environment and not clearly implicated in human disease (Eklund et al., 1988; Zhou et al., 1995), while most of the plasmids identified in *C. botulinum* strains in pioneering studies were cryptic and of relatively small sizes (Scott & Duncan 1978; Strom et al., 1984; Ferreira et al., 1987; Eklund et al., 1989). On the contrary, the BoNT-encoding plasmids and the β-lactamase-encoding plasmids more recently discovered in several botulinum neurotoxigenic clostridia strains are extremely large (> 100 kb); thus, they might have remained unrecognized due to past technical limitations in detecting extrachromosomal genetic elements of those sizes.

Carriage and maintenance of such large plasmids likely impose a high fitness cost on their clostridia hosts that must necessarily be compensated by the provision of advantageous traits for ecological adaptation. Specifically, the ability to produce BoNT could enhance the microorganism's chance of survival and proliferation, whereas the ability to produce β-lactamase could provide it with a selective advantage in a β-lactam-containing environment. Therefore, plasmids offering such benefits would be expected to easily spread among bacterial populations, with potentially serious public health implications. While the mobility of the β-lactamase-encoding megaplasmids of the BoNT-producing *C. butyricum* strains is still unknown, the transfer of some recently identified BoNT-encoding plasmids has been shown to occur in vitro through conjugation between phylogenetically unrelated clostridia (Marshall et al., 2010).

Thus, the above cited recent findings reemphasize the role of plasmids in botulinum neurotoxigenic clostridia and highlight the need to better characterize their properties. In this chapter, we describe the recent acquisitions on the novel BoNT-encoding plasmids and β-lactamase-encoding plasmids currently emerging in botulinum neurotoxigenic clostridia, and on their potential contribution to the remarkable diversity of these microbial pathogens.

Botulinum Neurotoxin (BoNT) and Its Antigenic Variants

Botulinum neurotoxin (BoNT) is the causative agent of the paralysis of botulism, a rare but severe disease affecting humans and animals. At least five different forms of botulism have been documented in humans, including foodborne botulism, intestinal toxemia botulism in infants and adults, wound botulism, iatrogenic botulism and inhalation botulism (Sobel, 2005) (Figure 1).

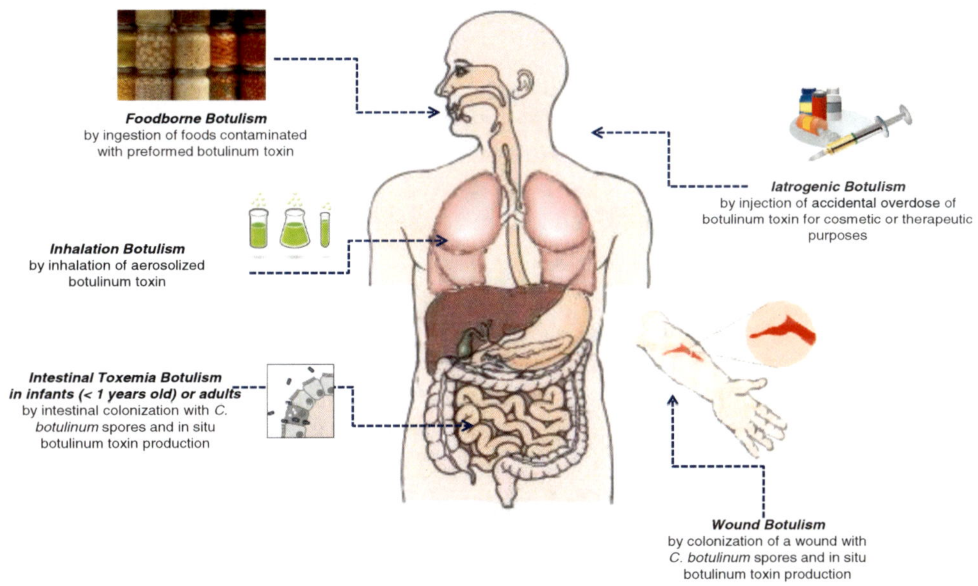

Figure 1. The different forms of human botulism.

The lethal dose of BoNT for humans has been estimated to be less than 1 ng/kg of body weight (Gill, 1982). Because of its high lethality and of the severity of the disease that it causes, BoNT is regarded as a high risk biothreat agent (Arnon et al., 2001). In contrast, when administered in extremely low doses, BoNT can provide effective clinical or even aesthetic benefits against a wide variety of conditions characterized by excessive muscle contraction, resulting in an improved quality of life for thousands of patients (Hallett et al., 2013; Shilpa et al., 2014).

Active BoNT is a 150 kDa protein consisting of two polypeptide chains linked by a disulphide bridge: a 100 kDa heavy chain containing two functional domains, one of which for nerve cell binding and the other one for translocation, and a 50 kDa light chain that includes a conserved zinc binding motif (HExxH) characteristic of zinc endopeptidases (Montecucco & Schiavo, 1993). The crystal structure of BoNT supports the existence of three distinct domains (Lacy et al., 1998; Swaminathan & Eswaramoorthy, 2000; Kumaran et al., 2009).

Once BoNT has entered the systemic circulation, either as a result of naturally occurring disease or after intentional local administration as a pharmacological or cosmetic agent, it is captured by specific ganglioside and protein receptors on the surface of the cholinergic nerve endings, and is then internalized within the neurons through receptor-mediated endocytosis (Simpson, 2013). The acidification inside the endocytic vesicles causes conformational modifications of BoNT that facilitate its membrane translocation. During this process, the disulphide bridge between the light and heavy chains is broken, and the light chain is released into the nerve terminal cytosol, where it selectively cleaves target synaptic SNARE (Soluble N-ethylmaleimide-sensitive factor Attachment Protein Receptor) proteins involved in the exocytosis of synaptic vesicles (Schiavo et al., 2000). By this mechanism, BoNT blocks the release of the acetylcholine neurotransmitter from the peripheral synapses, and ultimately impairs the nerve pulse transmission and muscle contraction. High-affinity binding to the peripheral cholinergic nerve endings and cleavage specificity of the synaptic protein targets are both responsible for BoNT high potency (Rossetto et al., 2013).

Historically, BoNT has been known to exist in seven (A to G) structurally and functionally similar serotypes displaying up to 70% differences in their amino acid contents, that result in variations in antigenicity, as well as in substrate binding abilities and catalytic properties. In fact, the seven BoNT serotypes are neutralized by different polyclonal antibodies (Hatheway & Ferreira, 1996); moreover, they bind diverse synaptic vesicle receptors (Brunger & Rummel, 2009), and recognize and cleave unique target peptides within one of three distinct SNARE proteins (Rossetto et al., 2013) (Table 1). Very recently, Barash & Arnon (2014) described a novel BoNT/H serotype based on the fact that no type A to G polyclonal antibodies could neutralize its toxicity, whereas complete neutralization was only achieved by using antibodies raised in rabbits challenged with the previously inactivated H toxin. While precise data on the binding and catalytic molecular targets of BoNT/H are not yet available, the authors reported that the catalytic domain of BoNT/H substantially differed from those of the previously known BoNT serotypes, although retaining the conserved zinc endopeptidase motif (Dover et al., 2014a). Among all BoNT serotypes, BoNT/A, /B, /E and /F are more frequently implicated in human botulism worldwide, while BoNT/C and /D mainly affect animals (Peck, 2009). So far, BoNT/G has never been proven to have caused human or animal botulism. The novel BoNT/H was found to be produced together with BoNT/B by a *C. botulinum* strain isolated from an infant with botulism (Barash & Arnon, 2014).

Based on epidemiological data on human botulism, BoNT/A and, to a lesser extent, BoNT/B are the only serotypes currently used for therapy. However, the possibility to exploit BoNT serotypes other than A and B for treating non-responder patients has been considered (Eleopra et al., 2009). Furthermore, the potential of modifying the BoNT binding and catalytic domains through genetic engineering has been explored, with the purpose to expand the number of therapeutic applications and/or to modify their duration of action (Pickett & Perrow, 2011).

Table 1. Cellular binding receptors and catalytic targets of the different BoNT serotypes

BoNT serotype	Synaptic vesicle receptors		Catalytic target protein	Target peptide and cleavage site
	Protein[1]	ganglioside		
A	SV2 A/B/C	GD1a, Gd1b, GT1b, GQ1b	SNAP-25	Glu-Ala-Asn-Gln /Arg-Ala-Thr-Lys
B	Syn I/II	GD1a, GD1b, GT1b	VAMP	Gly-Ala-Ser-Gln / Phe-Glu-Thr-Ser
C	SV2 A/B/C	GD1a, Gd1b, GT1b	SNAP-25	Ala-Asn-Gln-Arg / Ala-Thr-Lys-Met
			SYNTAXIN	Asp-Thr-Lys-Lys / Ala-Val-Lys-Phe
D	SV2 A/B/C	GD1b, GT1b	VAMP	Arg-Asp-Gln-Lys / Leu-Ser-Glu-Leu
E	SV2 A/B/C	GD1a, GD1b, GQ1b	SNAP-25	Gln-Ile-Asp-Arg / Ile-Met-Glu-Lys
F	SV2 A/B/C	GD1b	VAMP	Glu-Arg-Asp-Gln / Lys-Leu-Ser-Glu
F5			VAMP	Asp-Lys-Val-Leu / Glu-Arg-Asp-Gln
G	Syn I/II	GT1b, GD1b, GM1a, GD3	VAMP	Glu-Thr-Ser-Ala / Ala-Lys-Leu-Lys
H	n.a[2]	n.a.[2]	n.a.[2]	n.a.[2]

[1]SV2: Synaptic vesicle protein 2; Syn: synaptotagmin.
[2] n.a.: not available.

Within the BoNT serotypes A, B, E and F, subtypes exhibiting > 2.6% amino acid sequence variation have been established in order to generate diagnostic and therapeutic more specific antibodies (Smith et al., 2005).

To date, five different subtypes have been described for BoNT/A, eight for BoNT/B, nine for BoNT/E and seven for BoNT/F (Hill & Smith, 2013; Wangroongsarb et al., 2014), but the number of new BoNT subtypes is constantly increasing. Chimeric C/D and D/C variants also exist and they are considered as subtypes of BoNT/C and BoNT/D (Moriishi et al., 1996).

The BoNT subtypes might display different binding abilities to monoclonal and even polyclonal antibodies (Kalb et al., 2012a). In addition, some subtypes of BoNT/A and BoNT/F also differ in the binding affinities for their respective substrates (Kalb et al., 2011; Wang et al., 2013). Notably, the BoNT/F5 subtype has been shown to cut its target protein (i.e., the synaptobrevin-2) at a cleavage site that differs from the cleavage site recognized by the other BoNT/F subtypes (Kalb et al., 2012b) (Table 1). Recently, some BoNT/A subtypes have also been revealed to exert markedly different toxicities and to cause dissimilar symptoms in mice injected intraperitoneally and intravenously, respectively; furthermore, the activity levels of the BoNT/A subtypes varied when tested in several neuronal cell models, and their in vitro endopeptidase efficiencies also differed (Whitemarsh et al., 2013).

Although these data are currently available only for some BoNT/A and BoNT/F subtypes, it is expected that the amino acid sequence diversity of other BoNT subtypes can similarly affect their functionality and potency, besides modifying their antigenic epitopes, thus impacting both the production of specific antibodies against BoNTs and the development of new BoNT pharmaceuticals.

Nevertheless, given the general structure and function maintenance among all different BoNT serotypes and subtypes, it appears that their amino acid sequence divergence has evolved within a conserved framework. In fact, all antigenic variants of BoNT have retained the zinc-protease activity and attack the same eukaryotic complex involved in neuroexocytosis, even though at different sites, suggesting that they have derived from a common ancestral enzymatic toxin. The diversification among the BoNT serotypes and subtypes might have resulted from the need to counteract the host defences and/or to expand the range of eukaryotic hosts.

The Highly Heterogeneous Botulinum Neurotoxigenic Clostridia

BoNT is produced by Gram-positive, anaerobic, rod-shaped and spore-forming bacteria of the genus Clostridium that are common in the environment owing to their ability to adapt to a variety of niches (Hatheway, 1993). Under permissive conditions, these microorganisms multiply and synthesize BoNT as an inactive single polypeptide chain non-covalently linked to non-toxic accessory proteins that physically protect it (Gu & Jin, 2013): in fact, the BoNT complexes thus assembled may be released by the neurotoxigenic clostridia in terrestrial and aquatic environments, or in an improperly preserved contaminated food, or directly within the infected host intestine or wound (Figure 1), i.e. under all conditions where the protein toxin alone could easily be degraded. For instance, in the human gastrointestinal tract the non-toxic accessory proteins protect BoNT from low pH values and proteolytic cleavage; only at the alkaline pH values of the intestine, BoNT spontaneously dissociates from the accessory proteins of the complexes (Eisele et al., 2011) and is then proteolitycally cleaved to generate the active di-chain form, before entering the host systemic circulation (Simpson, 2013).

Most BoNT-producing clostridia strains have conventionally been classified within the *Clostridium botulinum* species and synthesize one BoNT serotype which defines the strain (e.g., *C. botulinum* type A, or type B, etc.); sometimes, the BoNT subtype produced by the strain is also specified as an Arabic number (e.g., *C. botulinum* type A1, or type A2, etc.). Certain rare *C. botulinum* strains can concomitantly produce two different BoNT serotypes, one in major amounts and the other one in minor amounts, that are indicated with an uppercase letter and a lowercase letter, respectively: the dual-toxin producing strains that have been identified so far are *C. botulinum* of types Ab, Ba, Af, Bf (Hatheway, 1993), and the most recently described type Bh (Barash & Arnon, 2014).

Given the high heterogeneity of the *C. botulinum* strains, they have been separated in three (I to III) physiologically and phylogenetically distinct Groups (Table 2). Group I includes all the *C. botulinum* strains that produce BoNT/A, and all known dual-toxin producing *C. botulinum* strains; Group II includes all *C. botulinum* type E strains; Group III includes all *C. botulinum* type C and D strains. The *C. botulinum* strains belonging to Groups I and II are also referred to as proteolytic and non-proteolytic strains, respectively, based on their ability to digest proteins; notably, the *C. botulinum* type B and type F strains can belong to either Group I or Group II (Hatheway, 1993).

Three (IV to VI) additional clostridia Groups include the botulinum neurotoxigenic clostridia assigned to species other than *C. botulinum* (Table 2). Of these, Group IV contains the BoNT/G-producing *C. argentinense* species that was early distinguished from the *C. botulinum* species because of the unique phenotypic, physiological and genetic characteristics of its members (Suen et al., 1988a); Groups V and VI comprise rare strains belonging to the *C. baratii* and *C. butyricum* species that are atypically capable of producing BoNT/F and BoNT/E, respectively (Suen et al., 1988b).

Remarkably, all six Groups also include closely related clostridia that do not synthesize BoNT (Table 2) (Hatheway, 1993).

Taxonomic clostridia Groups have been confirmed by a number of genetic approaches, including DNA/DNA homology (Lee & Riemann, 1970), 16S rRNA sequencing (Collins & East, 1998; Hill et al., 2007), pulsed-field gel electrophoresis (PFGE) (Hielm et al., 1998; Nevas et al., 2005), randomly-amplified polymorphic DNA analysis and repetitive element sequence-based PCR (Hyytiä et al., 1999), amplified fragment length polymorphism (AFLP) (Keto-Timonen et al., 2005), flagellin typing (Paul et al., 2007), and genome-wide comparative analyses (Carter et al., 2009; Hill et al., 2009; Stringer et al., 2013). All genetic studies support that each Group taxonomically corresponds to a distinct species, while currently different clostridia species are misleadingly included in each Group (Table 2). Although the proposals to reorganize the intricate taxonomy of the BoNT-producing clostridia and their non neurotoxigenic clostridia relatives are multiplying (Skarin & Segerman, 2011; Bradbury et al., 2012), it is unlikely that the *C. botulinum* designation will be abandoned.

Nevertheless, it is from taxonomy-related observations, such as evidences that clostridia strains of the same Group may or may not produce BoNT, or that strains of different Groups can produce the same BoNT serotype, that the ability to produce BoNT was initially suggested to be genetically transferable among clostridia. Further studies focusing on the genes coding for BoNTs and their flanking regions, and on their genomic locations within potentially mobile genetic elements, including plasmids, bacteriophages and remnants of transposons, have increasingly supported those early intuitions, as outlined in the following sections.

Table 2. Clostridia Groups and BoNT-encoding plasmids

Clostridia Group	*Clostridium spp.*	BoNT serotypes	BoNT subtypes	Plasmid-borne *bont* gene[4]
I (proteolytic)	BoNT-producing spp *C. botulinum*	A	A1, A2, A3, A4[1], A5	A2, A3, A4
		B	B1, B2, B3, B5[1,2], B6, B7, B8	B1, B2, B3, B5
		F	F1, F2[2], F3, F4[3], F5[3]	F2, F5
	Non- BoNT-producing spp *C. sporogenes*			
II (nonproteolytic)	BoNT-producing spp *C. botulinum*	B	B4	B4
		E	E1, E2, E3, E6, E7, E8, E9	E1
		F	F6	
	Non-BoNT-producing spp *undetermined*			
III	BoNT-producing spp *C. botulinum*	C	C, C/D, D/C	
		D	D, C/D, D/C	
	Non-BoNT-producing spp *C. novyi, C. haemolyticum*			
IV	BoNT-producing spp *C. argentinense*	G		G
	Non-BoNT-producing spp *C. subterminale, C. proteolyticus, C. schirmacherense*			
V	BoNT-producing spp *C. baratii* type F	F	F7	
	Non-BoNT-producing spp *C. baratii*			
VI	BoNT-producing spp *C. butyricum* type E	E	E4, E5	
	Non-BoNT-producing spp *C. butyricum*			

[1]BoNT/A4 and BoNT/B5 are concomitantly produced by the dual-toxin producing *C. botulinum* type Ba strain Ba657 (Hill & Smith, 2013)

[2]BoNT/F2 and BoNT/B5 are concomitantly produced by the dual-toxin producing *C. botulinum* type Bf (Raphael et al., 2010)

[3]BoNT/F4 and BoNT/F5 can be produced either alone or simultaneously by *C. botulinum* type Af strains (Dover et al., 2013; Raphael et al., 2014)

[4]*bont* subtype genes for which plasmid location has been reported (see text).

BoNT Genes and BoNT Gene Clusters

Each BoNT is encoded by a gene of about 3.8 kb. The *bont* genes coding for the different BoNT serotypes vary by about 25 to 45% at the nucleotide level, with the lowest nucleotide diversity observed between the *bont*/e and *bont*/f gene sequences, and the highest nucleotide diversity observed between the *bont*/a and *bont*/c gene sequences, respectively (Hill & Smith, 2013); within each serotype, the highest (24.1%) nucleotide divergence has been reported among the *bont/f* gene subtype sequences (Raphael et al., 2010).

Most BoNT-producing clostridia strains have a single *bont* gene in their genome and therefore only form the individual corresponding BoNT. The rare dual-toxin producing *C. botulinum* strains generally possess two different *bont* genes that are both expressed, although at different levels. Notably, the *C. botulinum* strain Af84, previously known to produce higher amounts of BoNT/A in addition to minor amounts of BoNT/F, has recently been shown to carry three distinct full-length *bont* genes: a *bont/a2* gene, a *bont*/f4 gene and a *bont/f5* gene (Dover et al., 2013). All three genes are functional: in fact, strain Af84 forms active BoNT/A2, BoNT/F5 and BoNT/F4 in a 88:11:1 ratio, respectively, as lately demonstrated by mass spectrometry (Kalb et al., 2014).

C. botulinum strains also exist that harbour a *bont/a* gene and a *bont/b* gene in their genome, but express only BoNT/A. These strains, designated as *C. botulinum* type A(B), appear to be more widespread than the rare dual-toxin producing strains (Franciosa et al., 1994; Cordoba et al., 1995; Hill et al., 2007; Wangroongsarb et al., 2014). Sequence analyses of the *bont/b* gene from some *C. botulinum* type A(B) strains have shown that the gene remains silent either because of point mutations that create premature stop codons leading to truncated inactive proteins (Hutson et al., 1996; Rodríguez Jovita et al., 1998), or because it lacks a large portion of the full-length *bont/b* gene sequence (Dover et al., 2009).

The *bont* genes lie at the 3' end of conserved clusters of genes of approximately 12-15 kb that include, among others, the genes coding for variable components of the BoNT complexes (Hill & Smith, 2013) (Figure 2). For instance, all *bont* gene clusters contain an *ntnh* gene located upstream of the *bont* gene and oriented in the same direction: both *bont* and *ntnh* genes are co-transcribed and form a single operon. The *ntnh* gene codes for a non-toxic non-haemagglutinin (NTNH) protein that has been shown to protect BoNT from degradation by adopting an architectural conformation very similar to that of the toxin in order to firmly bind and mask its more exposed molecular regions (Gu et al., 2012).

Based on the genes located upstream of the *ntnh* gene, two main *bont* gene clusters have been distinguished: the *ha* clusters and the *orfX* clusters (Figure 2).

In the *ha* clusters, three *ha* genes (*ha*33, *ha*17 and *ha*70) lay upstream of the *ntnh* gene; they code for as many hemagglutinins of different molecular weights that make part of the BoNT complexes in variable combinations and possibly enhance the BoNT rate of absorption across the intestinal epithelial cells, besides protecting it in the digestive tract (Fujinaga, 2010). A *bot*R gene coding for an alternative RNA polymerase sigma factor that positively regulates the *bont* gene transcription (Connan et al., 2013) can be located upstream either of the *ntnh* gene or of the *ha* genes, and in both cases is oriented in the same direction of the *bont* and *ntnh* genes (Figure 2).

Upstream of the *ntnh* gene and equally oriented, the *orfX* clusters have a *p47* gene whose role is still undefined. Upstream of the *p47* gene, the *orfX* clusters have three open reading frames (*orfX*1, *orfX*2 and *orfX*3 genes) that are oriented in an inverted direction: the functions of the *orf*X genes are also unknown. In addition, all *orf*X clusters, except for those containing the *bont*/e gene in all *C. botulinum* type E and *C. butyricum* type E strains, and those containing the *bont*/f gene in non-proteolytic *C. botulinum* type F and in neurotoxigenic *C. baratii* type F strains (Kubota et al., 1998; Dover et al., 2011; Dover et al., 2014b), possess a *botR* gene located between the *p47* and the *orfX1* genes and oriented in the same direction of the *orfX* genes (Figure 2).

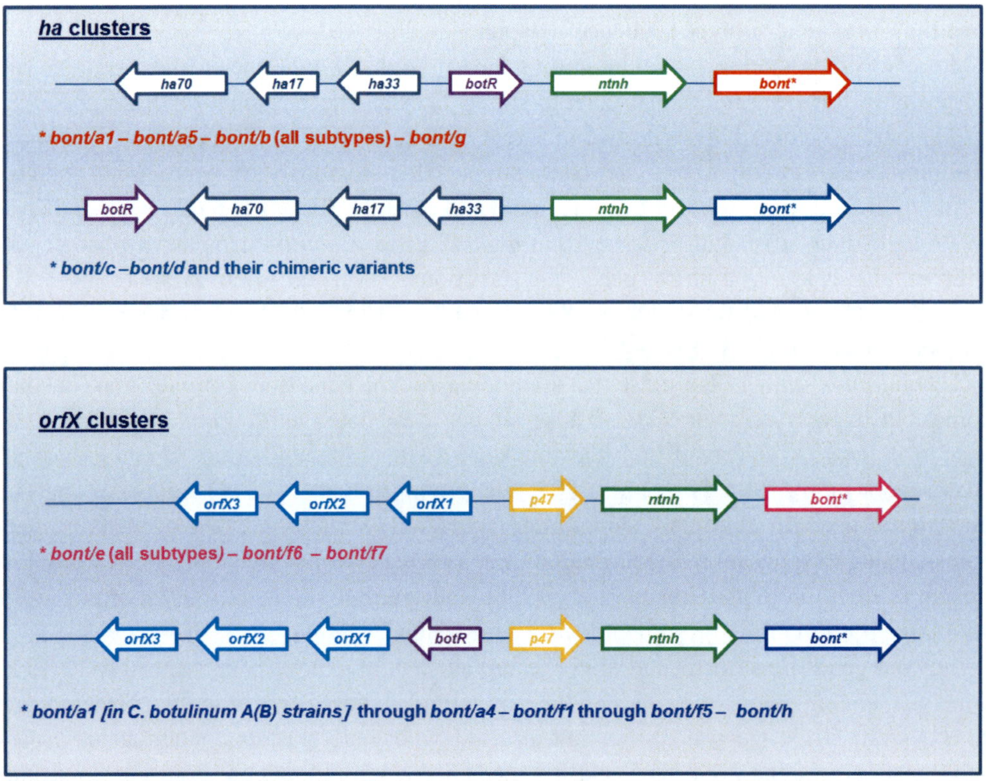

Figure 2. Schematic arrangements of the *ha* and *orf*X gene clusters of botulinum neurotoxigenic clostridia.

The *bont*/*b*, *bont*/*c*, *bont*/*d*, and *bont*/*g* genes are constantly present within an *ha* cluster; conversely, the *bont*/*e* and *bont*/*f* genes, as well as the novel *bont/h* gene recently identified in the *C. botulinum* type Bh strain, have so far been reported within an *orfX* cluster (Hill & Smith, et al., 2013; Dover et al., 2014). The different *bont/a* subtype genes can be located in a *ha* cluster (*bont*/*a1* and *bont*/*a5*) or in an *orfX* cluster (*bont*/*a1*, *bont*/*a2*, *bont*/*a3*, and *bont*/*a4*) (Hill & Smith, 2013; Dover et al., 2009). Hence, except for the *bont/a1* subtype gene that can reside in either *ha* or *orfX* clusters (Franciosa et al., 2004; Raphael et al., 2008), the associations between the other *bont* subtype genes and the *ha* or the *orfX* clusters appear to be conserved. It is worth noting that the *C. botulinum* strains possessing two *bont* genes usually contain one of them in the *ha* cluster and the other one in the *orfX* cluster, depending on the *bont* genes and consistently with the above reported associations: one notable exception is *C. botulinum* strain Af84 that carries all three *bont/a2*, *bont/f4*, and *bont/f5* genes within separate *orfX* clusters (Dover et al., 2013).

The overall conservation of the gene arrangement in the *ha* and *orfX* genetic clusters indicates that they have a common origin; however, the different associations between the distinct *bont* genes and either *ha* or *orf*X clusters suggest that the clusters have undergone divergent evolutionary pathways. Since chimeric sequences likely resulting from recombination and/or insertion events have been identified in both the *ntnh* and *bont* gene sequences of several *C. botulinum* strains, these genes have been proposed to be possible hot spots for the genetic rearrangements responsible of the *bont* gene clusters diversity (East et al., 1996; Moriishi et al., 1996; Hill et al., 2007).

Both *ha* and *orfX* genetic clusters contain partial or complete insertion sequence (IS) elements, i.e. the simplest known transposable genetic elements, at one or both ends, indicating possible mobile element-mediated horizontal gene transfer. Furthermore, a prophage gene (*lyc*A) coding for autolytic lysozyme (1,4-β-N-acetylmuramidase) is present at one end of some *bont* gene clusters, indicating that phage-mediated horizontal gene transfer might also have occurred during evolution (Popoff & Bouvet, 2013).

Thus, both evidence of genetic recombination and the presence of remnants of potentially mobile genetic elements flanking the *bont* gene clusters clearly indicate that multiple genetic mechanisms have likely contributed to their genetic diversity and to their variable positioning within the genomes of botulinum neurotoxigenic clostridia.

Comparative genomic analyses of these clostridia further support that the *bont* gene clusters have evolved independently in different genomic backgrounds (Jacobson et al., 2008; Hill et al., 2009; Carter et al., 2009; Fillo et al., 2011).

Locations of the BoNT Gene Clusters within the Genomes of Botulinum Neurotoxigenic Clostridia

The *bont* gene clusters are located either on the chromosome (clostridia Groups I, II, VI), a bacteriophage (clostridia Group III), or a plasmid (clostridia Groups I, II, IV). The location of the *bont*/f gene cluster in the genomes of Group V neurotoxigenic *C. baratii* type F strains is still undetermined.

1. Chromosome

In 2007, the first complete genome sequence of a representative Group I (proteolytic) *C. botulinum* type A strain (Hall, ATCC 3502) became available (Sebahia et al., 2007): it consisted of a chromosome of ~3.9 Mb and a plasmid of ~ 16 kb. The presence of a single *bont*/a1 gene within a *ha* cluster was confirmed in the chromosome of strain Hall 3502.

The increasing availability of genome sequences from BoNT-producing clostridia strains has revealed that the *bont* gene clusters are not randomly located in the chromosome, but they rather reside in one of the following fixed sites: i) the arsenic resistance (*ars*C) operon, that is associated to the *orfX* clusters containing either *bont*/a1 [in *C. botulinum* type A(B) strains], *bont*/a2 or *bont*/f genes; ii) the oligopeptide permease / branched chain amino acid transporter (*opp*A / *brn*Q) operon, that is associated to the *ha* clusters containing either *bont*/a1 or *bont*/b genes; iii) the type II secretion system protein E (*pul*E) operon, that was recently found to harbour the *orfX* cluster containing the *bont*/f4 gene in *C. botulinum* strain Af84; or iv) the resolvase (*rar*A) operon, that is associated to the *orfX* cluster containing the *bont*/e gene in both *C. botulinum* type E strains and neurotoxigenic *C. butyricum* type E strains (Hill et al., 2009; Dover et al., 2013). The latter *rarA* operon encodes a resolvase, i.e. a transposon-associated enzyme involved in the excision of transposons: therefore, it is conceivable that a pathogenicity island containing the *bont*/e gene cluster is responsible for the presence of *bont*/e in the two different *C. botulinum* and *C. butyricum* species.

Interestingly, two intact and almost identical transposase genes have been found to flank a large ~55 kb genetic region that includes the *orfX* cluster containing the *bont*/h gene in the chromosome of the recently described *C. botulinum* type Bh strain: sequence alignments indicate that the whole 55 kb genetic region is likely inserted into the chromosome by replacing a gene coding for an AraC family transcriptional regulator (Dover et al., 2014a). This observation implies that the *bont/*h gene cluster probably has been transferred within a pathogenicity island.

2. Bacteriophages

The *bont*/c and *bont*/d genes of Group III *C. botulinum* strains have been proven to be carried by distinct bacteriophages through experiments of phage curing and phage conversion, during the early 1970s: the non-neurotoxigenic strains resulting from phage curing could regain the ability to produce BoNT after infection with the bacteriophages induced from the parent strains. Interestingly, once the *C. botulinum* strains that originally produced BoNT/C or BoNT/D were cured of their phages, they could be re-infected by either BoNT/C- or BoNT/D-encoding phages, a phenomenon referred to as inter-conversion (Eklund et al., 1971; Inoue & Iida, 1971; Eklund et al., 1974). The bacteriophages have been isolated and characterized; their unstable lysogeny (also termed pseudolysogeny), i.e. the fact that they can easily be lost by Group III *C. botulinum* strains along with the ability to produce BoNT, has also historically been known, and has been attributed to the lack of integration of the bacteriophage into the host chromosome (Oguma, 1976; Eklund et al., 1989; Johnson, 2005).

More recently, the nucleotide sequences of an isolated BoNT/C-encoding bacteriophage and the whole genomic sequences of some Group III *C. botulinum* type C and D strains have been reported (Sakaguchi et al., 2005; Skarin et al., 2011; Hassan et al., 2013). The sequence data confirm that the bacteriophages harbouring the *bont*/c or the *bont*/d genes are not stably maintained as prophages into the host chromosome, as expected for temperate bacteriophages, but rather exist as circular extrachromosomal replicons of ~180–200 kb during lysogeny. Such BoNT-encoding plasmid-like prophages probably lack some specific factors required for their stable inheritance, which has been proposed to be the main cause of pseudolysogeny.

3. Plasmids

Eklund et al., (1988) first provided evidence that BoNT/G production by Group IV *C. argentinense* strains was related to the presence of a large plasmid**:** when those strains were cured of their plasmid, they ceased to produce BoNT/G. Zhou et al., (1995) later confirmed that the *bont/G* gene cluster was located in a 76 MDa plasmid in Group IV *C. argentinense* strains, by means of PFGE and Southern hybridization experiments using probes specific for the *bont*/g, *ha*, and *ntnh* genes.

Despite those findings, further advances in the area of the BoNT-encoding plasmids research were delayed because of the general assumption that the *bont* genes other that *bont/*c, /d and /g resided in the clostridia chromosome.

However, a fundamental change occurred in 2007 when Marshall et al., conducted a study to investigate the reason why the *C. botulinum* strains Loch Maree and Ba657, i.e. the only strains to date known to form BoNT/A3 and BoNT/A4, respectively, produced considerably lower amounts of BoNT/A compared to most *C. botulinum* type A strains. The initial hypothesis was that the *bont*/a gene of both strains might be located in an unstable bacteriophage similar to those encoding BoNT/C or BoNT/D in Group III *C. botulinum* strains: in fact, the loss of such a bacteriophage by part of the multiplying clostridia population could likely result in decreased toxin production. Using mitomycin C, a DNA-damaging compound, one and two bacteriophages were induced from the *C. botulinum* strains Loch Maree and Ba657, respectively: however, the induced bacteriophages were shown to lack any *bont* genes by PFGE and Southern hybridization with specific *bont* gene probes. Unexpectedly, the results revealed that both strains harboured the *bont*/a gene within a large (~ 280 kb) extrachromosomal genetic element that was deduced to be a plasmid because it lacked the essential ribosomal RNA genes typically found in the bacterial chromosomes. In addition, the *bont*/b5 gene of strain Ba657 was shown to reside within the same plasmid that carried the *bont*/a4 gene (Marshall et al., 2007).

Smith et al., (2007) independently reached similar conclusions by determining the whole genome sequence of the *C. botulinum* strains Loch Maree and Ba657. Specifically, a 267 kb plasmid (pCLK) harbouring the complete *bont*/a3 gene cluster was identified in the genome of strain Loch Maree, and a 270 kb plasmid (pCLJ) carrying both the *bont*/a4 gene cluster and the *bont*/b5 gene cluster, located at a distance of ~97 kb from each other, was revealed in the genome of strain Ba657: the distance between the two toxin gene clusters and the different location and orientation of the related *bot*R regulatory genes were suggested to account for the independent expression of BoNT/A and BoNT/B in strain Ba657. In the same study, the whole genome sequence of the *C. botulinum* strain OkraB was also obtained, and a new ~149 kb plasmid (pCLD) harbouring the *bont*/b1 gene cluster was identified in the genome of strain OkraB. The three BoNT-encoding plasmids pCLK (Loch Maree), pCLJ (Ba657), and pCLD (OkraB) were predicted to have a circular structure based on start-end overlapping sequences. Surprisingly, they all shared a high level of synteny (i.e., organization of the genetic loci) and large regions of sequence identity, suggesting evolutionary interrelationships. In contrast, plasmids pCLK, pCLJ, and pCLD did not show any similarities to the toxin-encoding plasmids of other clostridia *spp*, such as those of *C. tetani* or *C. perfringens,* nor to the bacteriophages carrying the *bont*/c gene or the *bont*/d gene in Group III *C. botulinum* strains.

The two above mentioned studies provided the first genetic evidences that the *bont*/a gene and the *bont*/b gene could be plasmid-borne in some *C. botulinum* strains, and that a single plasmid could simultaneously carry both the *bont*/a gene and the *bont*/b gene, as in strain Ba657.

The significant discovery of the novel BoNT-encoding plasmids in three peculiar *C. botulinum* strains prompted further research to investigate the genomic location of the *bont* genes in a larger number of *C. botulinum* strains.

Studies by our laboratory resulted in the surprising finding that the *bont*/b gene was plasmid-borne among *C. botulinum* strains much more frequently than previously known

(Franciosa et al., 2009). In fact, the PFGE and Southern blot analysis of 63 total *C. botulinum* type B, Ab, and A(B) strains isolated from different clinical and food sources in Italy and the United States in different periods of time, revealed that > 50% of the tested strains (specifically, 32 of 63 strains) harboured the *bont*/b gene within a plasmid. Of the strains with plasmid-borne *bont*/B gene, 29/54 were *C. botulinum* type B, and 3/4 were *C. botulinum* type Ab; notably, the *bont*/a gene of the latter three strains resided within the same plasmid containing the *bont*/b gene, similarly to the *bont*/a gene and the *bont*/b gene in plasmid pCLJ of *C. botulinum* Ba657. Conversely, no BoNT-encoding plasmid was detected in the 5 *C. botulinum* type A(B) strains included in the study.

Five *bont*/B subtype genes known at that time were identified among the strains, i.e. *bont*/b1, *bont*/b2, *bont*/b3, (non-proteolytic) *bont*/b4 and (bivalent) *bont*/b5. Except for (non-proteolytic) *bont*/b4 that was plasmid-located in the only two non-proteolytic *C. botulinum* type B strains included in the study, one of which was strain Eklund 17B, all other *bont*/b subtype genes resided on either plasmid or chromosome, with different frequencies: for instance, *bont*/b1 was most frequently plasmid-located (21/22 strains), while *bont*/b2 was prevalently chromosome-located (20/21 strains) (Table 3).

Table 3. Distribution of plasmid-borne *bont*/b subtype genes among 60 PFGE typeable *C. botulinum* strains (Franciosa et al., 2009)

bont/B subtype gene[1]	No. of strains with a plasmid-borne *bont*/b subtype gene / total No. of strains carrying the *bont*/b subtype gene	Plasmid size(s)[4] (No. of strains)
B1	21/22	245 kb (16) 217 kb (1) 139 kb (4)
B2	1/21	245 kb (1)
B3	1/2	245 kb (1)
B4 (np)[2]	2/2	55 kb (2)
B5 (bv)[3]	7/13	245 (5) 170 (2)

[1] *bont*/B subtype gene determined by PCR restriction fragment length polymorphisms (RFLP).
[2] np = nonproteolytic.
[3] bv = bivalent.
[4] The plasmid sizes were deduced by comparison with a molecular standard in PFGE experiments.

The newly identified BoNT-encoding plasmids were assumed to be circular, based on the recognition by PFGE of the structural isoforms typical of circular DNA molecules, i.e. closed supercoiled and open circular isoforms. Following enzymatic linearization, five plasmid sizes were estimated by comparison with a molecular marker, i.e. ~ 55 kb, 139 kb, 170 kb, 217 kb, and 245 kb; a *C. botulinum* strain could only carry a BoNT encoding plasmid of a single size. The largest (245 kb) plasmids prevailed, since they were identified in 24 of 32 plasmid-bearing strains, whereas the smallest (55 kb) plasmid was detected only in the 2 Group II (non-proteolytic) *C. botulinum* type B strains tested in the study. Among the different *bont*/b subtype genes, *bont/*b1 was localized within three differently sized plasmids of 245 kb, 217 kb and 139 kb, respectively; *bont*/b2 and *bont*/b3 were only found within plasmids of 245 kb; non-proteolytic *bont*/b4 was localized in the smallest (55 kb) plasmid; bivalent *bont*/b5 was localized within two differently sized plasmids of 245 kb and 170 kb, respectively (Table 3). The size (245 kb) of the plasmids carrying both *bont*/a and *bont*/b genes in the *C. botulinum* type Ab strains was similar to the size (270 kb) of plasmid pCLJ of *C. botulinum* Ba657.

Although no relationship between carriage of the BoNT-encoding plasmids and the clinical origin, food source or isolation date of the *C. botulinum* strains was evident, a geographic link could be established: in fact, plasmid-borne *bont*/b1 predominated among the *C. botulinum* strains from the United States, whereas chromosome-encoded *bont*/b2 prevailed among the *C. botulinum* strains from Italy, supporting that *C. botulinum* strains from diverse geographic areas have likely undergone different evolutionary pathways.

Plasmid-borne *bont*/b genes were confirmed in *C. botulinum* type B strains by a number of studies. Using PFGE and Southern blot analysis, Umeda et al., (2009) found that 5 of 15 Group I (proteolytic) *C. botulinum* type B strains, including strain OkraB, three strains from as many infant botulism cases occurred in Japan and the United States, respectively, and a food strain, carried their *bont*/b gene in a plasmid. Four plasmid sizes (i.e., approximately 150 kb, 260 kb, 275 kb and 280 kb) were estimated: the largest (280 kb) plasmid carried a newly identified *bont*/b6 subtype gene.

Hill et al. (2009) reported the partial genome sequence of a *C. botulinum* type Bf strain and the complete genome sequence of the Group II (nonproteolytic) *C. botulinum* type B strain Eklund 17B. Four contigs of the *C. botulinum* type Bf draft genome sequence exhibited extensive synteny with plasmid pCLJ of the *C. botulinum* strain Ba657, and high sequence homology to all three previously sequenced plasmids pCLJ (Ba657), pCLK (Loch Maree), and pCLD (OkraB): on these bases, the four contigs (170 kb in total) of the *C. botulinum* type Bf genome were deduced to correspond to part of a larger plasmid (pBf) harbouring the *bont*/b and *bont*/f genes.

Besides, the presence of a circular plasmid (pCLL, 48 kb) harbouring the nonproteolytic *bont*/b4 gene was confirmed in the genome sequence of the Group II (nonproteolytic) *C. botulinum* strain Eklund 17B: unlike plasmid pBf, plasmid pCLL of strain Eklund 17B was unique since it showed neither synteny nor sequence similarity to the other sequenced BoNT-encoding plasmids of Group I strains.

Stringer et al., (2013) carried out a comparative genomic analysis of 43 Group II (non-proteolytic) *C. botulinum* strains, including 14 type B, 24 type E, and 5 type F strains, by using DNA microarray. Fifty oligonucleotide probes of the microarray slide were designed from the coding sequences of plasmid pCLL (non-proteolytic Eklund 17B), including those of the *bont*/b gene cluster. The genomic DNAs of 3 of the 14 *C. botulinum* type B strains yielded hybridization signals with the pCLL-specific probes when tested by the microarray, indicating that they possessed the pCLL coding sequences in their genome. Another non-proteolytic *C. botulinum* type B strain which initially produced BoNT/B and had subsequently ceased to produce the toxin, failed to yield positive hybridization signals with the pCLL-specific probes: based on these observations, the authors concluded that a plasmid containing the *bont*/b gene had probably been lost by that strain.

Very recently, Wangroongsarb et al., (2014) reported that a *C. botulinum* type B strain implicated in a large foodborne botulism outbreak in Thailand harboured a 275 kb plasmid containing a novel *bont*/b subtype gene, designated as *bont*/b8.

While the above studies mainly focused on the genomic location of the *bont*/b gene, Zhang et al., (2013) analyzed the *bont*/e gene location in the genome of 36 *C. botulinum* type E strains isolated from fish products and aquatic environment in Finland and other geographic areas. By PFGE and Southern hybridization experiments the authors showed that 3 of the total tested 36 *C. botulinum* type E strains possessed an approximately 146 kb plasmid harbouring the *bont*/e1 gene.

As mentioned above, Dover et al., (2013) recently reported the draft genome sequence of *C. botulinum* strain Af84: the sequence data revealed that strain Af84 harboured three distinct *orf*X gene clusters, containing the *bont*/a2, *bont*/f4 and *bont*/f5 genes, respectively. Of these, the *orf*X gene cluster with the *bont*/f5 gene was identified within a 246 kb contig that showed extensive synteny and sequence homology to the sequenced plasmids pCLJ (Ba657), pCLK (Loch Maree), and pCLD (OkraB); moreover, the location of the *orf*X gene cluster containing the *bont*/f5 gene within the 246 kb contig was very similar to those of the *orf*X gene clusters containing the *bont*/a3 gene or the *bont*/a4 gene in plasmids pCLK (Loch Maree) or pCLJ (Ba657), respectively. Based on these observations, and because the ends of the 246 kb sequence contig of strain Af84 overlapped, it was concluded that the contig likely corresponded to a circular plasmid, designated as pCLQ.

Following the finding of plasmid-borne *bont*/F5 gene in strain Af84, Raphael et al., (2014) recently screened a panel of 15 Group I *C. botulinum* type F and Af strains previously shown to possess either the *bont*/f4 gene (8 strains) or the *bont*/f5 gene (7 strains) for the presence of BoNT-encoding plasmids. A PCR assay was designed in order to detect a plasmid-specific marker, i.e. a region of a putative primase gene that is conserved in all sequenced Group I BoNT-encoding plasmids, i.e. pCLJ (Ba657), pCLK (Loch Maree), pCLD (OkraB) and pCLQ (Af84). The 8 *C. botulinum* strains with the *bont*/f4 gene failed to amplify the plasmid-specific marker; conversely, the plasmid-specific marker was PCR amplified from all 7 strains possessing the *bont*/f5 gene, indicating that they likely harboured a BoNT/F5-encoding plasmid. Presence in the 7 strains of a ~ 242 kb plasmid containing the *bont*/f5 gene was demonstrated by PFGE and Southern hybridization with a *bont*/f5 specific gene probe. Three of the 7 strains with plasmid-borne *bont*/f5 gene were *C. botulinum* type Af strains: the *bont*/a gene of the three strains was located within their chromosome, as revealed by PFGE and Southern blot with a *bont*/a specific gene probe. Finally, the draft genome sequences generated for two of the *C. botulinum* type F5 strains confirmed that both genomes contained genetic regions sharing significant similarity with plasmid pCLQ of strain Af84.

Genetic Features of the Novel BoNT-Encoding Plasmids

The major genetic features of the five BoNT-encoding plasmids with complete nucleotide sequences available to date are listed in Table 4. The average %GC content of plasmids is about 25%, slightly lower than the average 28% GC content of the sequenced *Clostridium botulinum* genomes (http://www.ncbi.nlm.nih.gov/genome/?term=clostridium+botulinum). All plasmids are circular in structure and range in sizes from approximately 48 kb (pCLL of strain Eklund 17B) to 270 kb (pCLJ of strain Ba657). They contain 50 to 329 predicted coding sequences: of these, the most prominent are those coding for the BoNT complexes.

As detailed in the previous section, large regions of sequence homology and a high degree of synteny, including conserved locations of the *bont* gene clusters, have been identified among plasmids of the Group I *C. botulinum* strains, i.e. pCLK (Lock Maree), pCLJ (Ba657), pCLQ (Af84), and pCLD (OkraB) (Smith et al., 2007; Hill et al., 2009; Dover et al., 2013). Specifically, comparative analyses of the four Group I plasmid sequences revealed similar locations of the *orfX* clusters containing either *bont*/a3, *bont*/a4 or *bont*/f5 in plasmids pCLK (Loch Maree), pCLJ (Ba657), or pCLQ (Af84); and similar locations of the *ha* clusters containing either *bont*/b1 or *bont*/b5 in plasmids pCLD (OkraB) or pCLJ (Ba657), respectively. In addition, the analysis of the partial sequence of the BoNT-encoding plasmid pBf of Group I *C. botulinum* strain Bf indicates that both the *orfX* cluster containing the *bont*/f5 gene and the *ha* cluster containing the *bont*/b5 gene share similar locations within the plasmid as those of the related gene clusters in the other Group I plasmids (Hill et al., 2009).

Genetic regions upstream and downstream of the *bont* gene clusters are also conserved in the Group I plasmids. These regions do not display sequence similarity to the regions flanking the *bont* gene clusters in the chromosomes of *C. botulinum* strains, except for the presence of a prophage-related *lyc*A gene close to some *orfX* clusters; however, the *lyc*A gene is located upstream of the *orfX* clusters containing *bont*/a3, *bont*/a4, or *bont*/f5 in plasmids pCLK, pCLJ, pCLQ and pBf, respectively, whereas it is located downstream of the chromosomal *orfX* clusters containing *bont*/a1, *bont*/a2 or *bont*/f1 (Smith et al., 2007; Hill et al., 2009; Dover et al., 2013).

In contrast, there is neither synteny nor sequence similarity between the only sequenced Group II plasmid pCLL (Eklund 17B), which contains a *ha* cluster with *bont*/b4, and the Group I plasmids (Hill et al., 2009; Dover, 2013). A phylogenetic tree (Figure 3) generated from the five available BoNT-encoding plasmid sequences supports that the four Group I plasmids pCLK (Lock Maree), pCLJ (Ba657), pCLQ (Af84), and pCLD (OkraB) cluster together, whereas the Group II plasmid pCLL (Eklund 17B) forms a separate cluster: these data are consistent with the genetic diversity between the clostridia Groups I and II.

Significantly, a partial IS element and genes coding for transposon-associated recombinase and resolvase are located downstream of the *bont*/B4 gene cluster in plasmid pCLL, a clear indication that transposition could have been responsible for recombination/insertion of the *bont*/B gene cluster within this plasmid. Similar mechanisms have been proposed for the insertion of the *bont*/e gene cluster into the *rarA* operon within the chromosomes of *C. botulinum* type E and neurotoxigenic *C. butyricum* type E strains (Hill et al., 2009), and for the insertion of the *bont*/h gene cluster in the chromosome of *C. botulinum* type Bh (Dover et al., 2014a).

The occurrence of several partial or complete IS elements (Table 4), some of which flanking the *bont* gene clusters, in both Group I and Group II plasmids suggests that multiple transposition events leading to genetic rearrangements have likely contributed to the mobilization of the *bont* gene clusters and, more in general, to the overall genetic structure of plasmids, thus confirming the dynamic flexibility of these genetic elements.

While the *bont* gene clusters and their flanking regions have been well characterized in the BoNT-encoding plasmids, being their most significant characteristic, less than 25% of the remaining coding sequences of the plasmids have been assigned a functional role.

Table 4. Main characteristics of the BoNT-encoding plasmids whose nucleotide sequences are available. (http://www.ncbi.nlm.nih.gov/genome/?term=clostridium+botulinum)

Plasmid	Accession Number	*C. botulinum* strain	% GC	Size (bb)	Number of CDSs	*bont* subtype gene(s)	*bont* gene cluster	IS elements/ transposons	Reference
		Group I (proteolytic)							
pCLK	NC_010418.1	Loch Maree	25.6	266,785	329	*bont*/a3	*orf*X	5	Smith et al., 2007
pCLJ	NC_012654.1	657	25.6	270,022	306	*bont*/a4 *bont*/b5 (bivalent)	*orf*X *ha*	13	"
pCLD	NC_010379	OkraB	25.4	148,780	194	*bont*/b1	*ha*	3	"
pCLQ*	AOSX01000021.1	Af84	25.5	264,124	286	*bont*/f5	*orf*X	11	Dover et al., 2013
		Group II (nonproteolytic)							
pCLL	NC_010680.1	Eklund 17B	25	47,642	50	*bont*/b4 (nonproteolytic)	*ha*	1	Hill et al., 2009

*Draft sequence.

The CDSs with annotated functions include both putative metabolic genes and cell surface structural genes. Genes coding for enzymes potentially involved in plasmid replication and segregation, such as DNA polymerase III, primase, helicase, and ATPase have been identified in the Group I plasmids, suggesting that they are capable of self-replicating; however, since neither rolling-circle replication genes nor theta replication genes have been recognized, the exact replication mechanisms of Group I plasmids remain to be determined (Smith et al., 2007). Similar genes have not been recognized in the only sequenced Group II pCLL plasmid, which makes the Group II plasmids replication and partitioning even more obscure.

A

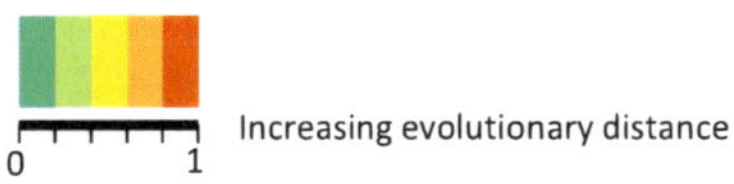

	pCLJ	pCLQ	pCLK	pCLD	pCLL
pCLJ	0,0000	0,0315	0,0448	0,4987	0,5181
pCLQ	0,0315	0,0000	0,0372	0,4996	0,5168
pCLK	0,0448	0,0372	0,0000	0,4992	0,5172
pCLD	0,4987	0,4996	0,4992	0,0000	0,6032
pCLL	0,5181	0,5168	0,5172	0,6032	0,0000

B

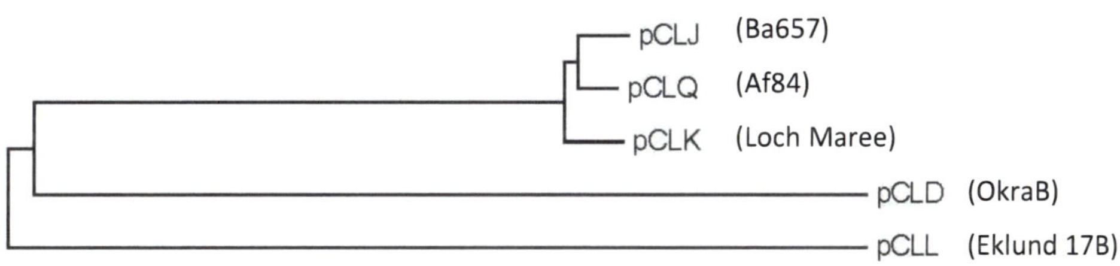

Figure 3. Phylogenomic analysis of the BoNT-encoding plasmids with known nucleotide sequences. A) Evolutionary distances between sequences performed with MEGA 5.2 by computing the proportion of nucleotide differences between each pair of sequences. B) Phylogenetic tree constructed with the neighbor joining method (NJ) using MEGA 5.2.

Conjugative Transfer of the BoNT-Encoding Plasmids

The conjugative transfer of two Group I plasmids and one Group II plasmid, respectively, to proteolytic *C. botulinum* strains has been demonstrated by Marshall et al., 2010.

The Group I plasmids used in the study were pCLJ (Ba657) and pBotCDC-A3, the latter likely similar to plasmid pCLK, as it carries the *bont*/a3 gene in a *C. botulinum* type A strain (CDC-A3) which is indistinguishable from strain Loch Maree (Jacobson et al., 2008); the Group II plasmid was pCLL (Eklund 17B). The plasmids were tagged by ClosTron mutagenesis (Kuehne & Minton, 2012): briefly, the *bont*/a3, *bont*/a4 and *bont*/b4 genes of plasmids pBotCDC-A3, pCLJ, and pCLL, respectively, were inactivated by the insertion of the group II intron containing the erythromycin B resistance (*erm*B) gene marker. The *C. botulinum* mutant strains carrying the tagged plasmids served as donors in mixed-plate mating experiments; the recipients were non-neurotoxigenic mutants of Group I *C. botulinum* type A strains with a deletion of the *bont*/a gene by the insertion of the tetracycline (*tet*) resistance transposon Tn916.

After mating, transconjugants growing in the presence of both erythromycin and tetracycline antibiotics were obtained, indicating that the tetracycline-resistant recipient strains had actually acquired the *erm*B gene from the donor strains; this was further confirmed by PFGE and Southern blot analyses with probes specific for *erm*B and the appropriate *bont* gene.

The fact that the *erm*B gene transfer had occurred through plasmid conjugation was indirectly confirmed by excluding i) phage-mediated transduction, because the filtered supernatants of the donor cultures, theoretically containing the bacteriophages, failed to yield transductans; and ii) transformation, since the addition of DNAseI, an enzyme degrading any crude DNA, did not inhibit the *erm*B transfer in the mating experiments. The conjugation frequency was determined to range from 1.7 x 10^{-5} to 1.5 x 10^{-8} transconjugants per donor cell.

Besides, Marshall et al., 2010 identified some genes coding for type II and type IV secretion system proteins such as pili in the Group I plasmid sequences, and regions of homology to conjugative plasmids of *C. perfringens*, including a *tcp* (namely, transfer of clostridium plasmids) locus essential for conjugative transfer (Li et al., 2013), in the Group II pCLL plasmid: these observations further supported that the transfer of plasmids had occurred through conjugation.

The demonstration of conjugative transfer of plasmid pCLL from a Group II (nonproteolytic) *C. botulinum* strain to Group I (proteolytic) *C. botulinum* strains was significant: in fact, given that the two clostridia Groups taxonomically correspond to distinct clostridia species, it clearly indicates that inter-species exchange is possible for the Group II plasmid. Whether the BoNT-encoding plasmids are narrow- or broad-host-range plasmids remains to be determined.

The β-Lactamase-Encoding Megaplasmids of Botulinum Neurotoxigenic C. butyricum Type E Strains

Strains belonging to the *C. butyricum* species are typically non-neurotoxigenic and are regular components of the gut microflora. In certain countries, *C. butyricum* strains with proven absence of toxicological effects are allowed as probiotics for human and animal use (Van Immerseel et al., 2010; Zhao et al., 2013; European Commission, 2014).

However, peculiar *C. butyricum* strains unusually capable of producing BoNT/E have been associated to human botulism in Italy, China, India, USA, Japan and UK (Aureli et al., 1986; Meng et al., 1997; Chaudry et al., 1998; Fenicia et al., 1999; Dykes et al., 2008; Abe et al., 2008; Grant et al., 2013). How the neurotoxigenic *C. butyricum* type E strains have acquired the ability of synthesizing BoNT/E is still unclear.

It has long been speculated that the *bont*/E gene was transferred from a progenitor strain to the neurotoxigenic *C. butyricum* type E strains through mobile genetic elements. The possibility that such a transfer was plasmid-mediated has been investigated in early studies leading to contradicting conclusions (Hauser et al., 1992; Zhou et al., 1993). More recently, the draft genome sequences generated for two *C. butyricum* type E strains and two *C.*

botulinum type E strains have shown that the *bont*/E gene cluster is located in the chromosome of those strains, within a *rarA* operon encoding the resolvase enzyme, which is indicative of mobilization of a genomic island (Hill et al., 2009). However, the genetic element(s) originally involved in the *bont*/e gene mobilization remain puzzling.

While re-screening by PFGE ten different neurotoxigenic *C. butyricum* type E strains isolated from clinical, food, and environmental sources in Italy and China for the presence of plasmids in their genomes, we unexpectedly found that each of them possessed a very large extrachromosomal genetic element (Franciosa et al., 2011). Four different sizes ranging approximately from 600 kb to 850 kb were estimated among the extrachromosomal genetic elements of strains by comparison with a molecular marker in PFGE experiments. Given that the genome sizes of two sequenced neurotoxigenic *C. butyricum* type E strains are approximately 4.5 Mb and 4.7 Mb (Hill et al., 2009), the extrachromosomal genetic elements were estimated to constitute 13 – 19% of the whole bacterial genome. The large sizes of the extrachromosomal genetic elements raised the question as to whether they were megaplasmids or second chromosomes; in fact, additional chromosomes that are stably inherited and carry genes for essential functions have been discovered in a number of bacterial genomes (Trucksis et al., 1998; Landeta et al., 2011). Classically, the presence of essential ribosomal RNA (rRNA) genes is indicative of a chromosome (Krawiec & Riley, 1990); since the extrachromosomal genetic elements identified in the neurotoxigenic *C. butyricum* type E strains lacked the 16S rRNA genes, it was concluded that they were megaplasmids rather than second chromosomes (Franciosa et al., 2011). However, no definite consensus currently exists on chromosome versus plasmid definition (Harrison et al., 2010), hence the question may need future reconsideration. If the extrachromosomal genetic elements of the *C. butyricum* strains were plasmids, then their molecular sizes would widely exceed those of the plasmids identified to date in the genomes of various *Clostridium* spp, including *C. botulinum, C. argentinense, C. tetani, C. perfringens, C. difficile* (Bruggeman, 2005).

The physical maps obtained for the megaplasmids of the *C. butyricum* type E strains from Italy provided evidence that they have a circular structure, thereby supplanting our earlier incorrect interpretation of linearity (Iacobino et al., 2013).

Attempts to associate the newly discovered megaplasmids of the neurotoxigenic *C. butyricum* type E strains to the *bont*/e gene were unsuccessful: in fact, the PFGE and Southern blot analyses with a gene probe specific for *bont*/e confirmed that the *bont*/e gene was a chromosome located in all ten strains. Instead, the megaplasmids of eight clinical neurotoxigenic *C. butyricum* type E strains were revealed to carry a β-lactamase gene, whereas this gene was absent from the megaplasmids of the two remaining environmental strains (Franciosa et al., 2011). These findings suggest that the β-lactamase-encoding megaplasmids provide their clostridia hosts with a selective advantage (i.e., increased resistance to β-lactam antibiotics, including penicillins and cephalosporins) in a clinical environment. Although antibiotic use is not specifically recommended for treating intestinal toxaemia botulism and foodborne botulism, i.e. the forms of botulism from which the neurotoxigenic *C. butyricum* type E strains with the β-lactamase-encoding megaplasmids had originally been isolated, their use may be necessary to treat secondary infections; β-lactams are among the most widely used antibiotics, hence the increased levels of exposure may result in selective pressure for resistance genes among bacteria. Whether these megaplasmids contribute to the dissemination of β-lactam resistance among bacteria is currently unknown, as their genetic transfer has not yet been demonstrated.

The ubiquitous presence of the megaplasmids in ten neurotoxigenic *C. butyricum* type E strains of variable sources, including the two environmental strains that lack the β-lactamase gene, would suggest that the megaplasmids have additional functions besides conferring antibiotic resistance. In order to identify other genetic regions that may play crucial roles, the available sequence of a megaplasmid of *C. butyricum* type E strain BL5262 isolated in Italy (http://www.ncbi.nlm.nih.gov/nuccore/NZ_ACOM01000001.1) was thoroughly analyzed. It consists of 757,653 bp, and, similarly to the BoNT-encoding plasmids of *C. botulinum* strains, it contains both structural and metabolic genes; interestingly, a number of transcriptional regulators and two-component regulatory systems are also present. Some two-component systems have recently been demonstrated to regulate the production of BoNT and the complex accessory proteins in *C. botulinum* type A (Connan et al., 2012); as no *bont*/E gene regulation mechanisms have so far been identified, the hypothesis that some megaplasmid-encoded putative regulators might influence the *bont*/E expression in neurotoxigenic *C. butyricum* type E strains is intriguing. Hopefully, the functional analysis of these megaplasmid genetic regions will provide answers to this question.

Other as yet undetermined issues include the mechanisms underlying the replication and partitioning of the *C. butyricum* type E megaplasmids.

CRISPR Loci in the BoNT-Encoding Plasmids and β-Lactamase-Encoding Plasmids of Botulinum Neurotoxigenic Clostridia

Bacteria use a variety of sophisticated defense systems against invading plasmids and bacteriophages. One of these systems rely on the CRISPR locus, consisting of clustered regularly interspaced short palindromic repeats (CRISPR) and a set of CRISPR-associated (*cas*) genes encoding the Cas proteins (Barrangou & Marraffini, 2014). The CRISPR-cas system of bacteria incorporates short sequences (spacers) from foreign DNA: the spacers are then transcribed and processed into small RNA molecules that, in conjunction with the Cas proteins, bind and cleave the invading DNA and similar DNA from subsequent invaders. Notably, recent studies have shown that the CRISPR-cas system may play a critical role in regulation of bacterial pathogenesis, in addition to its defensive role (Sampson et al., 2013; Louwen et al., 2014).

While most CRISPR loci are chromosome-encoded in bacteria, some have also been found within plasmids, suggesting that they may be subject to horizontal transfer (Godde & Bickerton, 2006; Horvath et al., 2009).

By using bioinformatics tools, we identified a CRISPR locus of 11,514 bp containing 39 distinct spacers in the available megaplasmid sequence of the neurotoxigenic *C. butyricum* type E strain BL5262 from Italy; a similar CRISPR locus was detected by PCR in the β-lactamase-encoding megaplasmids of two other Italian neurotoxigenic *C. butyricum* type E strains (Iacobino et al., 2013) (Figure 4a). Upon sequencing, the CRISPR loci of the two strains were shown to possess 39 spacers identical to those of strain BL5262 and in the same order; besides, they contained 2 and 6 additional distinct spacers, respectively. The sharing of 39 consecutive CRISPR spacers by the two *C. butyricum* type E strains and the fact that the strains were indistinguishable by PFGE indicated that they were likely clonally related; the presence of additional distinct spacers, attributable to a more prolonged activity of their respective CRISPR loci, suggested that they were independent isolates derived from a common recent ancestor.

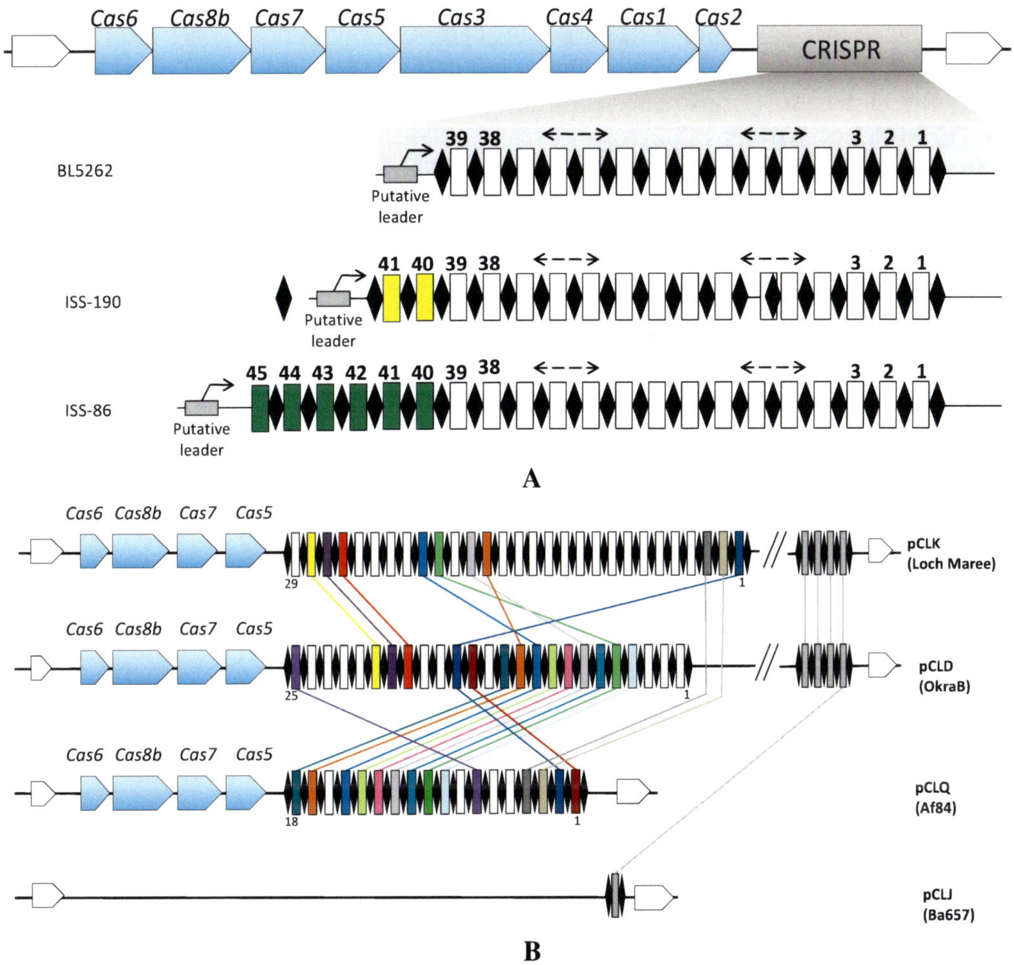

Figure 4. Organization of the CRISPR-cas loci within the plasmids of botulinum neurotoxigenic clostridia as determined by CRISPRFinder (http://crispr.u-psud.fr/Server/).

A. CRISPR-cas locus within the β-lactamase encoding plasmids of neurotoxigenic *C. butyricum* type E strains. The CRISPR spacers are numbered according to their acquisition order, with the more recently added spacers having the highest numbers. The additional distinct spacers of strains ISS-190 and ISS-86 are colored in yellow and green, respectively. ♦ 30 bp repeat consensus sequence (ATTTAAATACATCTAATGTTACTGTTAATC).

B. CRISPR-cas loci within the BoNT-encoding plasmids of Group I *C. botulinum* strains. pCLK and pCLD possess two CRISPR arrays, of which the second one is distant from the cas operon. All plasmids lack the *cas*1 and *cas*2 genes: pCLJ possesses no *cas* genes. White spacers are unique for each plasmid. Colored spacers represent shared spacers between two or three plasmids. ♦ repeat consensus sequence (GTTTAACTTCTACATTAGATGTATTTAAAC or AAAAAGTGTTTCACTTTTGTCGTGCACTTTT).

The hypervariable nature of CRISPRs makes their use suitable for differentiating between microorganisms, and the characterization of CRISPR spacers contents has been shown to be suitable for genotyping a number of bacteria (Zhang et al., 2010; Shariat et al., 2013). Most spacers of the CRISPR loci of the *C. butyricum* type E megaplasmids targeted bacteriophages and plasmids associated with the gut microbiome, indicating that they had been acquired upon exposure to those genetic elements in the intestinal environment. In fact, the two *C. butyricum* type E strains had originally been recovered from distinct cases of infant botulism and intestinal toxemia botulism in an adult: in both botulism forms, intestinal colonization by the BoNT-producing clostridia spores can last for months (Fenicia et al., 2007; Derman et al., 2014).

A 168 kb genetic region containing the CRISPR locus was absent from both megaplasmids and chromosomes of four additional *C. butyricum* type E strains from Italy that shared PFGE profile relatedness with the two strains possessing the CRISPR loci in their megaplasmid. The 168 kb genetic region is flanked by invertase and transposase genes in the megaplasmid sequence of the *C. butyricum* type E strain BL5262, which is indicative of horizontal transfer: whether the 168 kb region containing the CRISPR locus was acquired or lost by the megaplasmids of the Italian *C. butyricum* type E strains during evolution, however, is not clear (Iacobino et al., 2013).

In addition, partial CRISPR loci lacking one or more essential *cas* genes are present in the Group I BoNT-encoding plasmids pCLK (Loch Maree), pCLJ (Ba657), pCLD (OkraB), and pCLQ (Af84), as predicted by bioinformatics tools (unpublished data) (Figure 4b). The lack of some *cas* genes could indicate that the CRISPR loci in those plasmids have lost their functionality. Alternatively, since additional complete CRISPR loci with adjacent *cas* genes are also identifiable in the chromosomes of the same *C. botulinum* strains, except for the strain Loch Maree, the possibility exists that the Cas proteins encoded in the chromosome act in *trans* with the multiple chromosome and plasmid CRISPR loci (Bhaya et al., 2011). Interestingly, the CRISPR loci of some Group I BoNT-encoding plasmids share several spacer sequences (Figure 4b); however, they do not match sequences in known phages or plasmids (unpublished data). No CRISPR loci are recognizable within the Group II plasmid pCLL of *C. botulinum* strain Eklund 17B.

Collectively, these data indicate that complete or partial CRISPR loci may or may not be present within the plasmids and/or chromosomes of botulinum neurotoxigenic clostridia strains. While on one hand functional CRISPR-cas systems protect bacteria from bacteriophages infection or plasmid conjugation which could be detrimental to the host, on the other hand the limited presence and activity of CRISPR-cas systems in some bacteria has been related to a higher introduction of genes conferring advantageous traits, including virulence factors or antibiotic resistance, by bacteriophages and conjugative plasmids (Louwen et al., 2014). Further research will be necessary to clarify the role of the CRISPR loci in BoNT-producing clostridia genomes, particularly those carried by plasmids.

Conclusion

The discoveries of novel BoNT-encoding plasmids that are more widespread than expected among *C. botulinum* strains and the demonstration that some of those plasmids are conjugative between different clostridia Groups/species have provided experimental evidence that plasmid mobilization is one of the mechanisms accounting for the scattered distribution of the *bont* genes among taxonomically distinct clostridia.

Remarkably, all BoNT serotypes causing human botulism (i.e., serotypes A, B, E and F) have now been revealed to be plasmid encoded, though in some – not all – *C. botulinum* strains. Plasmids are generally regarded as vessels for the transport of accessory beneficial genetic traits allowing for the microbial host adaptation in specific ecological niches (Norman et al., 2009). Then, the presence of the *bont*/A, B, E and F genes within potentially spreadable plasmids would imply that those genes do confer some advantage to their clostridia hosts. Since the BoNTs of those serotypes can be lethal to humans and also mammalian animals, an hypothetical benefit to the clostridia host that has acquired the BoNT-encoding plasmids could derive from the supply of anaerobic substrate resulting from the necrotic human or animal organism affected by BoNT: this process would lead to massive clostridia proliferation and possibly to enhanced plasmid spread. It is worth noting that the *bont/*C and D genes, coding for the BoNT/C and D serotypes that are commonly associated with animal (mainly avian) botulism, are also carried within transferable genetic elements, i.e. the lysogenic bacteriophages. However, this speculation lacks experimental evidences.

According to evolutionary analyses, the *bont* genes and other genes within or adjacent to the *bont* gene clusters have divergently evolved from a precursor collagenase gene (Doxey et al., 2008), while the catalytic domain of BoNT has been proposed to derive from a viral protease (DasGupta, 2006). However, the reason why BoNT has developed the highly specific binding and catalytic domains which are responsible for its high lethality for humans and animals still remains elusive, as it is the reason why this deadly neurotoxin is only produced by some clostridia.

On the other side, the finding of β-lactamase-encoding plasmids in the botulinum neurotoxigenic *C. butyricum* type E strains associated to human botulism suggests that additional selective advantages besides the ability to produce BoNT might have been necessary to these atypical microorganisms in order to survive, for instance in a clinical setting, and emerge as human pathogens. Understanding whether the β-lactamase-encoding plasmids can be transferred from the botulinum neurotoxigenic *C. butyricum* type E strains to non-toxigenic *C. butyricum* strains – especially those that are used as probiotics - or to other clinically important clostridia *spp* will be of interest.

Based on the observation that both the BoNT-encoding plasmids and the β-lactamase-encoding plasmids newly identified in botulinum neurotoxigenic clostridia strains have very large sizes and contain a high number of IS elements and transposons, which are indicative of considerable genomic rearrangements, it is conceivable that they can serve as vectors of additional beneficial traits on their clostridia hosts. This hypothesis would be supported by the presence in both novel plasmids of CRISPR loci that have been demonstrated to play crucial defensive and regulatory roles in several microorganisms, even though their specific role(s) in botulinum neurotoxigenic clostridia are still undefined.

Clearly, more data on the frequency of occurrence of both the BoNT-encoding plasmids and the β-lactamase-encoding plasmids among bacterial populations, and the elucidation of their genetic contents, mobility properties, efficiency of transfer, and microbial host range are necessary in order to understand their function in the emergence and spread of their clostridia hosts.

References

Abe, Y., Negasawa, T., Monma, C., & Oka, A. (2008). Infantile botulism caused by *Clostridium butyricum* type E toxin. *Pediatr. Neurol.* 38: 55-57.

Arnon, S.S., Schechter, R., Inglesby, T.V, Henderson, D.A., Bartlett, J.G., Ascher, M.S., Eitzen, Ed., Fine, A.D., Haurer, J., Layton, M., Lillibridge, S., Osterholm, M.T., O'Toole, T., Parker, G., Perl, T.M., Russell, P.K., Swerdlow, D.L., Tonat, K., & Working Group on Civilian Biodefense. (2001). Botulinum toxin as a biological weapon: medical and public health management. *JAMA* 285, 1059-1070.

Aureli, P., Fenicia, L., Pasolini, B., Gianfranceschi, M., McCroskey, L.M., & Hatheway, C.L. (1986). Two cases of type E infant botulism caused by neurotoxigenic *Clostridium butyricum* in Italy. *J. Infect. Dis.* 154, 207-211.

Barash, J.R., & Arnon, S.S. (2014). A novel strain of *Clostridium botulinum* that produces type B and type H botulinum toxins. *J. Infect. Dis.* 209, 183-191.

Barrangou, R., & Marraffini, L.A. (2014). CRISPR-Cas systems: prokaryotes upgrade to adaptive immunity. *Mol. Cell.* 54, 234-244.

Bhaya, D., Davison, M., & Barrangou, R. (2011). CRISPR-Cas systems in bacteria and archaea: versatile small RNAs for adaptive defense and regulation. *Annu. Rev. Genet.* 45, 273-297.

Bradbury, M., Greenfield, P., Midgley, D., Li, D., Tran-Dinh, N., Vriesekoop, F., & Brown, J.L. (2012). Draft genome sequence of *Clostridium sporogenes* PA 3679, the common nontoxigenic surrogate for proteolytic *Clostridium botulinum*. *J. Bacteriol.* 194, 1631-1632.

Brüggemann, H. (2005). Genomics of clostridial pathogens: implication of extrachromosomal elements in pathogenicity. *Curr. Opin. Microbiol.* 8, 601-605.

Brunger, A.T., & Rummel, A. (2009). Receptor and substrate interactions of clostridial neurotoxins. *Toxicon* 54, 550-560.

Carattoli, A. (2013). Plasmids and the spread of resistance. *Int. J. Med. Microbiol.* 303, 298-304.

Carter, A.T., Paul, C.J., Mason, D.R., Twine, S.M., Alston, M.J., Logan, S.M., Austin, J.W., & Peck, M.W. (2009). Independent evolution of neurotoxin and flagellar genetic loci in proteolytic *Clostridium botulinum*. *BMC Genomics.* 10:115.

Chaudhry, R., Dhawan, B., Kumar, D., Bhatia, R., Gandhi, J.C., Patel, R.K., & Purohit, B.C. (1998). Outbreak of suspected *Clostridium butyricum* botulism in India. *Emerg. Infect. Dis.* 4: 506-507.

Collins, M.D., & East, A.K. (1998). Phylogeny and taxonomy of the food-borne pathogen *Clostridium botulinum* and its neurotoxins. *J. Appl. Microbiol.* 84, 5-17.

Connan, C., Brüggemann, H., Mazuet, C., Raffestin, S., Cayet, N., & Popoff, M.R. (2012). Two-component systems are involved in the regulation of botulinum neurotoxin synthesis in *Clostridium botulinum* type A strain Hall. *PLoS One* 7(7): e41848.

Connan, C., Denève, C., Mazuet, C., & Popoff, M.R. (2013) Regulation of toxin synthesis in *Clostridium botulinum* and *Clostridium tetani*. *Toxicon* 75, 90-100.

Cordoba, J.J., Collins, M.D., & East, A.K. (1995). Studies on the genes encoding botulinum neurotoxin type A of *Clostridium botulinum* from a variety of sources. *Syst. Appl. Microbiol.* 18, 13–22.

DasGupta, B.R. (2006). Botulinum neurotoxins: perspective on their existence and as polyproteins harboring viral proteases. *J. Gen. Appl. Microbiol.* 52, 1-8.

Derman, Y., Korkeala, H., Salo, E., Lönnqvist, T., Saxen, H., & Lindström, M. (2014). Infant botulism with prolonged faecal excretion of botulinum neurotoxin and *Clostridium botulinum* for 7 months. *Epidemiol. Infect.* 142, 335-339.

Dover, N., Barash, J.R., & Arnon S.S. (2009) Novel *Clostridium botulinum* toxin gene arrangement with subtype A5 and partial subtype B3 botulinum neurotoxin genes. *J. Clin. Microbiol.* 47, 2349-2350.

Dover, N., Barash, J.R., Hill, K.K., Detter, J.C., & Arnon, S.S. (2011). Novel structural elements within the nonproteolytic *Clostridium botulinum* type F toxin gene cluster. *Appl. Environ. Microbiol.* 77, 1904-1906.

Dover, N., Barash, J.R., Hill, K.K., Davenport, K.W., Teshima, H., Xie, G., & Arnon, S.S. (2013). *Clostridium botulinum* strain Af84 contains three neurotoxin gene clusters: *bont*/A2, *bont*/F4 and *bont*/F5. *PLoS One* 8(4):e61205.

Dover, N., Barash, J.R., Hill, K.K., Xie, G., & Arnon, S.S. (2014a). Molecular characterization of a novel botulinum neurotoxin type H gene. *J. Infect. Dis.* 209, 192-202.

Dover, N., Barash, J.R., Burke, J.N., Hill, K.K., Detter, J.C., & Arnon, S.S. (2014b). Arrangement of the *Clostridium baratii* F7 toxin gene cluster with identification of a σ factor that recognizes the botulinum toxin gene cluster promoters. *PLoS One* 9(5):e97983.

Doxey, A.C., Lynch, M.D., Müller, K.M., Meiering, E.M., & McConkey, B.J. (2008). Insights into the evolutionary origins of clostridial neurotoxins from analysis of the *Clostridium botulinum* strain A neurotoxin gene cluster. *BMC Evol Biol.* 8:316.

Dykes, J.K., Lùquez, C., Raphael, B.H., McCroskey, L.M., & Maslanka, S.E. (2008). Laboratory investigation of the first case of *Clostridium butyricum* type E infant botulism in the United States. In: Abstracts of the 45th Interagency Research Coordinating Committee Meeting, Philadelphia, PA, USA.

East, A.K., Bhandari, M., Stacey, J.M., Campbell, K.D., & Collins, M.D. (1996). Organization and phylogenetic interrelationships of genes encoding components of the botulinum toxin complex in proteolytic *Clostridium botulinum* types A, B, and F: evidence of chimeric sequences in the gene encoding the nontoxic nonhemagglutinin component. *Int. J. Syst. Bacteriol.* 46, 1105-1112.

Eisele, K.H., Fink, K., Vey, M., & Taylor, H.V. (2011). Studies on the dissociation of botulinum neurotoxin type A complexes. *Toxicon* 57, 555-565.

Eklund, M.W., & Poysky, F.T. (1974). Interconversion of type C and D strains of *Clostridium botulinum* by specific bacteriophages. Appl. Microbial. 27:251-258.

Eklund, M.W., Poysky, F.T., & Habig, W.H. (1989) Bacteriophages and plasmids in *Clostridium botulinum* and *Clostridium tetani* and their relationship to the production of toxin. In Simpson LL (ed): Botulinum Neurotoxin and Tetanus Toxin. Academic Press, San Diego, CA, pp 25-51.

Eklund, M.W., Poysky, F.T., & Habig, W.H. (1989). Bacteriophages and plasmids in *Clostridium botulinum* and *Clostridium tetani* and their relationship to the production of toxin. In Simpson, L.L. (ed.) Botulinum Neurotoxin and Tetanus Toxin, Academic Press San Diego, CA, pp 25-51.

Eklund, M.W., Poysky, F.T., Mseitif, L.M., & Strom, M.S. (1988). Evidence for plasmid-mediated toxin and bacteriocin production in *Clostridium botulinum* type G. *Appl. Environ. Microbiol.* 54, 1405-1408.

Eklund, M.W., Poysky, F.T., Reed, S.M., & Smith, C.A. (1971). Bacteriophage and the toxigenicity of *Clostridium botulinum* type C. *Science* 172, 480-482.

Eleopra, R., Tugnoli, V., & Montecucco, C. (2009). Biology and clinical pharmacology of botulinum neurotoxin type C and other non-A/non-B botulinum neurotoxins. In Jankovic J., Albanese A., Atassi Z., Dolly J.O., Hallett M., Mayer N.H. (eds): Botulinum Toxin, Therapeutic Clinical Practice & Science. New York, NY, Elsevier Health Sciences, pp 77-84.

European Commission. (2014). European Union Register of Feed Additives pursuant to Regulation (EC) No 1831/2003. Edition 185. Appendixes 3e, 4 – 12.05.2014.

Fenicia, L., Anniballi, F., & Aureli P. (2007). Intestinal toxemia botulism in Italy, 1984-2005. *Eur. J. Clin. Microbiol. Infect. Dis.* 26, 385-94.

Fenicia, L., Franciosa, G., Pourshaban, M., & Aureli, P. (1999). Intestinal toxaemia botulism in two young people, caused by *Clostridium butyricum* type E. *Clin. Infect. Dis.* 29, 1381-1387.

Ferreira, J.L., Hamdy, M.K., Zapatka, F.A., & McCay, S.G. (1987). Plasmids of *Clostridium botulinum* type A, B, E, F, and other clostridial species. *J. Food Safety* 8, 267-277.

Fillo, S., Giordani, F., Anniballi, F., Gorgé, O., Ramisse, V., Vergnaud, G., Riehm, J.M., Scholz, H.C., Splettstoesser, W.D., Kieboom, J., Olsen, J.S., Fenicia, L., & Lista, F. (2011). *Clostridium botulinum* group I strain genotyping by 15-locus multilocus variable-number tandem-repeat analysis. *J. Clin. Microbiol.* 49, 4252-4263.

Franciosa, G., Ferreira, J.L., & Hatheway, C.L. (1994). Detection of type A, B, and E botulism neurotoxin genes in *Clostridium botulinum* and other *Clostridium* species by PCR: evidence of unexpressed type B toxin genes in type A toxigenic organisms. *J. Clin. Microbiol.* 32, 1911-1917.

Franciosa, G., Floridi, F., Maugliani, A., & Aureli, P. (2004) Differentiation of the gene clusters encoding botulinum neurotoxin type A complexes in *Clostridium botulinum* type A, Ab, and A(B) strains. *Appl. Environ. Microbiol.* 70, 7192-7199.

Franciosa, G., Maugliani, A., Scalfaro, C., & Aureli, P. (2009). Evidence that plasmid-borne botulinum neurotoxin type B genes are widespread among *Clostridium botulinum* serotype B strains. *PLoS One* 4(3):e4829.

Franciosa, G., Scalfaro, C., Di Bonito, P., Vitale, M., & Aureli, P. (2011). Identification of novel linear megaplasmids carrying a ß-lactamase gene in neurotoxigenic *Clostridium butyricum* type E strains. *PLoS One* 6(6):e21706.

Frost, L.S., Leplae, R., Summers, A.O., & Toussaint, A. (2005). Mobile genetic elements: the agents of open source evolution. *Nat. Rev. Microbiol.* 3, 722-732.

Fujinaga, Y. (2010). Interaction of botulinum toxin with the epithelial barrier. *J. Biomed. Biotechnol.* 2010, Article ID 974943, 9 pages.

Gill, D.M. (1982). Bacterial toxins: a table of lethal amounts. *Microbiol. Rev.* 46, 86-94.

Godde, J.S. & Bickerton, A. (2006). The repetitive DNA elements called CRISPRs and their associated genes: evidence of horizontal transfer among prokaryotes. *J. Mol. Evol.* 62, 718-729.

Grant, K.A., McLauchlin, J., & Amar, C. (2013). Infant botulism: advice on avoiding feeding honey to babies and other possible risk factors. *Community Pract.* 86, 44-46.

Gu, S., & Jin, R. (2013). Assembly and function of the botulinum neurotoxin progenitor complex. *Curr. Top. Microbiol. Immunol.* 364, 21-44.

Gu, S., Rumpel, S., Zhou, J., Strotmeier, J., Bigalke, H., Perry, K., Shoemaker, C.B., Rummel, A., & Jin, R. (2012) Botulinum neurotoxin is shielded by NTNHA in an interlocked complex. *Science.* 335, 977-981.

Hallett, M., Albanese, A., Dressler, D., Segal, K.R., Simpson, D.M., Truong, D., & Jankovic, J. (2013). Evidence-based review and assessment of botulinum neurotoxin for the treatment of movement disorders. *Toxicon* 67, 94-114.

Harrison, P.W., Lower, R.P., Kim, N.K., & Young, J.P. (2010). Introducing the bacterial 'chromid': not a chromosome, not a plasmid. *Trends Microbiol.* 18, 141-148.

Hassan, K.A., Tetu, S.G., Elbourne, L.D., Johnson, E.A., & Paulsen, I.T. (2013). Genome sequence of the group III *Clostridium botulinum* strain Eklund-C. *Genome Announc.* 1(2):e0004413.

Hatheway, C.L. (1993). *Clostridium botulinum* and other clostridia that produce botulinum neurotoxin. In Hauschild, A.H.S., Dodds, K.L. (eds): *Clostridium botulinum*: ecology and control in foods. Marcel Dekker, New York, pp 3-20.

Hatheway, C.L., & Ferreira, J.L. (1996). Detection and identification of *Clostridium botulinum* neurotoxins. *Adv. Exp. Med. Biol.* 391, 481-98.

Hauser, D., Gilbert, M., Boquet, P., & Popoff, M.R. (1992). Plasmid localization of a type E botulinal neurotoxin gene homologue in toxigenic *Clostridium butyricum* strains, and absence of this gene in non-toxigenic *C. butyricum* strains. *FEMS Microbiol. Lett.* 99, 251-256.

Hielm, S., Björkroth, J., Hyytiä, E., & Korkeala, H. (1998). Genomic analysis of *Clostridium botulinum* group II by pulsed-field gel electrophoresis. *Appl. Environ. Microbiol.* 64, 703-708.

Hill, K.K., & Smith, T.J. (2013). Genetic diversity within *Clostridium botulinum* serotypes, botulinum neurotoxin gene clusters and toxin subtypes. *Curr. Top. Microbiol. Immunol.* 364, 1-20.

Hill, K.K., Smith, T.J., Helma, C.H., Ticknor, L.O., Foley, B.T., Svensson, R.T., Brown, J.L., Johnson, E.A., Smith, L.A., Okinaka, R.T., Jackson, P.J., & Marks, J.D. (2007). Genetic diversity among botulinum neurotoxin-producing clostridial strains. *J. Bacteriol.* 189, 818-832.

Hill, K.K., Xie, G., Foley, B.T., Smith, T.J., Munk, A.C., Bruce, D., Smith, L.A., Brettin, T.S., & Detter, J.C. (2009). Recombination and insertion events involving the botulinum neurotoxin complex genes in *Clostridium botulinum* types A, B, E and F and *Clostridium butyricum* type E strains. *BMC Biol.* 7:66.

Horvath, P., Coûté-Monvoisin, A.C., Romero, D.A., Boyaval, P., Fremaux, C., & Barrangou, R. (2009). Comparative analysis of CRISPR loci in lactic acid bacteria genomes. *Int. J. Food Microbiol.* 131, 62-70.

Hutson, R.A., Zhou, Y., Collins, M.D., Johnson, E.A., Hatheway, C.L., & Sugiyama, H. (1996). Genetic characterization of *Clostridium botulinum* type A containing silent type B neurotoxin gene sequences. *J. Biol. Chem.* 271, 10786-10792.

Hyytiä, E., Björkroth, J., Hielm, S., & Korkeala, H. (1999). Characterisation of *Clostridium botulinum* groups I and II by randomly amplified polymorphic DNA analysis and repetitive element sequence-based PCR. *Int. J. Food Microbiol.* 48, 179-189.

Iacobino, A., Scalfaro, C., & Franciosa, G. (2013). Structure and genetic content of the megaplasmids of neurotoxigenic *Clostridium butyricum* type E strains from Italy. *PLoS One* 8(8): e71324.

Inoue, K., & Iida, H. (1971). Phage-conversion of toxigenicity in *Clostridium botulinum* types C and D. *Jpn. J. Med. Sci. Biol.* 24, 53-56.

Jacobson, M.J., Lin, G., Whittam, T.S., & Johnson, E.A. (2008). Phylogenetic analysis of *Clostridium botulinum* type A by multi-locus sequence typing. *Microbiology.* 154(Pt 8), 2408-2415.

Johnson, E.A. (2005). Bacteriophages encoding botulinum and diptheria Toxins. In: Waldor MK, Friedman DI, Adhya SL, (eds): Phages: Their Role in Bacterial Pathogenesis and Biotechnology. ASM Press, Washington, D.C, pp. 280-296.

Kalb, S.R., Baudys, J., Egan, C., Smith, T.J., Smith, L.A., Pirkle, J.L., & Barr, J.R. (2011). Different substrate recognition requirements for cleavage of synaptobrevin-2 by *Clostridium baratii* and *Clostridium botulinum* type F neurotoxins. *Appl. Environ. Microbiol.* 77, 1301-1308.

Kalb, S.R., Baudys, J., Rees, J.C., Smith, T.J., Smith, L.A., Helma, C.H., Hill, K., Kull, S., Kirchner, S., Dorner, M.B., Dorner, B.G., Pirkle, J.L., & Barr, J.R. (2012a). De novo subtype and strain identification of botulinum neurotoxin type B through toxin proteomics. *Anal. Bioanal. Chem.* 403, 215-226.

Kalb, S.R., Baudys, J., Webb, R.P., Wright, P., Smith, T.J., Smith, L.A., Fernández, R., Raphael, B.H., Maslanka, S.E., Pirkle, J.L., & Barr, J.R. (2012b). Discovery of a novel enzymatic cleavage site for botulinum neurotoxin F5. *FEBS Lett.* 586, 109-115.

Kalb, S.R., Baudys, J., Smith, T.J., Smith, L.A., & Barr, J.R. (2014). Three enzymatically active neurotoxins of *Clostridium botulinum* strain Af84: BoNT/A2, /F4, and /F5. *Anal. Chem.* 86, 3254-3262.

Keto-Timonen, R., Nevas, M., & Korkeala, H. (2005). Efficient DNA fingerprinting of *Clostridium botulinum* types A, B, E, and F by amplified fragment length polymorphism analysis. *Appl. Environ. Microbiol.* 71, 1148-1154.

Koonin, E.V., & Wolf, Y.I. (2008). Genomics of bacteria and archaea: the emerging dynamic view of the prokaryotic world. *Nucleic Acids Res.* 36, 6688-6719.

Krawiec, S., & Riley, M. (1990). Organization of the bacterial chromosome. *Microbiol. Rev.* 54, 502-539.

Kubota, T., Yonekura, N., Hariya, Y., Isogai, E., Isogai, H., Amano, K., & Fujii, N. (1998). Gene arrangement in the upstream region of *Clostridium botulinum* type E and *Clostridium butyricum* BL6340 progenitor toxin genes is different from that of other types. *FEMS Microbiol. Lett.* 158, 215-221.

Kuehne, S.A., & Minton, N.P. (2012). ClosTron-mediated engineering of Clostridium. *Bioengineered.* 3, 247-254.

Kumaran, D., Eswaramoorthy, S., Furey, W., Navaza, J., Sax, M., & Swaminathan, S. (2009). Domain organization in *Clostridium botulinum* neurotoxin type E is unique: its implication in faster translocation. *J. Mol. Biol.* 386, 233-245.

Lacy, D.B., Tepp, W., Cohen, A.C., DasGupta, B.R., & Stevens, R.C. (1998). Crystal structure of botulinum neurotoxin type A and implications for toxicity. *Nat. Struct. Biol.* 5, 898-902.

Landeta, C., Dávalos, A., Cevallos, M.Á., Geiger, O., Brom, S., & Romero, D. (2011). Plasmids with a chromosome-like role in rhizobia. *J. Bacteriol.* 193, 1317-1326.

Lee, W.H., & Riemann, H. (1970). Correlation of toxic and non-toxic strains of *Clostridium botulinum* by DNA composition and homology. *J. Gen. Microbiol.* 60, 117-123.

Li, J., Adams, V., Bannam, T.L., Miyamoto, K., Garcia, J.P., Uzal, F.A., Rood, J.I., & McClane, B.A. (2013). Toxin plasmids of *Clostridium perfringens*. *Microbiol. Mol. Biol. Rev.* 77, 208-233.

Louwen, R., Staals, R.H.J., Endtz, H.P., van Baarlen, P., & van de Roost, J. (2014). The role of CRISPR-cas systems in virulence of pathogenic bacteria. *Microbiol. Mol. Biol. Rev.* 78, 74-88.

Marshall, K.M., Bradshaw, M., & Johnson, E.A. (2010). Conjugative botulinum neurotoxin-encoding plasmids in *Clostridium botulinum*. *PLoS One* 5(6):e11087.

Marshall, K.M., Bradshaw, M., Pellett, S., & Johnson, E.A. (2007). Plasmid encoded neurotoxin genes in *Clostridium botulinum* serotype A subtypes. *Biochem. Biophys. Res. Commun.* 361, 49-54.

McCroskey, L.M., Hatheway, C.L., Fenicia, L., Pasolini, B., & Aureli, P. (1986). Characterization of an organism that produces type E botulinal toxin but which resembles *Clostridium butyricum* from the feces of an infant with type E botulism. *J. Clin. Microbiol.* 23, 201-202.

Meng, X., Karasawa, T., Zou, K., Kuang, X., Wang, X., Lu, C., Wang, C., Yamakawa, K., & Nakamura, S. (1997). Characterization of a neurotoxigenic *Clostridium butyricum* strain isolated from the food implicated in an outbreak of food-borne type E botulism. *J. Clin. Microbiol.* 35, 2160-2162.

Montecucco, C., & Schiavo, G. (1993). Tetanus and botulism neurotoxins: a new group of zinc proteases. *Trends Biochem Sci.* 18, 324-327.

Moriishi, K, Koura, M., Fujii, N., Fujinaga, Y., Inoue, K., Syuto, B., & Oguma, K. (1996). Molecular cloning of the gene encoding the mosaic neurotoxin composed of parts of botulinum neurotoxin types C1 and D, and PCR detection of this gene from *Clostridium botulinum* type C organisms. *Appl. Environ. Microbiol.* 62, 662-667.

Nevas, M., Lindström, M., Hielm, S., Björkroth, K.J., Peck, M.W., & Korkeala, H. (2005). Diversity of proteolytic *Clostridium botulinum* strains, determined by a pulsed-field gel electrophoresis approach. *Appl. Environ. Microbiol.* 71, 1311-1317.

Norman, A., Hansen, L.H., & Sørensen, S.J. (2009). Conjugativc plasmids: vessels of the communal gene pool. *Philos. Trans. R. Soc. Lond. B. Biol. Sci.* 364, 2275-2289.

Oguma K. (1976). The stability of toxigenicity in *Clostridium botulinum* types C and D. *J. Gen. Microbiol.* 92, 67-75.

Paul, C.J., Twine, S.M., Tam, K.J., Mullen, J.A., Kelly, J.F., Austin, J.W., & Logan SM. (2007). Flagellin diversity in *Clostridium botulinum* groups I and II: a new strategy for strain identification. *Appl. Environ. Microbiol.* 73, 2963-2975.

Peck, M.W. (2009). Biology and genomic analysis of *Clostridium botulinum*. *Adv. Microb. Physiol.* 55, 183-265.

Pickett, A., & Perrow, K. (2011). Towards new uses of botulinum toxin as a novel therapeutic tool. *Toxins* 3, 63-81.

Popoff, M.R., & Bouvet, P. (2013). Genetic characteristics of toxigenic Clostridia and toxin gene evolution. *Toxicon* 75, 63-89.

Raphael, B.H., Bradshaw, M., Kalb, S.R., Joseph, L.A., Lúquez, C., Barr, J.R., Johnson, E.A., & Maslanka, S.E. (2014). Characterization of *Clostridium botulinum* strains producing BoNT/F4 or BoNT/F5. *Appl. Environ. Microbiol.* 80, 3250-3257.

Raphael, B.H., Choudoir, M.J., Lúquez, C., Fernández, R., & Maslanka, S.E. (2010). Sequence diversity of genes encoding botulinum neurotoxin type F. *Appl. Environ. Microbiol.* 76, 4805-4812.

Rodríguez Jovita, M., Collins, M.D., & East, A.K. (1998) Gene organization and sequence determination of the two botulinum neurotoxin gene clusters in *Clostridium botulinum* type A(B) strain NCTC 2916. *Curr. Microbiol.* 36, 226-231.

Rossetto, O., Megighian, A., Scorzeto, M., & Montecucco, C. (2013). Botulinum neurotoxins. *Toxicon* 67, 31-36.

Sakaguchi, Y., Hayashi, T., Kurokawa, K., Nakayama, K., Oshima, K., Fujinaga, Y., Ohnishi, M., Ohtsubo, E., Hattori, M., & Oguma, K. (2005) The genome sequence of *Clostridium botulinum* type C neurotoxin-converting phage and the molecular mechanisms of unstable lysogeny. *Proc. Natl. Acad. Sci. U S A.* 102, 17472-17477.

Sampson, T.R., Saroj, S.D., Llewellyn, A.C., Tzeng, Y.L., & Weiss, D.S. (2013). A CRISPR/Cas system mediates bacterial innate immune evasion and virulence. *Nature* 497, 254-257.

Schiavo, G., Matteoli, M., & Montecucco, C. (2000). Neurotoxins affecting neuroexocytosis. *Physiol. Rev.* 80, 717-766.

Scott, V.N., & Duncan, C.L. (1978). Cryptic plasmids in *Clostridium botulinum* and *C. botulinum*-like organisms. *Fems Microbiol. Lett.* 4, 55-58.

Sebaihia, M., Peck, M.W., Minton, N.P., Thomson, N.R., Holden, M.T., Mitchell, W.J., Carter, A.T., Bentley, S.D., Mason, D.R., Crossman, L., Paul, C.J., Ivens, A., Wells-Bennik, M.H., Davis, I.J., Cerdeño-Tárraga, A.M., Churcher, C., Quail, M.A., Chillingworth, T., Feltwell, T., Fraser, A., Goodhead, I., Hance, Z., Jagels, K., Larke, N., Maddison, M., Moule, S., Mungall, K., Norbertczak, H., Rabbinowitsch, E., Sanders, M., Simmonds, M., White, B., Whithead, S., & Parkhill, J. (2007). Genome sequence of a proteolytic (Group I) *Clostridium botulinum* strain Hall A and comparative analysis of the clostridial genomes. *Genome Res.* 17, 1082-1092.

Shariat, N., Kirchner, M.K., Sandt, C.H., Trees, E., Barrangou, R., & Dudley, E.G. (2013). Subtyping of *Salmonella enterica* serovar Newport outbreak isolates by CRISPR-MVLST and determination of the relationship between CRISPR-MVLST and PFGE results. *J. Clin. Microbiol.* 51, 2328-2336.

Shilpa, P.S., Kaul, R., Sultana, N., Bhat, S. (2014). Botulinum toxin: the Midas touch. *J. Nat. Sci. Biol. Med.* 5, 8-14.

Simpson, L. (2013). The life history of a botulinum toxin molecule. *Toxicon* 68, 40-59.

Skarin, H., & Segerman, B. (2011). Horizontal gene transfer of toxin genes in *Clostridium botulinum*: involvement of mobile elements and plasmids. *Mob. Genet. Elements.* 1, 213-215.

Skarin, H., Håfström, T., Westerberg, J., & Segerman, B. (2011). *Clostridium botulinum* group III: a group with dual identity shaped by plasmids, phages and mobile elements. *BMC Genomics.* 12, 185.

Smith, T.J., Hill, K.K., Foley, B.T., Detter, J.C., Munk, A.C., Bruce, D.C., Doggett, N.A., Smith, L.A., Marks, J.D., Xie, G., & Brettin, T.S. (2007). Analysis of the neurotoxin complex genes in *Clostridium botulinum* A1-A4 and B1 strains: BoNT/A3, /Ba4 and /B1 clusters are located within plasmids. *PLoS One* 2(12): e1271.

Smith, T.J., Lou, J., Geren, I.N., Forsyth, C.M., Tsai, R., Laporte, S.L., Tepp, W.H., Bradshaw, M., Johnson, E.A., Smith, L.A., & Marks, J.D. (2005). Sequence variation within botulinum neurotoxin serotypes impacts antibody binding and neutralization. *Infect. Immun.* 73, 5450-5457.

Sobel, J. (2005). Botulism. *Clin. Infect. Dis.* 41, 1167-1173.

Stringer, S.C., Carter, A.T., Webb, M.D., Wachnicka, E., Crossman, L.C., Sebaihia, M., & Peck, M.W. (2013). Genomic and physiological variability within Group II (non-proteolytic) *Clostridium botulinum. BMC Genomics.* 14, 333.

Strom, M.S., Eklund, M.W., Poysky, F.T. (1984) Plasmids in *Clostridium botulinum* and related Clostridium species. *Appl. Environ. Microbiol.* 48, 956-963.

Suen, J.C., Hatheway, C.L., Steigerwalt, A.G., & Brenner, D.J. (1988a). *Clostridium argentinense* sp. nov.: a genetically homogeneous group composed of all strains of *Clostridium botulinum* toxin type G and some nontoxigenic strains previously identified as *Clostridium subterminale* or *Clostridium hastiforme*. *Int. J. Syst. Bacteriol.* 38, 375-381.

Suen, J.C., Hatheway, C.L., Steigerwalt, A.G., & Brenner, D.J. (1988b). Genetic confirmation of identities of neurotoxigenic *Clostridium baratii* and *Clostridium butyricum* implicated as agents of infant botulism. *J. Clin. Microbiol.* 26, 2191-2192.

Swaminathan, S., & Eswaramoorthy, S. (2000). Structural analysis of the catalytic and binding sites of *Clostridium botulinum* neurotoxin B. *Nat. Struct. Biol.* 7, 693-699.

Trucksis, M., Michalski, J., Deng, Y.K., & Kaper, J.B. (1998). The *Vibrio cholerae* genome contains two unique circular chromosomes. *Proc. Natl. Acad. Sci. USA.* 95,14464-14469.

Umeda, K., Seto, Y., Kohda, T., Mukamoto, M., & Kozaki, S. (2009). Genetic characterization of *Clostridium botulinum* associated with type B infant botulism in Japan. *J. Clin. Microbiol.* 47, 2720-2728.

Van Immerseel, F., Ducatelle, R., De Vos, M., Boon, N., Van De Wiele, T., Verbeke, K., Rutgeerts, P., Sas, B., Louis, P., & Flint, H.J. (2010). Butyric acid-producing anaerobic bacteria as a novel probiotic treatment approach for inflammatory bowel disease. *J Med Microbiol* 59 (Pt 2), 141-143.

Wang, D., Krilich, J., Pellett, S., Baudys, J., Tepp, W.H., Barr, J.R., Johnson, E.A., & Kalb, S.R. (2013). Comparison of the catalytic properties of the botulinum neurotoxin subtypes A1 and A5. *Biochim. Biophys. Acta* 1834, 2722-2728.

Wang, X., Maegawa, T., Karasawa, T., Kozaki, S., Tsukamoto, K., Gyobu, Y., Yamakawa, K., Oguma, K., Sakaguchi, Y., & Nakamura, S. (2000). Genetic analysis of type E botulinum toxin-producing *Clostridium butyricum* strains. *Appl. Environ. Microbiol.* 66, 4992-4997.

Wangroongsarb, P., Kohda, T., Jittaprasartsin, C., Suthivarakom, K., Kamthalang, T., Umeda, K., Sawanpanyalert, P., Kozaki, S., Ikuta, K. (2014). Molecular characterization of *Clostridium botulinum* isolates from foodborne outbreaks in Thailand, 2010. *PLoS One* 9(1):e77792.

Whitemarsh, R.C., Tepp, W.H., Bradshaw, M., Lin, G., Pier, C.L., Scherf, J.M., Johnson, E.A., & Pellett, S. (2013). Characterization of botulinum neurotoxin A subtypes 1 through 5 by investigation of activities in mice, in neuronal cell cultures, and in vitro. *Infect. Immun.* 81, 3894-3902.

Zhang, J., Abadia, E., Refregier, G., Tafaj, S., Boschiroli, M.L., Guillard, B., Andremont, A., Ruimy, R., & Sola, C. (2010). *Mycobacterium tuberculosis* complex CRISPR genotyping: improving efficiency, throughput and discriminative power of 'spoligotyping' with new spacers and a microbead-based hybridization assay. *J. Med. Microbiol.* 59(Pt 3), 285-294.

Zhang, Z., Hintsa, H., Chen, Y., Korkeala, H., & Lindström, M. (2013). Plasmid-borne type E neurotoxin gene clusters in *Clostridium botulinum* strains. *Appl. Environ. Microbiol.* 79, 3856–3859.

Zhao, X., Guo, Y., Guo, S., & Tan J. (2013). Effects of *Clostridium butyricum* and *Enterococcus faecium* on growth performance, lipid metabolism, and cecal microbiota of broiler chickens. *Appl. Microbiol. Biotechnol.* 97, 6477-6488.

Zhou, Y., Sugiyama, H., & Johnson, E.A. (1993). Transfer of neurotoxigenicity from *Clostridium butyricum* to a nontoxigenic *Clostridium botulinum* type E-like strain. *Appl. Environ. Microbiol.* 59, 3825-3831.

Zhou, Y., Sugiyama, H., Nakano, H., & Johnson, E.A. (1995). The genes for the *Clostridium botulinum* type G toxin complex are on a plasmid. *Infect. Immun.* 63, 2087-2091.

In: Prokaryotes
Editor: Marina Nisnevitch

ISBN: 978-1-63321-592-4

Chapter 9

Sensitivity of Bacteria to Photodynamic Chemotherapy

Marina Nisnevitch [*] ***and Faina Nakonechny***
Department of Chemical Engineering, Biotechnology and Materials,
Ariel University, Ariel, Israel

Abstract

Increasing widespread resistance of microbial cells to antibiotics raises a need to develop novel antibacterial technologies. One of the approaches that may lead to improved antimicrobial treatment and replace or least significantly reduce the use of antibiotics is photodynamic chemotherapy. This technique includes visible light excitation of nontoxic components – photosensitizers, which transfer energy to molecular oxygen, resulting in the production of reactive oxygen species that in turn cause irreversible damage to cellular components of bacteria. Photodynamic chemotherapy has been extensively studied towards both Gram-positive and Gram-negative bacteria, including antibiotic-resistant strains. Gram-positive and Gram-negative bacteria react differently to photodynamic chemotherapy, at that the former exhibit high sensitivity to the treatment. Gram-negative bacteria are less susceptible to photodynamic chemotherapy due to the presence of phospholipids, complex lipoproteins and polysaccharides in the additional outer cell envelope. From the numerous studies on the effect of photodynamic chemotherapy on different microorganisms no development of bacterial resistance to photosensitizers has been reported so far. Excitation of photosensitizers in the process of bacterial eradication can be performed by several methods, including illumination with external light sources, chemiluminescent illumination and sonodynamic activation. The two latter techniques enable to deliver photodynamic treatment without external illumination in the dark, which opens prospects for combating internal infections that are difficult or not available for treatment by traditional photodynamic chemotherapy. This property turns these techniques into promising novel antimicrobial therapeutic strategies.

[*] Corresponding author: Email: marinan@ariel.ac.il.

Keywords: Photodynamic antimicrobial chemotheraphy; photosensitizer; Gram-positive bacteria; Gram-negative bacteria; chemiluminescent illumination; sonodynamic activation

Introduction

The twentieth century witnessed zigzags in the history of treatment of bacterial infections. The year 1900 opened an epoch of photodynamic inactivation of cells when Oscar Raab (Raab, 1900) discovered and described the inhibiting effect of the dye acridine orange on the microorganism *Paramecium caudatum* under illumination. This effect was soon implemented by von Tappeiner and Jesionek (1903) for the practical development of a treatment for skin diseases, but was actually soon forgotten for several decades. The discovery of penicillin by Sir Alexander Fleming in 1928 (Fleming, 1929) led to great expectations and hopes that at last, humankind was afforded a magic means for combating bacterial infections. Unfortunately, the Golden Age of antibiotics did not last very long, and hopes for a victory over microorganisms were dashed by an obstacle raised by Mother Nature – the development of bacterial resistance to antibiotics. The need to find new approaches for eradication of pathogenic bacteria led to renewed interest in photodynamic killing of bacteria, thus opening a new era of photodynamic antimicrobial chemotheraphy (PACT) targeted against bacteria and bacterial spores (Hamblin and Hasan, 2004; Nitzan and Pechatnikov, 2011; Malik et al., 2010; Daia et al., 2009; Zerdin et al., 2009; Pereira et al., 2012).

PACT is based on a combined action of non-toxic or low-toxicity dyes – photosensitizers (PSs) – and harmless visible light. PSs that exhibit antibacterial activity are represented by uncharged, cationic and anionic molecules which belong to several chemical groups, such as phenothiazine dyes, porphyrins, porphycenes, halogenated xanthenes, chlorin derivatives and fullerenes. A far from complete list of PSs and their abbreviations is presented in Table 1. All PS groups have one common denominator – all the molecules contain conjugated double bonds which are responsible for the absorption of visible light.

Table 1. Examples of photosensitizers used in PACT

Name	Structure	References *
Methylene blue (MB)	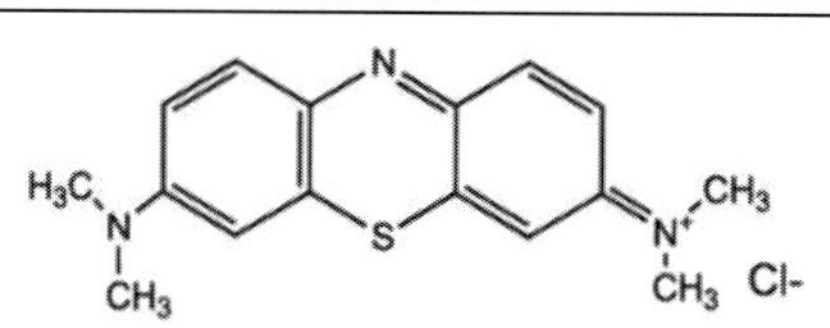	Rolim et al., 2012; Donnelly et al., 2009; Choi et al., 2010; Cassidi et al., 2011; Cassidi et al., 2012; Kashef et al., 2012; Vilela et al., 2012; Nisnevitch et al., 2010; Nakonechny et al., 2011, 2012 ; Sahu et al., 2009

Name	Structure	References *
Toluidine blue O (TBO)	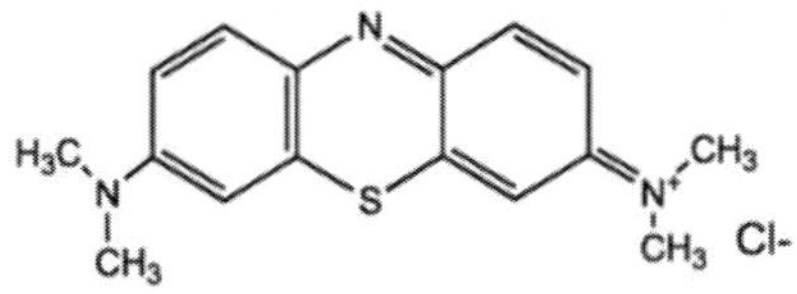	Rolim et al., 2012; Kashef et al., 2012; Vilela et al., 2012; Nakonechny et al., 2011; Sahu et al., 2009
Neutral Red (NR)	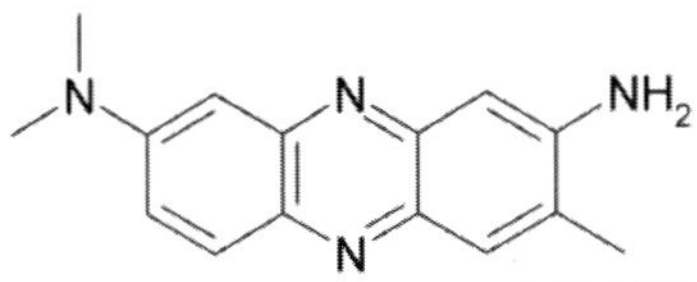	Nisnevitch et al., 2010
Rose Bengal (RB)	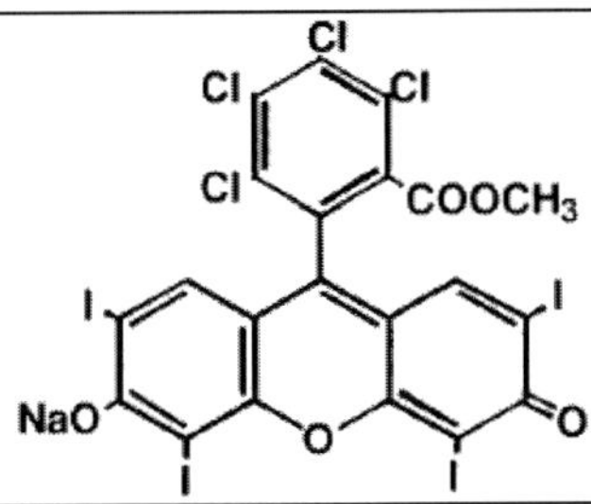	Rolim et al., 2012; Bolean et al., 2010; Nisnevitch et al., 2010; Nakonechny et al., 2012 a and b)
Malachite green (MG)	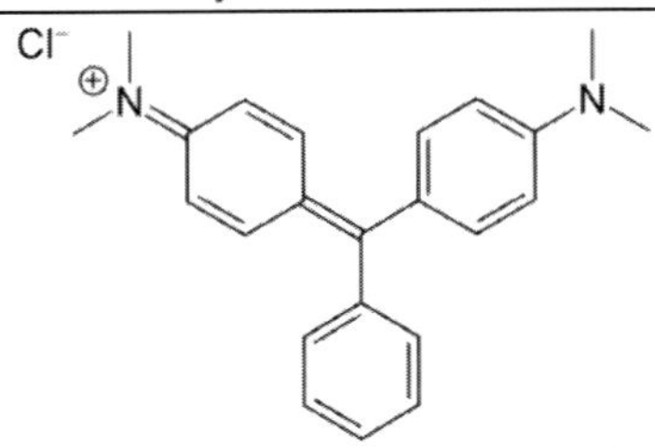	Rolim et al., 2012; Vilela et al., 2012
Eosin (EOS)	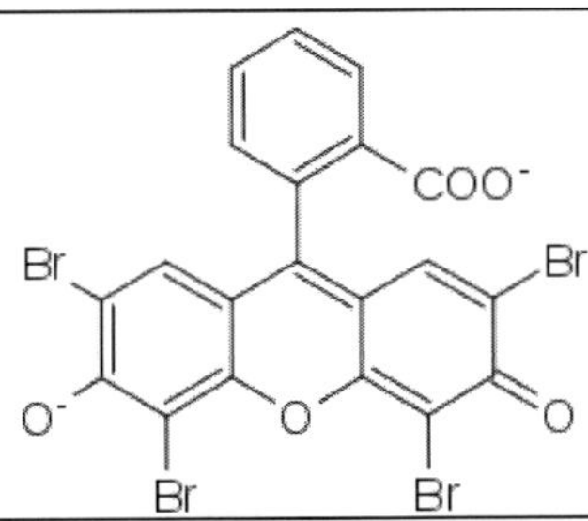	Rolim et al., 2012
Erythrosine (ERI)	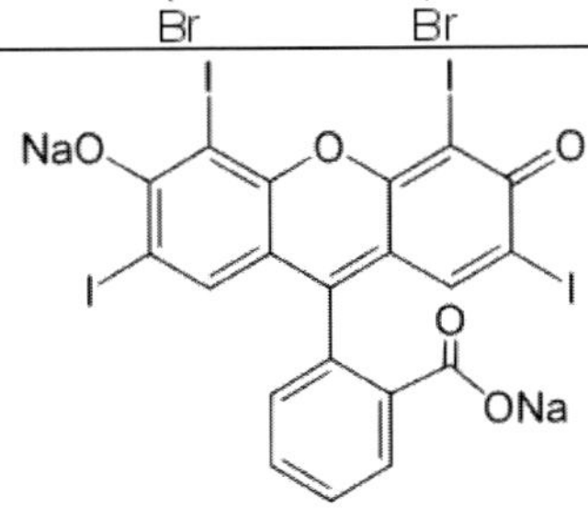	Rolim et al., 2012
δ-Aminolaevulinic acid (ALA)	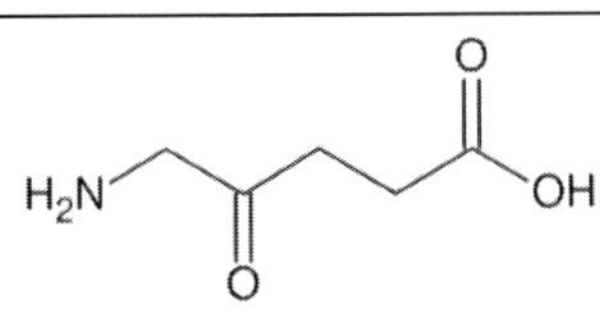	Lee et al., 2004

Table 1. (Continued)

Name	Structure	References *
Photoporphyrin IX	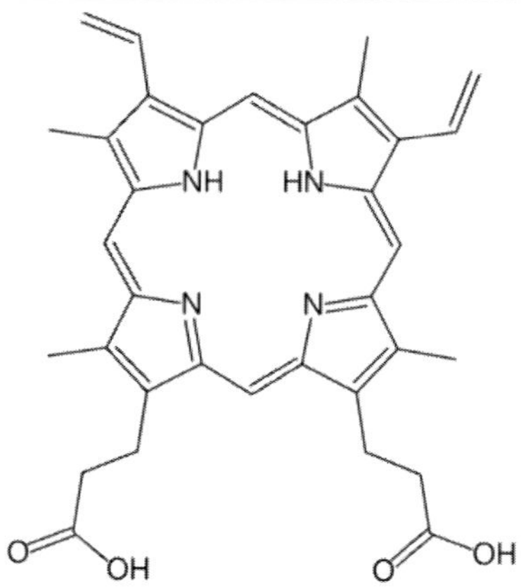	Lee et al., 2004
Meso-tetra (N-methyl-4-pyridyl) porphine tetra tosylate (TMP)	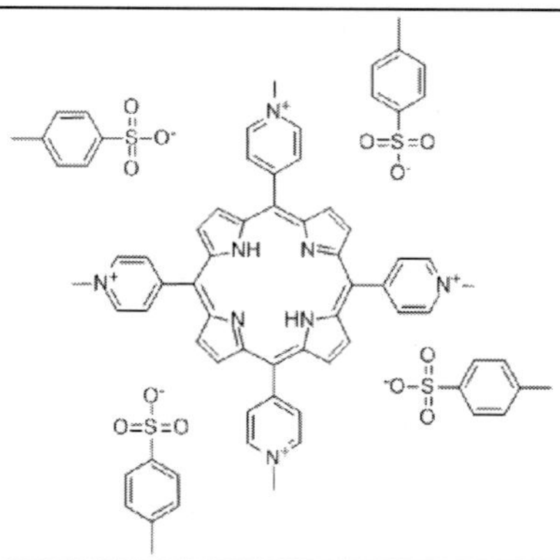	Donnelly et al., 2009; Cassidi et al., 2010 and 2012
Tetrahydroporphyrin-tetratosylat (THPTS)	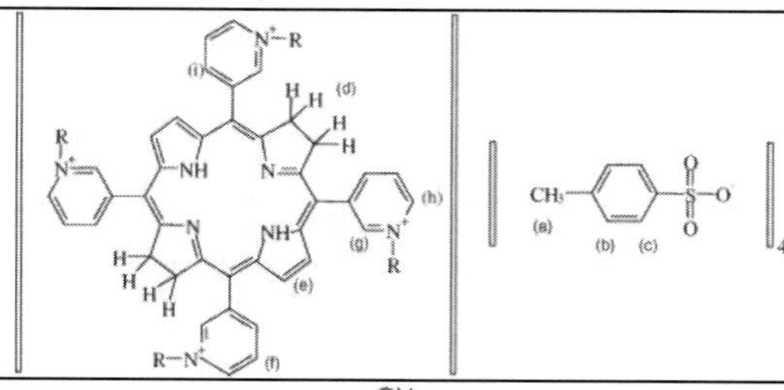	Schastak et al., 2010
5,10,15,20-tetrakis (N-methyl-4-pyridyl)-21H,23H-porphine (TMPyP)	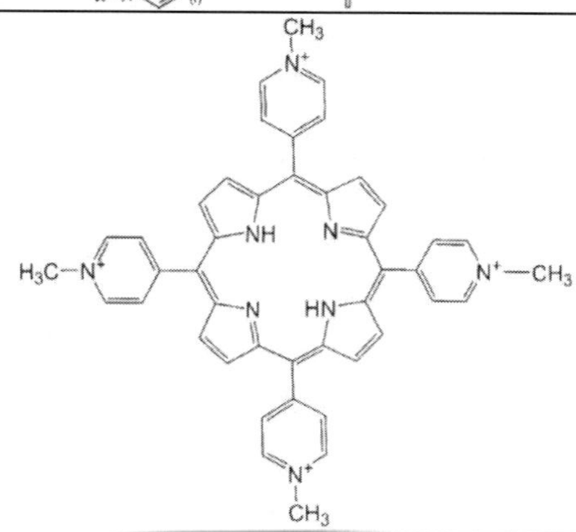	Ragàs and Nonell, 2010; Hanakova et al., 2012

Name	Structure	References *
5-(pentafluorophenyl)-10,15,20-tris(1-methylpiridinium-4-yl)porphyrin tri-iodide (Tri-Py$^+$-Me-PF)	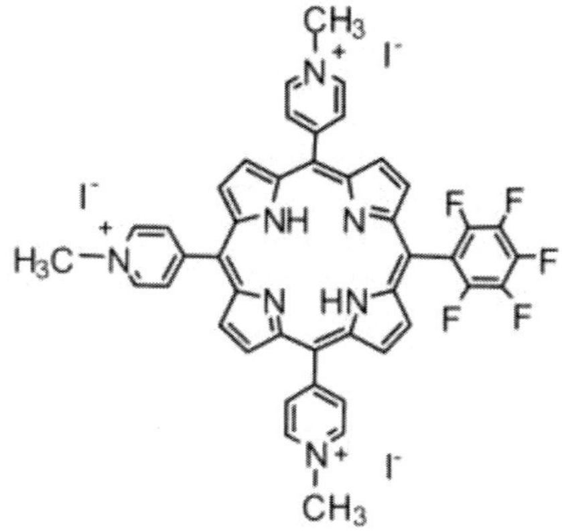	Alves et al., 2009
Porphycenepyridyl tribromide (Py$_3$MeO-TBPo·Br$_3$)	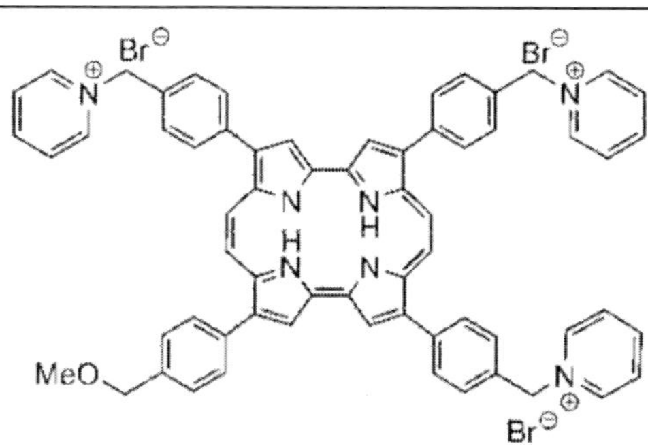	Ragàs et al., 2010
5-mono(N-decyl-4-pyridyl)-10,15,20-tri(N-methyl-4-pyridyl)-21H,23H-porphine (MDPyTMPyP).	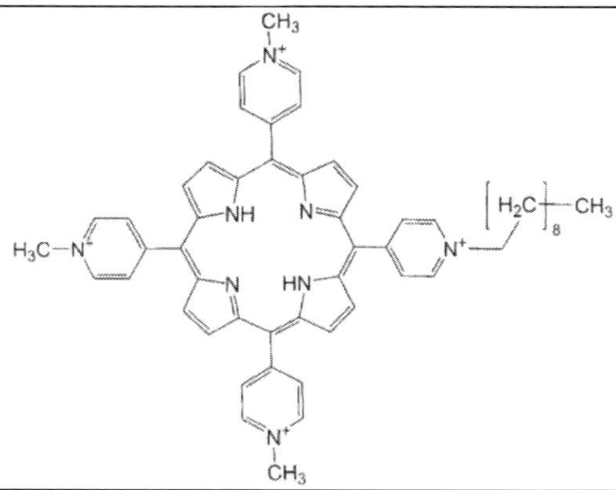	Ragàs and Nonell, 2010
Meso-tetra(3-hydroxyphenyl)chlorin (mTHPC)	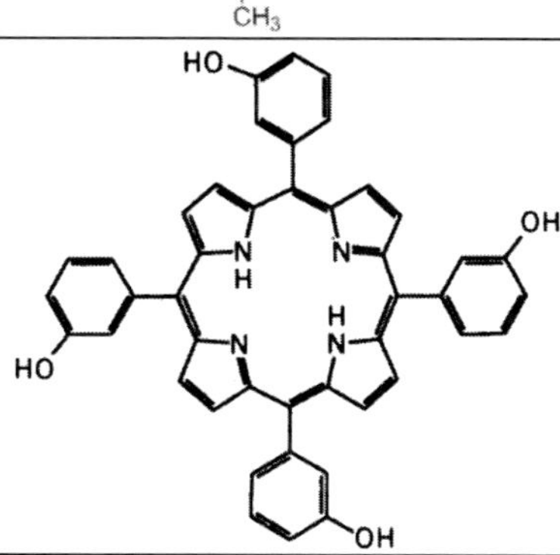	Peèkaitytë et al., 2005; Kranz et al., 2011

Table 1. (Continued)

Name	Structure	References *
Photogem®	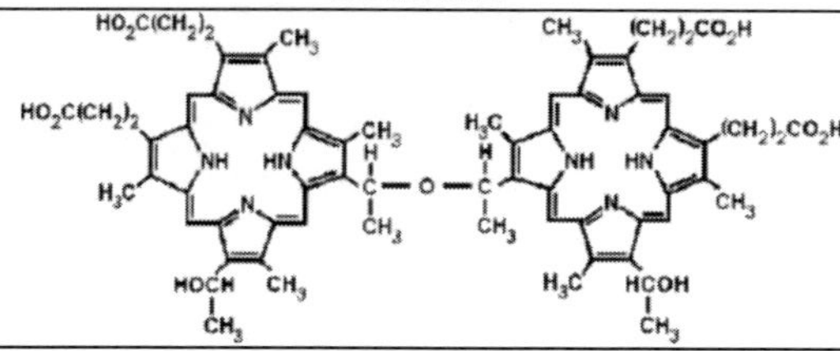	Ferraz et al., 2011
Chlorin e_6 (Ce_6)	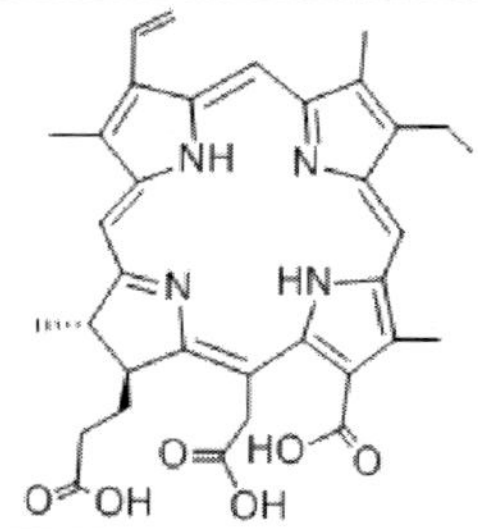	Hamblin et al., 2002; Park et al., 2010; Ferraz et al., 2011
Tetra-3-S-mercaptopyridinyl-5,10,15,20-tetrakis (pentafluorophenyl)chlorin -penta-iodide	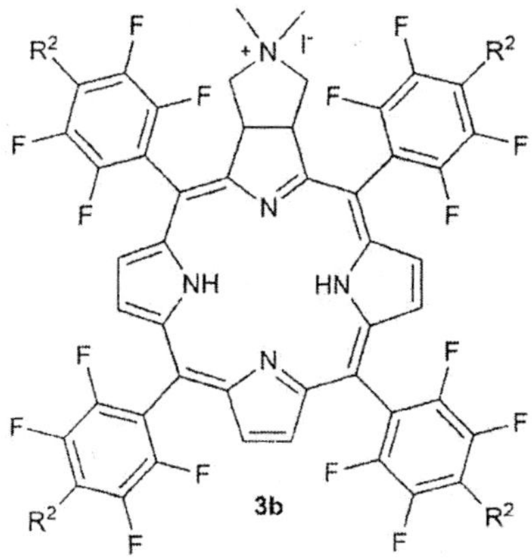 3b where: 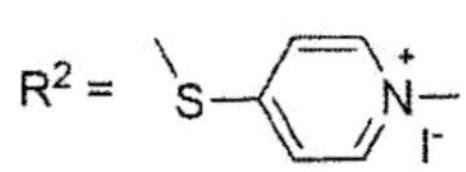	Costa et al., 2012
Zinc(II) 2,9,16,23-tetrakis[4-(N-methylpyridyloxy)]phthalocyanine ($ZnPPc^{+4}$)	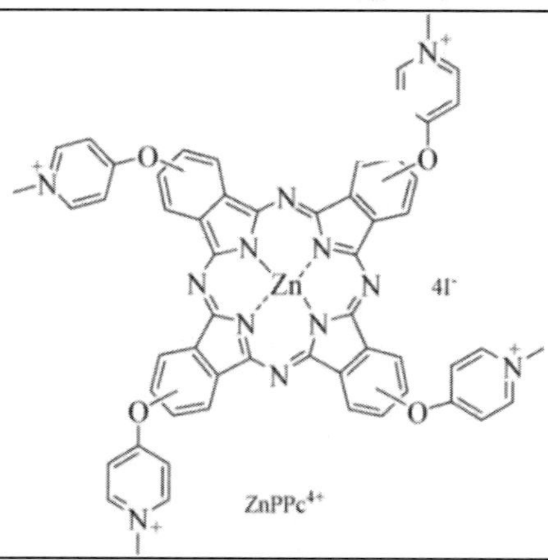	Spesia et al., 2009

Name	Structure	References *
Octa-methylpyridyloxy-gallium(III) phthalocyanine	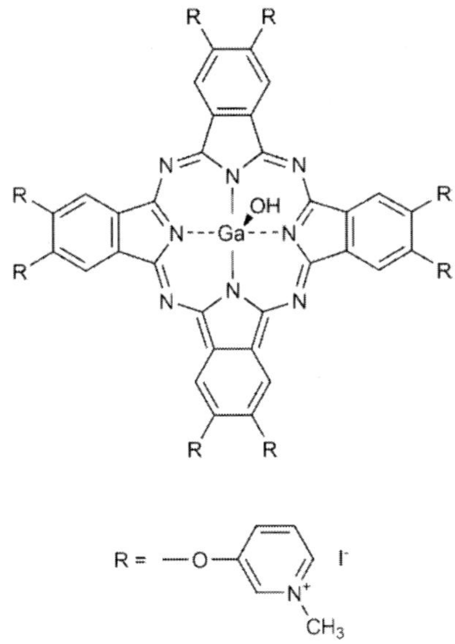	Mantareva et al., 2011
Tetra-methylpyridyloxy-gallium(III) phthalocyanines	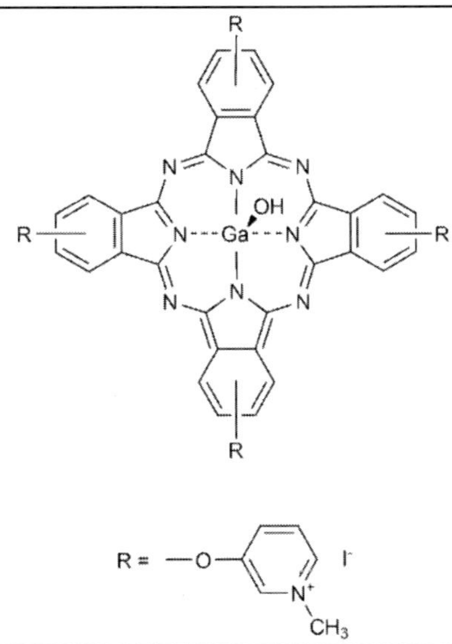	Mantareva et al., 2011
2,6-diiodo-1,3,5,7-tetramethyl-8-(N-methyl-4-pyridyl)-4,4'-difluoroboradiazaindacene	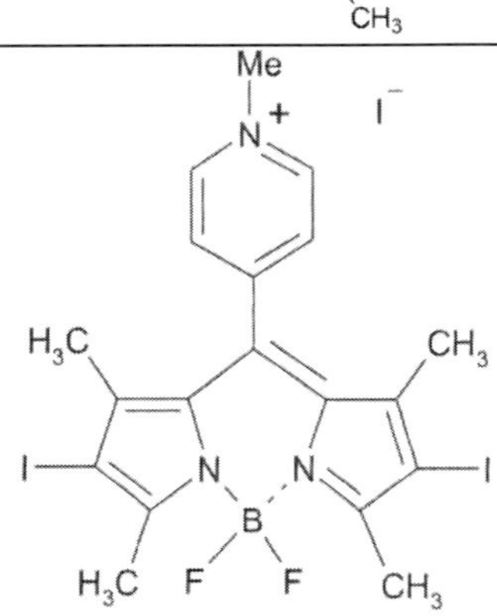	Caruso et al., 2012

Table 1. (Continued)

Name	Structure	References *
N,N-Dimethyl-2-(40-N,N,N-trimethylaminophenyl)fulleropyrrolidinium iodide (DTC_{60}^{2+}).	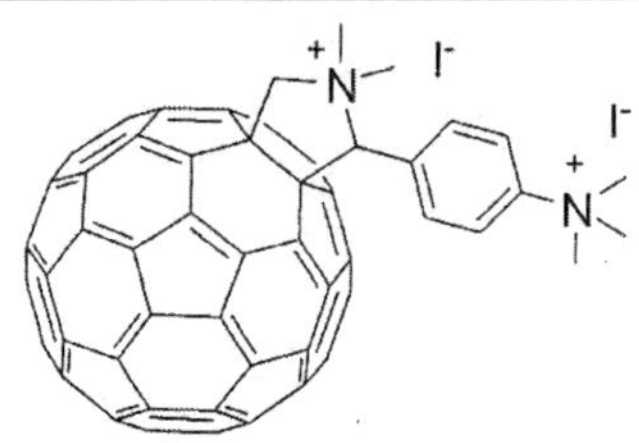	Spesia et al., 2008
N,N-Dimethyl-2-(40-acetamidophenyl)fulleropyrrolidinium iodide (DAC_{60}^{+})	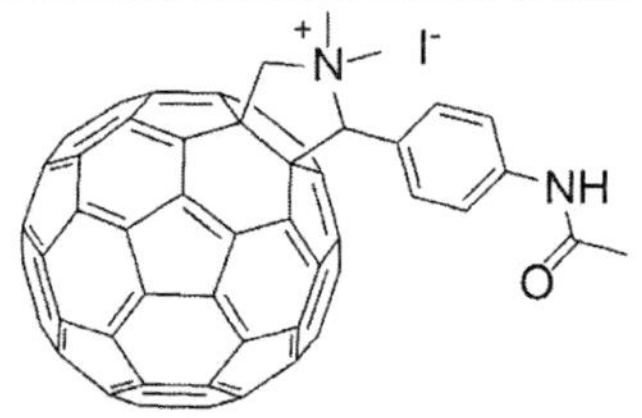	Spesia et al., 2008

* This table relates only to selected references of the past ten years.

PACT Mechanisms

When PS molecules are illuminated by visible light at an appropriate wavelength, they absorb the light energy and transfer from a ground state to an excited triplet state (PS*). Further transformations of the PS* can occur by two different paths, called Type I and Type II reactions (Martin and Logsdon, 1987; Moan and Peng, 2003). In the Type I reaction, PS* can interact with an organic substrate (such as components of a cell membrane), producing free organic radicals which may react with dissolved oxygen, leading to the formation of reactive oxygen species (ROS). A PS* can also transfer an electron to molecular oxygen, thus promoting superoxide ion (O_2^{-}) production, which may in turn be involved in the generation of hydrogen peroxide (H_2O_2) or highly reactive hydroxyl radicals ($HO^{\cdot}$). In the Type II reaction, the PS* can pass energy to dissolved oxygen molecules directly and transfer them to an excited state. The excited oxygen as well as other ROS cause irreversible oxidative damage to biological molecules (lipids, fatty acids, proteins, amino acids, sugars or DNA) and disorder the cell wall structure, thus leading to cell death (Maisch, 2007). A more detailed description of the photosensitization process can be found in the review of Castano et al. (2004). Some of the PSs, for example methylene blue (MB), can be excited by either Type I or Type II mechanisms or by a combination of these processes (Tardivo et al., 2005).

Effect of PACT on Gram-Positive and Gram-Negative Bacteria

The sensitivity of Gram-positive bacteria to PACT is very different to that of Gram-negative strains. Cells of Gram-negative bacteria possess two lipid bilayers, making their cell envelope less permeable to PSs, whereas Gram-positive bacteria have only one lipid bilayer and a relatively permeable outer envelope. These differences are responsible for the higher concentrations of PSs required for inactivation of Gram-negative compared to Gram-positive bacteria (Sahu et al., 2009).

In early studies, Gram-negative bacteria were even considered to be resistant to photodynamic treatment, since phospholipids, complex lipoproteins and polysaccharides present in the additional outer envelope of these bacterial cells inhibited the binding of anionic PS molecules (Nitzan et al., 1992). Several alternative strategies have been developed in order to overcome the barrier which prevents the binding and permeation of PSs into these cells. These strategies include the use of uncharged (e.g., deuteroporphyrin, prochlorphyllide) or positively charged PSs (e.g., tetra-cationic porphyrin substitutes, cationic phthalocyanines, phenothiazine derivatives, porphycenes), particularly when coupled with membrane-penetrating peptides or conjugated to polycations (Alves et al., 2009; Nitzan and Pechatnikov, 2011; Ragàs et al., 2010; Caruso et al., 2012). In addition to their ability to inactivate both Gram-positive and Gram-negative bacteria, cationic PSs demonstrate a higher rate of uptake into bacterial cells than into mammalian cells (Rovaldi et al., 2000; Hamblin et al., 2002). This feature seems to afford an enormous advantage to PSs in a clinical perspective, with reference to their use as a treatment for intrinsic bacterial infections.

Sahu et al. (2009) studied morphological alterations in *Escherichia coli* and in *Staphylococcus aureus* caused by photodynamic action induced by toluidine blue O (TBO) and found different changes in the two bacterial species. The photobactericidal effect of TBO was higher for the Gram-positive *S. aureus*. Positively charged TBO molecules readily bound to the negatively charged residues of teichuronic acid which are present on the cell wall. Photodynamic treatment caused an increase in the cell surface roughness, disappearance of slime, bleb formation and leakage of cellular contents. TBO probably also penetrated into the cytoplasm and caused damage to cytoplasmic components such as DNA (Sahu et al., 2009). In the case of the Gram-negative *E. coli*, TBO bound to the lipopolysaccharide layer of the bacterial envelope due to the highly anionic nature of the outer membrane provided by the presence of phosphate groups, acidic sugars and fatty acyl groups. At low photosensitization doses the bacterial cells underwent structural perturbations in outer membrane components which resulted in an increased roughness of the outer surface. The damage to the outer membrane promoted enhanced permeation of PS molecules into the inner membrane, causing damage to its constituents. At a higher TBO concentration, the cell surface demonstrated blebbing and the cell shape became distorted. This process caused leakage of intracellular contents. Fluid leakage was more noticeable and occurred at a faster rate in *E. coli* than in *S. aureus*. The authors explain this phenomenon by a difference in the peptidoglycan layer of these bacteria, which is much thinner (2-3 nm) in *E. coli* than the multilayered peptidoglycan layer in Gram-positive cells (20-80 nm), and could facilitate faster leakage of intracellular contents (Sahu et al., 2009).

A combination of photodynamic and antibiotic treatment has been reported in addition to the sole application of PACT. Poto et al. (2009) tested exposure of *S. aureus* biofilms to vancomycin after photodynamic treatment with meso-tetra (N-methyl-4-pyridyl) porphine tetra tosylate (TMP) under visible light and achieved a drastic reduction in staphylococcal survival. A similar idea was realized in the work of Cassidi et al. (2012), who combined MB and TMP-mediated PACT with several antibiotics (tobramycin, ceftazidime, chloramphenicol or ciprofloxacin) for treatment of both planktonic and biofilm cultures of *Burkholderia cepacia* complex infection. PACT increased the bactericidal activity of the antibiotics against these bacteria.

PS Excitation Techniques

Excitation of PSs in the process of bacterial eradication can be performed by several methods, including illumination with external light sources, chemiluminescent illumination and sonodynamic activation.

PS Excitation by External Illumination

The first group of external electrical light sources for PS excitation includes lasers, light-emitting diodes (LED) and halogen lamps, all of which have a single specified wavelength or narrow emission regions which match the maximal light absorption regions of the PS. For instance, a He-Ne laser emitting at 633 nm was used for excitation of Radachlorin®, composed mostly of chlorin e_6 with maximum absorption at λ_{max}=662 nm (Fekrazad et al., 2011). This system was shown to be effective in killing *Streptococcus mutans*. The same source of irradiation was used for excitation of MB (λ_{max}=665 nm) in the process of photodynamic inactivation of pulmonary pathogens (Cassidi et al., 2011). Diode lasers emitting at 660 and 630 nm were used for excitation of MB and TBO (λ_{max}=630 nm) with the aim of eradicating methicillin-resistant *S. aureus* (MRSA) and multi-drug resistant *E. coli* (Kashef et al., 2012). A red LED emitting in the 633-645 nm region (predominantly at 636 nm) excited MB, TBO and malachite green (MG) (λ_{max}=675 nm) (Rolim et al., 2012). The system was used for eradication of *S. mutans*. A LED array lamp with major emittance at 630 nm was applied for killing *Pseudomonas aeruginosa* as planktonic and biofilm cultures by means of the endogenous PS photoporphyrin IX synthesized upon administration of δ-aminolaevulinic acid (ALA) (Lee et al., 2004). In the work of Papastamou et al. (2011), a blue-band halogen lamp emitting at 450-485 nm was applied for eradication of *Fusobacterium nucleatum* and *Porphyromonas gingivalis* by excited ruthenium-based PSs, having absorbance maxima at 463, 452 and 462 nm.

These examples demonstrate that even partial overlapping of the source emission wavelengths with the absorption spectra of PSs was sufficient for effective PACT. Donnelly et al. (2007) reported using TBO and TMP excited by a xenon/ceramic lamp with a filtered output of 635±25 nm for a high rate of killing *P. aeruginosa* growing planktonically or in biofilms. It should be noted that the maximal absorption of TBO lies at 630 nm and that of TMP lies at 450, 540 and 570 nm (Mitra et al., 2011).

Use of light sources with specific emittance wavelengths for PS excitation provides the most effective utilization of the light source. However, use of these sources is limited to excitation of a small group of PSs, which absorb light at the appropriate wavelengths and therefore cannot be used for activation of other PSs, which absorb in other regions.

Sources of wide-band light, such as luminescent or xenon lamps, do not have this limitation and are widely used in PACT by applying the entire range of the lamp's emission region or by filtering a narrow region from the emission spectrum.

Our group used a luminescent white lamp emitting at 400-700 nm for photosensitization of MB, TBO, Rose Bengal (RB) and neutral red (NR). MB demonstrated high efficacy in inhibiting Gram-positive *Sarcina lutea*, *S. aureus* (methicillin-sensitive, MSSA, and MRSA) strains and *Staphylococcus epidermidis*, as well as Gram-negative *E. coli* and *Shigella flexneri*. NR was effective against *S. lutea*, *S. aureus*, *E. coli* and *Salmonella para B*. RB demonstrated high ability to eradicate *S. lutea*, *S. aureus* and *S. epidermidis*, *P. aeruginosa* and *S. flexneri*. TBO was tested against *S. aureus* and *E. coli* and exhibited high antibacterial activity (Nisnevitch et al., 2010; Nakonechny et al., 2010; Nakonechny et al., 2011; Nakonechny et al., 2012). It is interesting that the MRSA strain was even more sensitive to the PACT treatment in the presence of MB than the MSSA strain (Nakonechny et al., 2011). Figure 1 illustrates this phenomenon: both *S. aureus* strains exhibited susceptibility to MB excited by white light, and what is more, after an hour of treatment the viability of the MSSA strain decreased by 4 orders of magnitude of colony forming units (CFU) and the viability of the MRSA strain decreased by 4.5 orders.

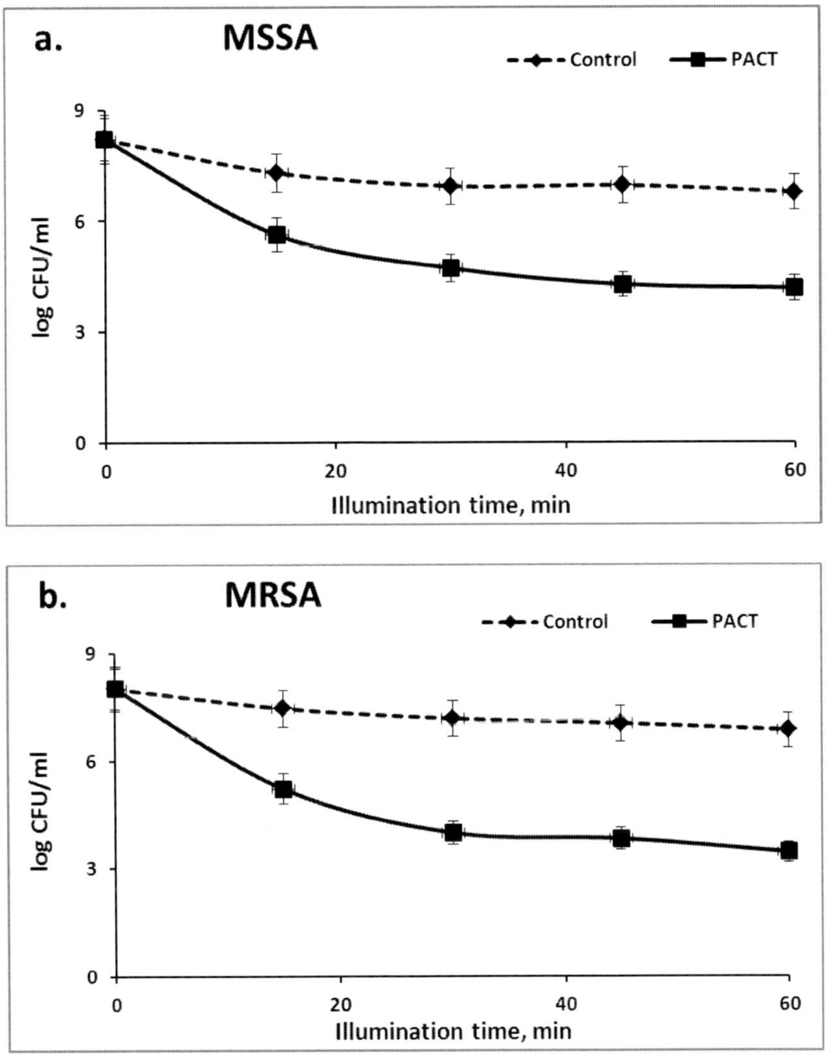

Figure 1. PACT effect on the viability of a. MSSA and b. MRSA. Cells with initial concentration of 10^8 cells mL^{-1} were incubated with 25 μM of MB for 20 min in the dark and then illuminated with a white luminescent lamp with a fluence rate of 1.6 mW cm^{-2} at 25°C under temperature control. After treatment, cells were diluted in 10-fold dilutions and evenly spread over BH-agar plate. Plates were incubated at 37°C overnight and CFU were counted.

Poor penetration of visible light does not enable treating internal body tissues. The light range of 380-480 nm cannot penetrate more than 1 mm into tissue. Possible clinical applications of PACT are therefore usually limited to dermatology and stomatology (Ibbotson, 2012; Vilela et al., 2012; Lee and Baron, 2011; Séguier et al., 2010; Sakamoto et al., 2010).

PSs are classified as belonging to the first and second generation, according to their ability to absorb visible light in various regions (Costa et al., 2012). PSs of the first generation are excited by blue or white light which exhibits very limited skin penetration, whereas representatives of the second generation can be excited by red light (650-800 nm), which is capable of penetrating up to several cm into the depth of the tissues (Costa et al., 2012). A group of chlorin-based PSs was found to be an ideal choice for photosensitization in this light region.

Schastak et al. (2010) used near infrared (NIR) activation of Tetrahydroporphyrin-tetratosylat (THPTS) at 761±3 nm provided by a diode laser in order to abolish, or at least diminish, the limitations of PACT. The wavelength chosen for the photodynamic treatment allowed activating THPTS in tissues with a thickness of up to 10 mm. The scheme using NIR was shown to be effective for photokilling of Gram-positive *S. aureus* (both MSSA and MRSA strains) and Gram-negative *E. coli* and *P. aeruginosa*.

Additional information on a range of electrical light sources can be found in the review of Nagata et al. (2012).

Chemiluminescent Excitation of PSs

A new approach for overcoming restrictions of PACT related to limited penetration of visible light into tissues is based on a replacement of the external light source by chemiluminescent light emitted during the course of a chemical reaction.

Our group used the chemiluminescent oxidation of luminol (LM) by hydrogen peroxide in the presence of Fe^{+2} ions, which is characterized by a release of light energy. The reaction includes several stages, including formation of a number of intermediates and is accompanied by detachment of nitrogen and light emission (Ferreira and Rossi, 2002). The spectrum of the emitted light lies in the visible region (λ<550 nm) and is characterized by maximal emission at 430-450 nm.

The technology of chemiluminescent excitation of PSs, which we called Chemiluminescent Photodynamic Antimicrobial Therapy (CPAT), was effective against both Gram-positive and Gram-negative bacteria. Our data showed that CPAT applied against bacteria in the presence of MB or TBO and LM in the dark effectively reduced the viability of *S. aureus* and *E. coli*. CPAT treatment of the cells with MB in the presence of LM led to a significant reduction (2-3 $\log_{10}$) in bacterial viability.

It was interesting to compare the effects of CPAT and PACT on MRSA and MSSA strains. Both *S. aureus* strains were incubated for 1 h in the presence of MB and LM in the dark or exposed to external illumination. Both CPAT and PACT were more effective in eradicating MRSA than MSSA (Figure 2). Separate incubation of the cells with MB or with LM in the mixture with H_2O_2 and a catalyst did not affect cell viability, demonstrating that CPAT, similarly to PACT, requires the presence of both components - a light source and a

PS. As seen from the Figure 2, when CPAT and PACT were applied for the same 1 h time period, CPAT was less effective than PACT. However, when treatment time was extended to 2 h, the efficacy of CPAT was comparable to that of PACT (Nakonechny et al., 2010 and 2011).

The obtained results show that CPAT can become a practical and effective alternative to traditional PACT in killing and inhibiting the growth of both Gram-positive and Gram-negative bacteria, including antibiotic-resistant strains. The important advantage of CPAT is the possibility of administering photodynamic treatment without external illumination, which opens prospects for combating internal infections that are difficult or not available for treatment by traditional PDT. This property turns CPAT into a promising novel antimicrobial therapeutic strategy.

Another idea of using chemiluminescent light was suggested by Ferraz et al. (2011), who proposed exploiting a chemiluminescent reaction of bis(2,4,6-trichlorophenyl) oxalate in the presence of hydrogen peroxide and an activator, 9,10-bis(phenylethynyl) anthracene, in a solution of imidazole and ethyl acetate. The produced chemiluminescent light was characterized by an emission spectrum between 450 and 600 nm and was applied for external activation of Photogem® with the aim of reducing the concentration of *S. aureus* cells. The advantage of this system lies in a possibility of constructing a flexible and portable non-wired device based on chemiluminescence as a potential external light source for PACT.

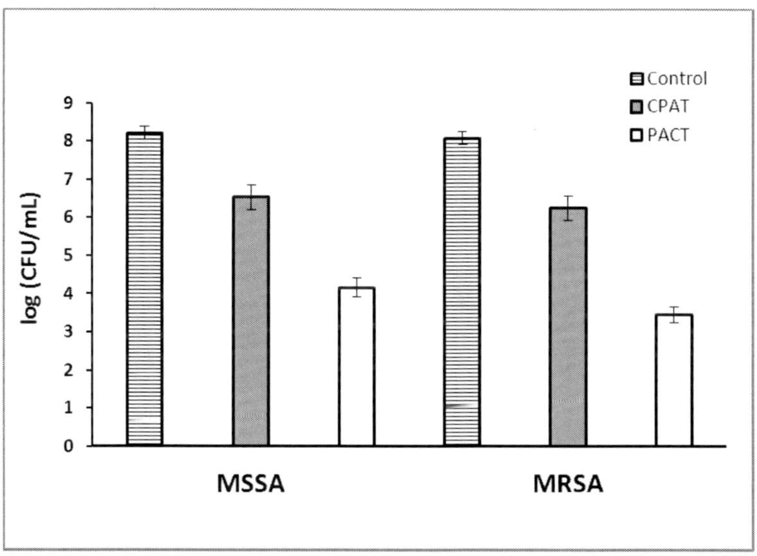

Figure 2. PACT and CPAT effect on the viability of MSSA and MRSA. In CPAT experiments 10^8 cells mL^{-1} were incubated with 25 μM MB for 1 h at 25°C together with 0.15 mM of luminol in the presence of 4 μM $FeSO_4$ and 3μM H_2O_2. In control experiments cells were incubated separately with each of the components (MB and LM together with $FeSO_4$ and H_2O_2). Strict precautions were taken to avoid external illumination of the system during addition of components and further incubation. In PACT experiments 10^8 cells mL^{-1} were incubated with 25 μM MB for 20 min in the dark and then illuminated with a white luminescent lamp with a fluence rate of 1.6 mW cm^{-2} for 1 h at 25°C under temperature control. After the treatments aliquots of mixtures were diluted in 10-fold dilutions and evenly spread over BH-agar plate. Plates were then incubated at 37°C overnight and CFU were counted taking dilutions into account.

Sonodynamic Excitation of PSs

Another interesting way for PS activation which does not require an external light source is sonodynamic excitation of PSs by ultrasound. Sonodynamic therapy (SDT) is based on ultrasound-induced cytotoxicity of compounds called sonosensitizers and this mode has already been studied for inhibition of cancer cells.

The term SDT is used for all non-thermally related therapeutic ultrasound applications, from induction of apoptosis when combined with chemotherapy to ultrasound therapy. However, this term is usually used for ultrasonic activation of drugs for cancer therapy (Tachibana et al., 2008). The exact mechanism of SDT has not yet been clarified, and it is assumed that there is no universal mechanism of synergism between ultrasound and drugs, and that different classes of sonosensitizers are activated in the dark by different mechanisms (Rosenthal et al., 2004; Mišík and Riesz, 2000). It is thought that SDT can affect biological tissues by one of three different mechanisms: heat, mechanical effects and acoustic cavitation, which result in structural and/or functional changes in the cells (Yu et al., 2004).

Mišík and Riesz (2000) believe that PSs (for example porphyrins) can be sonosensitized by the following scheme: a drug undergoes pyrolysis inside collapsing cavitation bubbles or in the heated gas-liquid interface, forming free radical intermediates which react with dissolved oxygen to form peroxyl radicals, and the latter attack cellular sites due to their ability to diffuse to significant distances.

Other authors explain the sonodynamic effect of PSs by electronic excitation of the drug molecules by sonoluminescence. This phenomenon is well-known since 1934, and is expressed by light flashes produced during acoustic ultrasonic cavitation in liquids without any application of external illumination (Putterman, 1995). The range of sonoluminescent emission depends on the type of liquid phase and on dissolved gases. In an aqueous medium, sonoluminescence exhibits a broadband emission spectrum from 200 to 700 nm, with maximum emission intensity between 250 and 600 nm (Hatanaka et al., 2002). Sonoluminescence is considered to induce a chain of photochemical processes leading to production of cytotoxic singlet oxygen (Rosenthal et al., 2004; Yumita et al., 1990 and1997) .

The STD technique has been well studied in the field of cancer cell inhibition. However, the hypothetical possibility of applying this technique to bacterial eradication was first proposed only in 2009 by Ma et al. (2009), who suggested a new concept of sonodynamic antimicrobial chemotherapy (SACT) as a promising novel antimicrobial strategy.

This hypothesis was proven experimentally by our group. Two types of cells, *E. coli* and *S. aureus*, were chosen for primary evaluation of sonodynamic activation for eradication of bacteria (Nakonechny et al., 2013). RB and MB, which are sonosensitizers well-known for their photodynamic activity, were chosen. The sonodynamic activity of RB against cancer cells has been reported previously (Umemura et al., 1999; Hiraoka et al., 2006). To the best of our knowledge, the sonodynamic activity of MB has not been studied to date, but PSs are believed to be good candidates for exhibiting sonodynamic properties (Rosenthal et al., 2004; He et al., 2011). Furthermore, RB and MB are water-soluble, which enabled performing the experiments in aqueous solutions

Our experiments for testing sonodynamic antibacterial properties of RB and MB were held in an ultrasonic bath with cold water under an ultrasound frequency of 28 kHz and an intensity of 0.84 W cm^{-2}. All experiments were performed under strict dark conditions in order to exclude any possible photodynamic excitation of the examined compounds. SACT activated RB exhibited a high cytotoxic effect against both *E. coli* and *S. aureus* tested at concentrations of 10^6-10^9 CFU mL^{-1}. The *E. coli* concentration decreased by 2-4.7 log_{10} and the highest *E. coli* eradication rate was achieved when 10^6 CFU mL^{-1} of cells were treated by 15 μM RB, when the remaining *E. coli* concentration dropped to only 20 CFU mL^{-1} (Figure 3a). *S. aureus* was even more sensitive to the sonodynamic treatment. The bacteria were suppressed by 3.5-6 log_{10} compared to the untreated samples at all tested cell concentrations, and treatment of *S. aureus* at the lowest concentration of 10^6 CFU mL^{-1} by 5 μM RB actually caused total eradication of the cells (Figure 3b). The SACT effect presumably depends on the ratio between the PS dose and the bacterial concentration (Nakonechny et al., 2013). An analogous phenomenon of different effects of PSs on bacterial cells at various concentration ratios was demonstrated previously in PACT experiments, and it was shown that bacteria exhibited stronger susceptibility for PACT treatment at low concentrations (Nakonechny et al., 2012). Prolonged sonication led to total bacterial suppression even at high initial cell concentrations: when treated at an initial concentration of 10^7 CFU mL^{-1}, *S. aureus* cells were effectively eradicated after 2 h sonication in the presence of 5 μM RB, whereas after 1 h treatment, the residual concentration of viable cells was 50 CFU mL^{-1} (Nakonechny et al., 2013).

In contradistinction to RB, MB showed very poor activity under SACT conditions. At an initial bacterial concentration of 10^9 CFU mL^{-1}, sonication in the absence of MB caused a 1.4 log_{10} decrease in CFU, and only a 1.7 log_{10} reduction in the *S. aureus* CFU in the presence of MB, such that the difference between these data was not significant. The inefficiency of MB in the SACT experiment was quite surprising. It can be assumed that not all photosensitizing compounds are sensitive to sonodynamic activation. The sensitivity of PSs to sonication probably depends on the structure of the molecules, and especially their hydrophobicity and electrical charge. The PACT technique is generally more effective in eradicating bacteria than SACT, but the efficiency of the latter can probably be increased by application of ultrasound at higher intensities or other frequencies, thus transforming this technique into a practical and convenient tool for bacterial suppression under dark conditions. Additional studies should undoubtedly be carried out for more profound investigation of the SACT method, including use of different PSs, different concentrations and application of a wide range of ultrasound intensities (Nakonechny et al., 2013).

Combating bacterial infections *in vivo* can be more effective and can diminish side effects of the treatment when targeted to the site of infection. In addition to known methods of targeting drugs to cells, use of ultrasound may afford direct and exact focusing on the treated site. Since non-activated PSs do not harm healthy tissues, excitation of the PS only after it reaches the treatment site can prevent systemic toxic effects during drug delivery to the treatment site. Such targeted treatment was carried out on mouse cancer cells and resulted in cessation of tumor growth and even in destruction of tumor tissue (Yumita et al., 1997; Umemura et al., 1996). The same targeting scheme can most probably be applied for antibacterial treatment by sonosensitized PSs.

A method for cold sterilization of medical instruments and surgical accessories can be proposed, in addition to future potential therapeutic applications of SACT. A simple addition of sonosensitizers, for example RB, to the aqueous phase in the ultrasonic bath will significantly improve and increase the effectiveness of sterilization processes (Nakonechny et al., 2013).

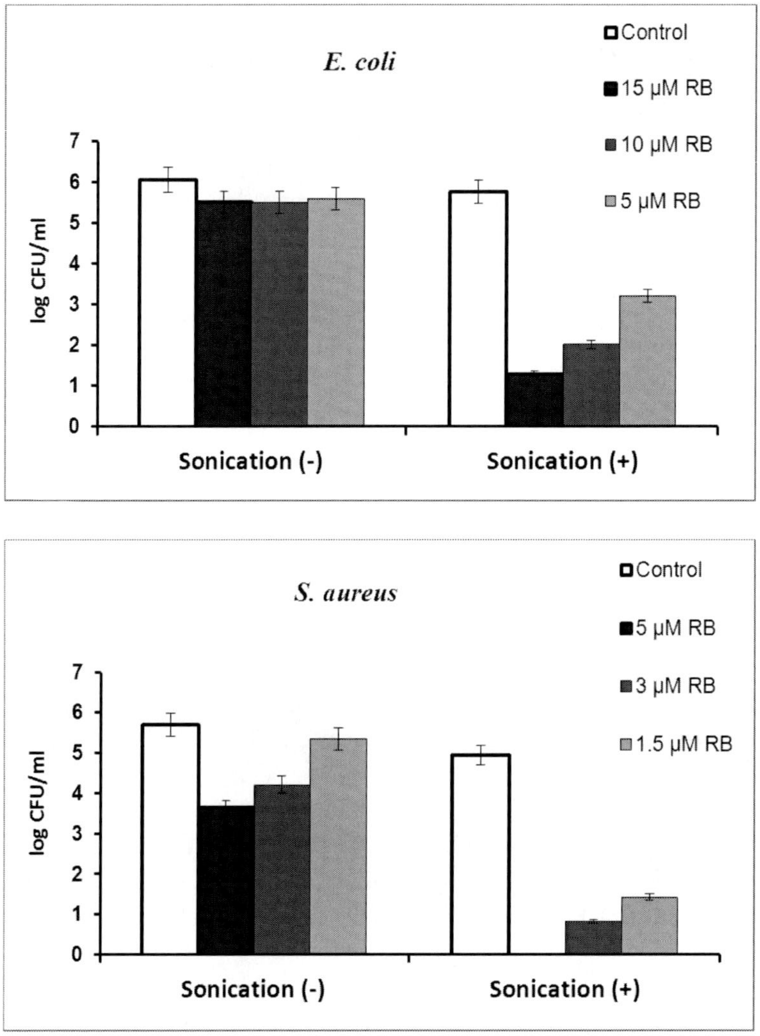

Figure 3. SACT effect of RB on Gram-negative and Gram-positive bacteria. Cells of a. *E. coli* and b. *S. aureus* at initial concentration of 10^6 CFU mL^{-1} were incubated with 0-15 μM RB (*E. coli*) or 0-5 μM (*S. aureus*) under sonication in an ultrasonic bath or without sonication for 1 h in the dark. Controls: sonicated or not sonicated bacterial cultures without RB. After the treatment, bacterial samples were serially diluted in 10-fold dilutions and evenly spread over BHA plates with a Drigalsky spreader. Plates were incubated at 37°C overnight and CFU were counted.

Hypothetical Resistance of Bacteria to PACT

In contradistinction to antibiotics, no development of bacterial resistance to PSs has been reported, although several possible mechanisms of such hypothetical resistance are discussed in the literature (Hamblin, 2012; Peèkaitytë et al., 2005). The first mechanism considers an action of efflux pump systems in bacterial cells, which are known to be responsible for drug resistance phenomena in bacteria (Poole, 2005; Nikaido, 1998). Peèkaitytë et al. (2005) discussed the possibility of such a mechanism and Tegos and Hamblin (2006) reported that phenothiazinium-based PSs served as substrates for microbial efflux pumps, such that the possibility of this mechanism cannot be excluded. Another mechanism is related to an expression of GroEL and DnaK heat shock proteins in bacterial cells. This phenomenon was described as a bacterial response to a stress caused by heat or by antibiotics, indicating a general mechanism of bacterial cell reactions to external stress conditions (Neidhardt and Van Bogelen, 1987; Tran et al., 2011; Cardoso et al., 2010). Expression of the same proteins was reported upon photodynamic treatment of bacteria (Bolean et al., 2010; Hamblin et al., 2011). This fact does not enable neglecting the probability of bacterial resistance to PSs, although no experimental evidence for the resistance phenomenon has been found to date, and in practice, multiple repeated re-growth of bacterial cells that survived photodynamic treatment did not lead to development of bacterial resistance to PSs (Lauro et al., 2002; Pedigo et al., 2009; Tavares et al., 2010). Cassidy et al. (2010) studied the response of *S. aureus* and *P. aeruginosa* to treatment by PACT and by a range of antibiotics used in a standard procedure against infections caused by these bacteria. The results of this study demonstrated that exposure of *P. aeruginosa* and *S. aureus* to MB and TMP under sub-lethal PACT conditions did not decrease the susceptibility of the cells to a range of antibiotics. *P. aeruginosa* also exhibited no reduction in susceptibility to PACT after habituation under sub-lethal PACT conditions. Some of the clinical isolates of *S. aureus* demonstrated a decreased response to MB and TMP, but the $\log_{10}$ reduction in CFU was still greater than four. PSs are still generally considered to be bactericidal agents that do not cause bacterial resistance, because, unlike other agents, PSs in combination with light produce ROS in a controlled time-dependent manner and not upon exposure to the PS alone (Cassidy et al., 2010). This phenomenon is presumably not limited to microorganisms but is also related to cancer cells, which except for rare situations, also do not develop resistance to photodynamic treatment (Laptev et al., 2006; Hornung et al., 1998). The question of the difference between PSs and other cytotoxic agents such as antibiotics and anti-cancer drugs which are known to lead to the development of multi-drug resistance after repeated exposure to the drug, remains open (Davies and Davies, 2010; Tan et al., 2010). This subject needs further investigation, since understanding the mechanisms for preserving the cells' sensitivity to the photodynamic treatment could promote development of advanced generations of cytotoxic drugs.

Conclusion

Photodynamic treatment was proved to be a powerful tool for eradication of both Gram-positive and Gram-negative bacteria, including antibiotic-resistant strains. Excitation of PSs can be performed by illumination with external light sources or by chemiluminescent illumination and sonodynamic activation in the dark.

Acknowledgments

The preparation of this chapter was supported by the Research Authority of the Ariel University, Israel.

References

Alves, E., Costa, L., Carvalho, C. M., Tomé, J. P., Faustino, M. A., Neves, M. G., Tomé, A. C., Cavaleiro, J. A., Cunha, A. & Almeida, A. (2009). Charge effect on the photoinactivation of Gram-negative and Gram-positive bacteria by cationic meso-substituted porphyrins. *BMC Microbiol.*, *9*, 70-82.

Bolean, M., de Paiva Paulino, T., Thedei Jr., G. & Ciancaglini, P. (2010). Photodynamic Therapy with Rose Bengal Induces GroEL Expression in *Streptococcus mutans*. *Photomed Laser Surg.*, *28*(S1), S79-S84.

Cardoso, K., Gandra, R. F., Wisniewski, E. S., Osaku, C. A., Kadowaki, M. K., Felipach-Neto, V., Haus, L. F. A.-A., de Cássia, R. & Simão, G. (2010). DnaK and GroEL are induced in response to antibiotic and heat shock in *Acinetobacter baumannii*. *J. Med. Microbiol.*, *59*, 1061-1068.

Caruso, E., Banfi, S., Barbieri, P., Leva, B. & Orlandi, V. T. (2012). Synthesis and antibacterial activity of novel cationic BODIPY photosensitizers. *J. Photochem. Photobiol. B–Biol.*, *114*, 44-51.

Cassidy, C. M., Donnelly, R. F. & Tunney, M. M. (2010). Effect of sub-lethal challenge with Photodynamic Antimicrobial Chemotherapy (PACT) on the antibiotic susceptibility of clinical bacterial isolates. *J. Photochem. Photobiol. B–Biol.*, *99*, 62-66.

Cassidy, C. M., Tunney, M. M., Magee, N. D., Elborn, J. S., Bell, S., Singha, T. R. R. & Donnelly, R. F. (2011). Drug and light delivery strategies for photodynamic antimicrobial chemotherapy (PACT) of pulmonary pathogens: A pilot study. *Photodiagn Photodyn*, *8*, 1-6.

Cassidy, C. M., Donnelly, R. F., Elborn, J. S., Magee, N. D. & Tunney, M. M. (2012). Photodynamic Antimicrobial Chemotherapy (PACT) in combination with antibiotics for treatment of *Burkholderia cepacia* complex infection. *J. Photochem. Photobiol. B–Biol.*, 106, 95-100.

Castano, A. P., Demidova, T. N. & Hamblin, M. R. (2004). Mechanisms in photodynamic therapy: part one – photosensitizers, photochemistry and cellular localization. *Photodiagn Photodyn*, *1*, 279-293.

Choi, S. S., Lee, H. K. & Chae, H. S. (2010). In vitro photodynamic antimicrobial activity of methylene blue and endoscopic white light against *Helicobacter pylori* 26695. *J. Photochem. Photobiol. B–Biol.*, *101*, 206-209.

Costa, D. C. S., Gomes, M. C., Faustino, M. A. F., Neves, M. G. P. M. S., Cunha, Â., Cavaleiro, J. A. S., Almeida, A. & Tomé, J. P. C. (2012). Comparative photodynamic inactivation of antibiotic resistant bacteria by first and second generation cationic photosensitizers. *Photochem Photobiol Sci*, *11*, 1905-1913.

Daia, T., Huanga, Y.-Y. & Hamblin, M. R. (2009). Photodynamic therapy for localized infections – State of the art. *Photodiagn Photodyn*, 6, 170-188.

Davies, J. & Davies, D. (2010). Origins and evolution of antibiotic resistance. *Microb. Mol. Biol. Reviews*, *74*, 417-427 .

Donnelly, R. F., McCarron, P. A., Cassidy, C. M., Elborn, J. S. & Tunney, M. M. (2007). Delivery of photosensitisers and light through mucus: Investigations into the potential use of photodynamic therapy for treatment of *Pseudomonas aeruginosa* cystic fibrosis pulmonary infection. *J Control Release*, *117*, 217-226

Donnelly, R. F., Cassidy, C. M., Loughlin, R. G., Brown, A., Tunney, M. M., Jenkins, M. G., McCarron, P. A. (2009). Delivery of methylene blue and meso-tetra (N-methyl-4-pyridyl) porphine tetra tosylate from cross-linked poly(vinyl alcohol) hydrogels: A potential means of photodynamic therapy of infected wounds. *J. Photochem. Photobiol. B–Biol.*, 96, 223-231.

Fekrazad, R., Bargrizan, M., Sajadi, S. & Sajadi, S. (2011). Evaluation of the effect of photoactivated disinfection with Radachlorin® against *Streptococcus mutans* (an in vitro study). *Photodiagn Photodyn.*, *8*, 249-253.

Ferraz, R. C. M. C., Fontana, C. R., de Ribeiro, A. P., Trindade, F. Z., Bartoloni, F. H., Baader, J. W., Lins, E. C., Bagnato, V. S. & Kurachi, C. (2011). Chemiluminescence as a PDT light source for microbial control. *J. Photochem. Photobiol. B–Biol.*, *103*, 87-92.

Ferreira, E. C. & Rossi, A. V. A. (2002). Quimioiluminescncia como ferramenta analtica: do mecanismo a aplicaes da reao do luminol em mtodos cinticos de anlise. *Quim. Nova*, *25*, 1003-1011.

Flemming, A. (1929). On the antibacterial action of cultures of a penicillium with special reference to their use in the isolation of *B. influenza*. *Brit J Exp Pathol*, *10*, 226-236.

Hamblin, M. R. (2012). Antimicrobial photodynamic therapy and photodynamic inactivation, or killing bugs with dyes and light - a symposium-in-print. *Photochem Photobiol*, *88*, 496-498.

Hamblin, M. R., O'Donnell, D. A., Murthy, N., Rajagopalan, K., Michaud, N., Sherwood, M. E. & Hasan, T. (2002). Polycationic photosensitizer conjugates: effects of chain length and Gram classification on the photodynamic inactivation of bacteria. *JAC*, *49*, 941-951.

Hamblin, M. R. & Hasan, T. (2004). Photodynamic therapy: a new antimicrobial approach to infectious disease? *Photochem Photobiol* Sci, *3*, 436-450.

Hamblin, M., Tegos, G., Denis, T. St. & Huang, L. (2011). Antimicrobial photodynamic therapy: Can resistance develop? Proceedings of the13th IPA World Congress, *Photodiagnosis and Photodynamic Therapy*, *8*, 178.

Hanakova, A., Bogdanova, K., Tomankova, K., Binder, S., Bajgar, R., Langova, K., Kolar, M., Mosinger, J. & Kolarova, H. (2012). Study of photodynamic effects on NIH 3T3 cell line and bacteria. *Biomed Pap Med Fac Univ Palacky Olomouc Czech Repub.*, *156*, 1-8.

Hatanaka, S., Mitome, H., Yasui, K. & Hayashi, S. (2002). Single-bubble sonochemiluminescence in aqueous luminol solutions. *J Am Chem Soc*, *124*, 10250-10251.

He, L. L., Wang, X., Liu, B., Wang, J., Sun, Y.-G. & Xu, S.-K. (2011). Study on the sonodynamic activity and mechanism of promethazine hydrochloride by multi-spectroscopic techniques. *Spectrochim Acta A-M*, *81*, 698- 705.

Hiraoka, W., Honda, H., Feril, Jr. L. B., Kudo, N. & Kondo, T. (2006). Comparison between sonodynamic effect and photodynamic effect with photosensitizers on free radical formation and cell killing. *Ultrason Sonochem*, *13*, 535–542.

Hornung, R., Walt, H., Crompton, N. E., Keefe, K. A., Jentsch, B., Perewusnyk, G., Haller, U. & Köchli, O. R. (1998). m-THPC-mediated photodynamic therapy (PDT) does not induce resistance to chemotherapy, radiotherapy or PDT on human breast cancer cells in vitro. *Photochem Photobiol*, *68*, 569-574.

Ibbotson, S. (2012). What is the role of photodynamic therapy in the treatment of acne vulgaris? *Photodiagn Photodyn*, *9*, 2-4.

Kashef, N., Abadia, G. R. S. & Djavid, G. E. (2012). Phototoxicity of phenothiazinium dyes against methicillin-resistant *Staphylococcus aureus* and multi-drug resistant *Escherichia coli*. *Photodiagn Photodyn*, *9*, 11-15.

Kranz, S., Guellmar, A., Völpel, A., Gitter, B., Albrecht, V. & Sigusch, B. W. (2011). Photodynamic suppression of *Enterococcus faecalis* using the photosensitizer mTHPC. *Lasers Surg. Med.*, *43*, 241-248.

Laptev, R., Nisnevitch, M., Siboni, G., Malik, Z. & Firer, M. A. (2006). Intracellular chemiluminescent activation of ligand- photosensitizer conjugate for the targeted destruction of cancer cells. *Br. J. Cancer*, *95*, 189-196 .

Lauro, F. M., Pretto, P., Covolo, L., Jori, G. & Bertoloni, G. (2002). Photoinactivation of bacterial strains involved in periodontal diseases sensitized by porphycene-polylysine conjugates. *Photochem Photobiol Sci*, *1*, 468–470 .

Lee, C.-F., Lee, C.-J., Chen, C.-T. & Huang, C.-T. (2004). δ-Aminolaevulinic acid mediated photodynamic antimicrobial chemotherapy on *Pseudomonas aeruginosa* planktonic and biofilm cultures. *J. Photochem. Photobiol. B–Biol.*, *75*, 21–25.

Lee, Y. & Baron, E. D. (2011). Photodynamic therapy: current evidence and applications in dermatology. *Semin Cutans Med Surg*, *30*, 199-209.

Ma, X., Pan, H., Wu, G., Yang, Z. & Yi, J. (2009). Ultrasound may be exploited for the treatment of microbial diseases. *Med Hypotheses*, *73*, 18–19.

Maisch, T. (2007). Anti-microbial photodynamic therapy: useful in the future? *Lasers Med Sci*, *22*, 83–91.

Malik, R., Manocha, A. & Suresh, D. K. (2010). Photodynamic therapy – a strategic review. *Indian J of Dental Research*, *21*, 285-291.

Mantareva, V., Kussovski, V., Angelov, I., Wöhrle, D., Dimitrov, R., Popova, E. & Dimitrov, S. (2011). Non-aggregated Ga(III)-phthalocyanines in the photodynamic inactivation of planktonic and biofilm cultures of pathogenic microorganisms. *Photochem Photobiol* Sci, *10*, 91-102.

Martin, J. P. & Logsdon, N. (1987). Oxygen radicals are generated by dye-mediated intracellular photo-oxidations: a role for superoxide in photodynamic effects. *Arch Biochem Biophys*, *256*, 39-49.

Mišík, V. & Riesz, P. (2000). Free radical intermediates in sonodynamic therapy. *Ann NY Acad Sci*, *899*, 335-348.

Mitra, S., Haidaris, C. G., Snell, S. B., Giesselman, B. R., Hupcher, S. M. & Foster, T. H. (2011). Effective photosensitization and selectivity in vivo of *Candida albicans* by meso-tetra (N-methyl-4-pyridyl) porphine tetra tosylate. *Lasers Surg Med*, *43*, 324-332.

Moan, J. & Peng, Q. (2003). An outline of the history of PDT. In: Patrice T, editor. *Photodynamic Therapy, Comprehensive Series in Photochemistry & Photobioligy,* V. *2*, RSC, Cambridge, UK, 1-18.

Nagata, J. Y., Hioka, N., Kimura, E., Batistela, V. R., Terada, R. S. S., Graciano, A. X., Baesso, M. L. & Hayacibara, M. F. (2012). Antibacterial photodynamic therapy for dental caries: Evaluation of the photosensitizers used and light source properties. *Photodiagn Photodyn*, *9*, 122-131.

Nakonechny, F., Firer, M. A., Nitzan, Y. & Nisnevitch, M. (2010). Intracellular antimicrobial photodynamic therapy: a novel technique for efficient eradication of pathogenic bacteria. *Photochem Photobiol*, *86*, 1350-1355

Nakonechny, F., Nisnevitch, M., Nitzan, Y. & Firer, M. A. (2011). New techniques in antimicrobial photodynamic therapy: scope of application and overcoming drug resistance in nosocomial infections. In: Méndez-Vilas A, editor. *Science against microbial pathogens: communicating current research and technological advances*, Vol. *1*. Formatex Research Center Publisher, Badajoz, Spain, 684-691.

Nakonechny, F., Nitzan, Y. & Nisnevitch, M. (2012). Olive oil based delivery of photosensitizes for bacterial eradication. In: Boskou D, editor. Olive oil – Constituents, Quality, Health Properties and Bioconversions, InTech Open Access Publisher, Rijeka, Croatia, 471-492.

Nakonechny, F., Nisnevitch, M. M., Nitzan, Y. & Nisnevitch, M. (2013). Sonodynamic excitation of Rose Bengal for eradication of Gram-positive and Gram-negative bacteria. *BioMed Research International*, 2013, Article ID 684930, 1-7.

Neidhardt, F. C. & VanBogelen, R. A. (1987). Heat shock response .In: Neidhardt FC, editor. *Escherichia coli* and *Salmonella typhimurium*: cellular and molecular biology. American Society for Microbiology, Washington, DC, 1334–1345 .

Nikaido, H. (1998). Multiple antibiotic resistance and efflux. *Curr. Opin. Microbiol*, *1*, 516-523.

Nisnevitch, M., Nakonechny, F. & Nitzan, Y. (2010). Photodynamic antimicrobial chemotherapy by liposome-encapsulated water-soluble photosensitizers. *Russ J Bioorg Chem*, *36*, 396-402.

Nitzan, Y. & Pechatnikov, I. (2011). Approaches to kill Gram-negative bacteria by photosensitized process. In: Hamblin MR, Jori G, editors. Photodynamic inactivation of microbial pathogens: medical and environmental applications, RSC Publishing, UK, 47-67.

Nitzan, Y., Gutterman, M., Malik, Z. & Ehrenberg, B. (1992). Inactivation of Gram-negative bacteria by photosensitized porphyrins. *Photochem. Photobiol.*, *55*, 89-96 .

Papastamou, V., Nietzsch, T., Staudte, H., Orellana, G. & Sigusch, B. W. (2011). Photoinactivation of *F. nucleatum* and *P .gingivalis* using the ruthenium-based RD3 sensitizer and a conventional halogen lamp. *Arch Oral Biol*, *56*, 264-268.

Park, J. H., Moon, Y. S., Bang, I. S., Kim, Y. C., Kim, S. A., Ahn, S. G. & Yoon, J. H. (2010). Antimicrobial effect of photodynamic therapy using a highly pure chlorin e6. *Laser Med Sci*, *25*, 705-710.

Pedigo, L. A., Gibbs, A. J., Scott, R. J. & Street, C. N. (2009). Absence of bacterial resistance following repeat exposure to photodynamic therapy. Proc. SPIE 7380, 73803H-73803H-7.

Peèkaitytë, A., Daugelavièius, R., Sadauskaitë, A., Kirvelienë, V., Bonnett, R. & Bakienë, E. (2005). Comparative analysis of the photobactericidal action of selected tetrapyrrole compounds on Gram-positive and Gram-negative bacteria. *Biologija*, *1*, 41–46.

Pereira, J. B., Carvalho, E. F., Faustino, M. A., Fernandes, R., Neves, M. G., Cavaleiro, J. A., Gomes, N. C., Cunha, A., Almeida, A. & Tomé, J. P. (2012). Phthalocyanine thio-pyridinium derivatives as antibacterial photosensitizers .*Photochem Photobiol*, *88*, 537-547.

Poole, K. (2005). Efflux-mediated antimicrobial resistance. *J. Antimicrob. Chemother.*, *56*, 20–51.

Poto, A. D., Sbarra, M. S., Provenza, G., Visai, L. & Speziale, P. (2009). The effect of photodynamic treatment combined with antibiotic action or host defense mechanisms on *Staphylococcus aureus* biofilms. *Biomaterials*, *30*, 3158–3166.

Putterman, S. J. (1995). Sonoluminescence: sound into light. *Scientific American*, Febr., 32-37.

Raab, O. (1900). Über die Wirkung, fluorezcierender Stoffe auf infusorien. *Z. Biol.*, *39*, 524-546.

Ragàs, X., Sánchez-García, D., Ruiz-González, R., Dai, T., Agut, M., Hamblin, M. R. & Nonell, S. (2010). Cationic porphycenes as potential photosensitizers for antimicrobial photodynamic therapy. *J Med Chem*, *53*, 7796-7803.

Ragàs, X. & Nonell, S. (2010). Singlet oxygen in *Escherichia coli*: New insights for antimicrobial photodynamic therapy, *Free Radical Bio Med*, *49*, 770–776.

Rolim, J. P. M. L., de-Melo, M. A. S., Guedes, S. F., Albuquerque-Filho, F. B., de Souza, J. R., Nogueira, N. A. P., Zanin, I. C. J. & Rodrigues, L. K. A. (2012). The antimicrobial activity of photodynamic therapy against *Streptococcus mutans* using different photosensitizers. *J. Photochem. Photobiol. B–Biol.*, *106*, 40–46.

Rosenthal, I., Sostaric, J. & Riesz, P. (2004). Sonodynamic therapy – a review of the synergistic effects of drugs and ultrasound. *Ultrason Sonochem*, *11*, 349–363.

Rovaldi, C. R., Pievsky, A., Sole, N. A., Friden, P. M., Rothstein, D. M. & Spacciapoli, P. (2000). Photoactive porphyrin derivative with broad-spectrum activity against oral pathogens in vitro. *Antimicrobial Agents and Chemotherapy*, *44*, 3364-3367.

Sahu, K., Bansal, H., Mukherjee, C., Sharma, M. & Gupta, P. K. (2009). Atomic force microscopic study on morphological alterations induced by photodynamic action of Toluidine Blue O in *Staphylococcus aureus* and *Escherichia coli*. *J. Photochem. Photobiol. B–Biol.*, *96*, 9-16.

Sakamoto, F. H., Torezan, L. & Anderson, R. R. (2010). Photodynamic therapy for acne vulgaris: A critical review from basics to clinical practice. *J Am Acad Dermatol*, *63*, 195-211.

Schastak, S., Ziganshyna, S., Gitter, B., Wiedemann, P. & Claudepierre, T. (2010). Efficient photodynamic therapy against Gram-positive and Gram-negative bacteria using THPTS, a cationic photosensitizer excited by unfrared wavelength. *PLoS ONE*, *5*, e11674, 1-8.

Séguier, S., Souza, S. L. S., Sverzut, A. C. V., Simioni, A. R., Primo, F. L., Bodineau, A., Corrêa, V. M. A., Coulomb, B. & Tedesco, A. C. (2010). Impact of photodynamic therapy on inflammatory cells during human chronic periodontitis. *J. Photochem. Photobiol. B–Biol.*, *101*, 348–354.

Spesia, M. B., Milanesio, M. E. & Durantini, E. N. (2008). Synthesis, properties and photodynamic inactivation of *Escherichia coli* by novel cationic fullerene C_{60} derivatives. European *J Med Chem*, *43*, 853-861.

Spesia, M. B., Caminos, D. A., Pons, P. & Durantini, E. N. (2009). Mechanistic insight of the photodynamic inactivation of *Escherichia coli* by a tetracationic zinc(II) phthalocyanine derivative. *Photodiagn Photodyn*, *6*, 52-61.

Tachibana, K., Feril Jr., L. B. & Ikeda-Dantsuji, Y. (2008). *Sonodynamic therapy, Ultrasonics*, *48*, 253–259.

Tan, D. S. W., Gerlinger, M., Teh, B. T. & Swanton, C. (2010). Anti-cancer drug resistance: Understanding the mechanisms through the use of integrative genomics and functional RNA interference. *Eur. J. Cancer*, *46*, 2166-2177.

von Tappeiner, H. & Jesionek, A. (1903). Therapeutische Versuche mit fluorescierenden Stoffen. *Munch. Med. Wochenschr.*, *47*, 2042-2044.

Tardivo, J. P., Giglio, A. D., de Oliveira, C. S., Gabrielli, D. S., Junqueira, H. C., Tada, D. B., Severino, D., de Fátima Turchiello, R. & Baptista, M. S. (2005). Methylene blue in photodynamic therapy: From basic mechanisms to clinical applications. *Photodiagn Photodyn*, *2*, 175-191.

Tavares, A., Carvalho, C. M. B., Faustino, M. A., Neves, M. G. P. M. S., Tomé, J. P. C., Tomé, A. C., Cavaleiro, J. A. S., Cunha, Â., Gomes, N. C. M., Alves, E. & Almeida, A. (2010). Antimicrobial photodynamic therapy: study of bacterial recovery viability and potential development of resistance after treatment. *Mar. Drugs*, *8*, 91-105.

Tegos, G. P. & Hamblin, M. R. (2006). Phenothiazinium antimicrobial photosensitizers are substrates of bacterial multidrug resistance pumps. Antimicrob. *Agents Chemother.*, *50*, 196-203.

Tran, T. D.-H., Kwon, H.-Y., Kim, E.-H., Kim, K.-W., Briles, D. E., Pyo, S. & Rhee, D.-K. (2011). Decrease in penicillin susceptibility due to heat shock protein ClpL in *Streptococcus pneumonia*. *Antimicrob. Agents Chemother.*, *55*, 2714-2728.

Umemura, K., Yumita, N., Nishigaki, R. & Umemura, S. (1996). Sonodynamically induced antitumor effect of pheophorbide a. *Cancer Lett*, 102, 151-157.

Umemura, S., Yumita, N., Umemura, K. & Nishigaki, R. (1999). Sonodynamically induced effect of Rose Bengal on isolated sarcoma 180 cells. *Cancer Chemoth Pharm*, 43, 389-393.

Vilela, S. F. G., Junqueira, J. C., Barbosa, J. O., Majewski, M., Munin, E. & Jorge, A. O. C. (2012). Photodynamic inactivation of *Staphylococcus aureus* and *Escherichia coli* biofilms by malachite green and phenothiazine dyes: An in vitro study. *Archives of Oral Biology*, *57*, 704-710.

Yu, T., Wang, Z. & Mason, T. J. (2004). A review of research into the uses of low-level ultrasound in cancer therapy. *Ultrason Sonochem*, *11*, 95–103.

Yumita, N., Nishigaki, R., Umemura, K. & Umemura, S. (1990). Synergetic effect of ultrasound and hematoporphyrin on Sarcoma 180. *Jpn J Cancer Res*, *81*, 304-308.

Yumita, N., Sasaki, K., Umemura S., Yukawa, A. & Nishigaki, R. (1997). Sonodynamically induced antitumor effect of gallium-porphyrin complex by focused ultrasound on experimental kidney tumor. *Cancer Lett*, *112*, 79-86.

Zerdin, K., Horsham, M. A., Durham, R., Wormell, P. & Scully, A. D. (2009). Photodynamic inactivation of bacterial spores on the surface of a photoactive polymer. *React Funct Polym*, *69*, 821–827.

In: Prokaryotes
Editor: Marina Nisnevitch
ISBN: 978-1-63321-592-4

Chapter 10

Monod's Bacterial Growth Model. Kinetic Arguments for "Simple, but Not Too Simple"

Vlad Tudor Popa*
Romanian Academy, "Ilie Murgulescu" Institute of Physical Chemistry,
Bucharest , Romania

Abstract

The Monod-type model was analyzed from the chemical kinetics side (Hinshelwood) in "Langmuir adsorption" terms, or from the microbiology side (Button) in "Michaelis-Menten" enzyme kinetics terms, but it seems that a general kinetic derivation of the Monod growth equation is missing. Also missing is a more or less general answer to the question: Why the model still holds, within surprisingly broad limits concerning bacterial strains, nutrient and environment conditions? The present study proves that the actual form of the Monod equation may be derived in strictly kinetic terms within assumptions / approximations / methods borrowed from heterogeneous catalysis (Horiuti, Temkin, Happel and Boudart): (quasi)steady state approximation, most abundant reaction intermediate, kinetically significant step, reaction route or path. The main approximation involves considering bacterial individuals (in various stages of their growth) as some quasi-chemical species, the reactivity of which is subject to mass action law. The growth is amenable to a two-step sequence where the term "step" stands for some complex, possibly single-route, and kinetically significant process. Consequently, the growth may be approached in quasi-chemical terms and the derivation of the Monod equation is straightforward for two limiting cases: 1) "starving conditions", with substrate binding as a limiting step; 2) "normal" laboratory conditions with abundant nutrients (excepting possibly the growth controlling or limiting one) and bacterial fission as limiting step. Using the hypothesis of a most abundant growth intermediate, *magi*, as a kinetic key, several other mechanistic alternatives are presented, involving multiple-substrate binding before and/or after the formation of this key intermediate. Several experimental

* Corresponding author: Email: vtpopa@icf.ro.

observations are discussed in connection with the mechanistic expressions obtained for the Monod growth parameters. Application to microcalorimetric growth data in batch conditions involves the decomposition of the complex thermogram via Peakfit processing of the thermogram. The thermal version of the Monod equation describes component, non-steady growth processes surprisingly well. A possible explanation would be the conservation of the *magi* (and thus of the rate equation) over the whole extension of the growth process.

Keywords: Bacterial growth; quasi-chemical approach; complex reaction theory; most abundant growth intermediate; steady state

Introduction

Even before Jacques Monod first advanced his equation for bacterial growth in 1941-1942 (Monod, 1942), Sir Cyril Norman Hinshelwood and his co-workers were examining their results on bacterial growth kinetics in "Langmuir terms". Hinshelwood's book (Hinshelwood, 1946), published in 1946, was promptly (and in some respect critically) reviewed by Monod (Monod, 1946): "This book is not a treatise or a monograph. It is an essay, in which the author has expanded the main ideas which led to his own and his group's important work on the kinetics of bacterial growth."

Hinshelwood's approach consisted in focusing on various (experimentally proven or carefully inferred) details of the growth followed by extrapolation to the whole bacterial cell / process. Its origins were to be found in the author's vast experience in heterogeneous catalysis kinetics leading to the extensively utilized correspondence:

adsorbent-adsorbate (adsorption) ↔ enzyme-substrate (binding)

20 years later this approach was comprehensively developed (Dean and Hinshelwood, 1966) around the concept of "total integration" that made use of new experimental evidence concerning the adaptive mechanisms involved in bacteria – medium interactions. Monod's main objection (Monod, 1946) to Hinshelwood's 1946 book was concerned with the insuficient handling of the role of mutations in the bacterial evolution subject to various stress (environment) factors: "True, the occurrence of spontaneous mutations and the importance of selection as factors of bacterial variability are discussed at some length, the general conclusion being that it seems more 'profitable' and less 'arbitrary' to consider most changes as resulting from adaptive reactions, rather than from spontaneous mutations. This type of discussion is not very convincing, however." Hinshelwood's response in his 1966 book (Dean and Hinshelwood, 1966) consisted in a whole chapter "Mutants and Selection" where his former views on the mater were reiterated with some fresh arguments: "Furthermore, it is questionable wether substrate-adaptation and drug-resistance should be described as 'acquired characters'. They represent quantitative enhancement of capacities already present, and there would seem to be no reason to postulate separate and specific genetic information for each possible quantitative level of every existing property."

On the other hand, Monod's approach to growth kinetics was purely empiric. In his 1949 contribution (Monod, 1949), while acknowledging the formal resemblance of his growth equation to adsorption and enzyme kinetics, he resumed his intuitive, "common sense" arguments:

"Several mathematical formulations could be made to fit the data. But it is both convenient and logical to adopt a hyperbolic equation:

$$R = R_K \frac{C}{C_1 + C} \tag{1}$$

similar to an adsorption isotherm or to the Michaelis equation. In the above equation C stands for the concentration of the nutrient. R_K is the rate limit for increasing concentrations of C. C_1 is the concentration of nutrient at which the rate is half the maximum." (Boldface stress belongs to the present contribution.)

This hyperbolic equation (or various of its modifications) was and still is widely used in bacterial growth kinetics in either Langmuir (Langmuir, 1918) or Michaelis-Menten (Michaelis and Menten, 1913) terms. Several authors gradually developed and emphasised the original idea (Monod, 1949) that formal resemblance between Monod's equation and Langmuir or Michaelis-Menten ones has no general mechanistic grounds. D. K. Button's most important involvement in this matter deserves a special attention: it started with his former contributions concerning continuous cultures of *Torulopsis utilis*, where Michaelis constant was associated with the Monod one (Button and Garver, 1966), followed by his first important review published some 20 years later (Button, 1985), where a clear distinction was made between "Michaelis or substrate concentration at half-maximal growth rate" and "Michaelis or substrate concentration at half-maximal transport rate", continued with a deeper insight concerning the mechanistic expression of the Michaelis constant (Button, 1991), and a second review (Button, 1998) where his "structural model of substrate uptake via cellular membrane permeation" is further developed, up to more recent articles, where an experimentally demonstrated multiple-nutrient collection mechanism was showed to "supercede the Michaelis-Menten paradigm" (Button et al., 2004), and where "specific affinity theory" was applied (Button, 2010) to "predict the identity and concentrations of some naturally occurring hydrocarbons". However, in spite of a previous explicite statement of doubt concerning the Michaelis-Menten concept as representing a situation of a single (master) growth determining / limiting enzyme (Button, 2000), Button noted the fact that "yet the paradigm persists" and did not completely abandon its use in the evaluation bacterial growth kinetics.

An excellent review marked half a century use of Monod-type kinetics (Kovarova-Kovar and Egli, 1998) and examined the alternative of "mixed-substrate kinetics"; a special attention was devoted to the biological meaning of the model parameters with emphasis on the difference between growth and enzyme kinetic parameters. Hinshelwood's "zooming approach", systematically developed in chemical engineering, offered the basis of the so-called "structured models" of growth, presented in a detailed review devoted to bacterial and fungal growth models (Nielsen and Villadsen, 1992). These models "describe microbial kinetics by means of selected cell components rather than by the undifferential biomass" making use of "Monod kinetics" as a building block of more detailed mechanistic schemes, addressing even the "morphologic heterogeneity" of the bacterial/fungal population. Evidence concerning this type of approach was more recently reviewed (Dubnau and Losick, 2006):

adaptive efforts to various stress factors, leading to "nonuniform gene expression" and thus nonuniform bacterial reactivity, ultimately results in bistable cultures. This is illustrated for *Escherichia coli* (that forms persister cells) and *Bacillus subtillis* (that may develop genetic competence, spore formation/cannibalism and swimming/chaining). Recent results on *Paenibacillus vortex* (Roth et al., 2013) demonstrate the generation of two morphotypes, "builders" and "explorers" that express the adaptation to antibiotics. A recent comprehensive review (Ryall et al., 2012), concerning experimental evidence on the complex bacterial response to environmental stress, introduced the concept of "adaptomics" and expressed the hope that "the *adaptomic* description of **regulatory** and **mutational** responses to a single environmental change is a sensible step toward modeling the outcomes of multistep evolutionary change". (One may recognise in boldface fonts the 1940s points of view of both Hinshelwood and Monod!). Recent important contributions assessing links between growth and gene expression (Scott et al., 2010) and bistability and antibiotic resistence (Derris et al., 2013), containing a vast supplementary material, continue the use of Monod-type equations in the modeling of complex experimental data. However, phenomena such as "heterogeneous reactivity" and "bistability", although fully acknowledged, are beyond the reach of the following approach.

A Simple Quasi-Chemical Growth Model with Uniform Reactivity

The basic question we are trying to answer in this section is why the complex phenomena involved in bacterial growth ***appear to be suitably described***, in certain (but frequent) circumstances, by the simple Monod equation (2), where standard terminology (Kovarova-Kovar and Egli, 1998) was adopted: **B**(t) is the total bacterial population (concentration), μ is the specific growth rate, μ_m its maximum model-allowed value, K_S is the "substrate affinity constant" and $S(t)$ is the substrate concentration (at rigorous steady state growth the time dependence may be omitted).

$$\frac{d\mathbf{B}(t)/dt}{\mathbf{B}(t)} = \mu = \frac{\mu_m S(t)}{K_S + S(t)} \tag{2}$$

The initial intention of the present contribution was not only to "revisit" Monod's growth model but also to state that from the very title. In fact, this title was used before (Lobry et al., 1992) in an attempt to improve the model in "mechanistic terms". Actually, this attempt was a merely mathematical revisit, not of the Monod's model, but of the (100 years older) logistic one (Verhulst, 1845); with no specific reference to (new) mechanistic features, except the introduction of a "lag time in one variable only because there is a delay for biomass production and not for substrate consumption". Taylor expansion and keeping of the first term (which points to low values of this lag) results in the usual hyperbolic equation (2) with respect to substrate concentration. The modified model predicts "an unexpected increase of parameter K_S with μ_m" and the authors review various experimental contributions in which such a dependence was actually observed.

Several derivations and/or interpretations of the Monod equation were advanced, in thermodynamic (Liu et al., 2003), (Liu, 2006), kinetic and/or nonequilibrium thermodynamic terms. These were reviewed by Liu (Liu, 2007) with a comprehensive analysis concerning the most debated and problematic physical meaning of the Monod constant, K_S. The majority of these derivations are focused on the more or less mechanistically detailed substrate evolution. The closest approach to the one we are going to advance here seems to be that of Jin and Bethke: they proposed a model concerning electron transfer via the respiratory chain (Jin and Bethke, 2002), improved it with obtaining of a new rate law "that would apply to respiration by prokaryotic cells" (Jin and Bethke, 2003) and later generalized it to microbial metabolism (Jin and Bethke, 2007). However, the main actors of the growth process, i.e. individuals of the bacterial culture are not specifically involved in the Jin-Bethke mechanistic picture.

The present contribution originates in the so-called Horiuti-Temkin theory of complex (catalytic) reactions: (Horiuti, 1948; 1957; 1973; Horiuti and Nakamura, 1967; Temkin, 1963; 1968; 1971; 1979); this theory was developed by Happel (Happel, 1968; 1981; 1986; Csuha and Happel, 1971; Happel and Sellers, 1992; 1995); and successfully refined, popularized and advocated by Boudart (Boudart, 1968; 1975; Boudart and Djéga-Mariadassou, 1984). It involves bacterial individuals contained in the growth medium (suspension) as **quasi-chemical reactants/products** involved in a sequence (network) of complex reactions, the detailed chemistry of which is not specified, with the exception of those involving the limiting substrate. The term "step" designates in general a complex set of reactions and will be used in the sense of "kinetically significant step", following Boudart's definition (Boudart, 1975), i.e. as a sum of elementary reactions with **no individual contribution** to the **observed rate law**. This approach will be extended to the general mechanistic description of the bacterial growth: "complex steps" with no individual kinetic significance may sum up to give "kinetically significant complex steps". For convenience, the term "complex" will be in general omitted in the following argument, but it has to be implicitly assumed whenever the term "step" occurs.

Various stages in bacterial cell growth will be treated as **reaction intermediates** and termed "growth intermediates" or "species" (in a **quasi-chemical, not biological sense**). The growth process is represented as a sequence of mechanistic steps where all chemicals involved, with the exception of the limiting one, are assumed in excess and thus omitted. Their concentrations are embodied in the rate constants, or rather "rate coefficients" (Vlad et al., 2005), k_j. For convenience, the term "rate constant" will be utilized throughout this chapter, although we acknowledge and, when needed, even reveal the complex character of these growth parameters. An important review (Chebotareva et al., 2004) examined the influence of "macromolecular crowding" and introduced the concept of "excluded volume" in dealing with intracellular chemical kinetic effects; the contribution of both jamming, i.e. overcrowding and fractal effects on the rate coefficients pertaining to heterogeneous, spatially confined enzymatic catalysis made the object of a special kinetic model in (Xu and Ding, 2007). In principle, "step" rate constants will be termed "mechanistic", while more complex, kinetically significant rate constants will be termed "apparent". (In fact, all "rate constants" or "rate coefficients" utilized in the following argument are complex, apparent quantities.)

Thus these steps, denoted by *R(j)*, are not elementary, but rather kinetically significant complex (possibly single-route) reactions (Horiuti and Nakamura, 1967). Mass action law and (quasi)steady state equations will be assumed to approximate the rate of the mechanistic (step) reactions and the rate of the overall growth, respectively. While the (quasi)steady state approximation, *qssa*, is generally accepted, at least in the logarithmic growth phase of bacterial cultures, mass action law may be regarded as a first-order approximation when dealing with bacterial cells in various phases of their growth (Vlad et al., 2005).

In the mechanistic scheme (3) all steps were taken as "one-way" (Boudart and Tamaru, 1991), i.e. "irreversible". We will show later that the reverse steps do alter the actual form of the rate equations, but not the whole argument. The main reason for taking one-way steps is the fact that the proposed mechanism accounts for cell division (this is not the case for many growth models in which "biomass formation" does not involve division in a detailed manner). With a chemical system of such complexity, involving hundreds of species connected by thousands of reactions, following the time evolution of the **actual bacterial cell** resembles following an "atomic species transfer from a reactant to a product" (Happel, 1981). Such a "chemical trajectory" was termed "reaction path" by Happel (Happel, 1968). For (heterogeneous) catalytic reactions a "potential factor" may be factored out the overall rate equation (Csuha and Happel, 1971) "when all slow steps for a reaction are present in a single path" (Happel, 1981). This is the source of the so-called "thermodynamic potential factor" present in the contributions of Jin and Bethke (Jin and Bethke, 2002; 2003; 2007).

In principle, the general mechanism proposed here is an extension of the reaction path concept to the complex bacterial growth process, where the adjective "complex" will be, as much as possible, omitted in connection with "step", "path", etc. For convenience, in eq.(3) and throughout the following, B_j denotes both the growth intermediate and its concentration. The intermediate B_1 stands for the "new-born" individual. It is a kinetically significant, or determining "growth species", as it depicts the very first stage in bacterial growth and accounts for its "autocatalytic" nature.* It is, in some respect, the analogous of the "free catalytic site" in homogeneous and heterogeneous catalysis, with the important difference that its concentration doubles at the end of the catalytic cycle (on each "overall reaction" turnover) for batch cultures. In the case of "continuous cultures", also pioneered by Monod (Monod, 1950) within a gradientless setup and developed into various versions of modern chemostats, (Zhang et al., 2006; de Crécy et al., 2007), a rigorous steadiness of the bacterial species' concentrations is obtained.

$$
\begin{array}{ll}
R(1) & B_1 + S \xrightarrow{k_1} B_2 \\
R(2) & B_2 \xrightarrow{k_2} B_3 \\
\text{-----------------} & \\
R(j-1) & B_{j-1} \xrightarrow{k_{j-1}} B_j \\
R(j) & B_j \xrightarrow{k_j} B_{j+1} \\
\text{-----------------} & \\
R(n-1) & B_{n-1} \xrightarrow{k_{n-1}} B_n \\
R(n) & B_n \xrightarrow{k_n} 2B_1 \\
\hline
 & B_1 + S = 2B_1
\end{array}
\qquad (3)
$$

* François Jacob, 1965: Le rêve d'une bactérie doit devenir deux bactéries (The dream of a bacterium is to become two bacteria).

A normal consistency requirement for eqs. (2) and (3) is **biomass balance**:

$$\sum_{j=1}^{n} B_j = \mathbf{B} \tag{4}$$

where the time dependence of species' concentrations has been omitted.

The general mechanism (3) may be reduced to simpler, two-step variants by means of steady state (catalytic) kinetics approximations.

The first one, (I), involves the assumption of a most abundant reaction intermediate (Boudart and Djéga-Mariadassou, 1984), *mari*. Suppose that in eq. (3) new-born species are these "most abundant growth intermediates", *magi*, and growth limiting substrate binding is the very first step of the whole process. In this case the above scheme reduces to two kinetically significant steps, the first one and a sum of all subsequent ones, where all intermediates except B_1 and B_2 cancel out. Only these intermediates matter in the reduced, two-step mechanism, and they will be termed **growth determining intermediates**, *gdi*:

$$\left\{ R(1) \quad B_1 + S \xrightarrow{k_1} B_2 \right\} \xrightarrow{B_1 = magi} (\text{I.1})\ B_1 + S \xrightarrow{\kappa_1 = k_1} B_2$$

$$\left\{ \begin{array}{ll} R(2) & B_2 \xrightarrow{k_2} B_3 \\ ------------- & \\ R(j-1) & B_{j-1} \xrightarrow{k_{j-1}} B_j \\ R(j) & B_j \xrightarrow{k_j} B_{j+1} \\ ------------- & \\ R(n-1) & B_{n-1} \xrightarrow{k_{n-1}} B_n \\ R(n) & B_n \xrightarrow{k_n} 2B_1 \end{array} \right\} \xrightarrow{\sum_{j=2}^{n} R(j)} (\text{I.2})\ B_2 \xrightarrow{\kappa_2 = \prod_{j=2}^{n} k_j} 2B_1 \tag{5}$$

The second approximation, (II), assumes B_n as *magi*. In this case, the simplified two-step sequence involves a sum of the first *(n-1)* steps and the last one, where all intermediates cancel out, except B_1 and B_n which are the growth determining intermediates:

$$\left\{ \begin{array}{ll} R(1) & B_1 + S \xrightarrow{k_1} B_2 \\ R(2) & B_2 \xrightarrow{k_2} B_3 \\ ------------- & \\ R(j-1) & B_{j-1} \xrightarrow{k_{j-1}} B_j \\ R(j) & B_j \xrightarrow{k_j} B_{j+1} \\ ------------- & \\ R(n-1) & B_{n-1} \xrightarrow{k_{n-1}} B_n \end{array} \right\} \xrightarrow{\sum_{i=1}^{n-1} R(j)} (\text{II.1})\ B_1 + S \xrightarrow{\kappa_1 = \prod_{j=1}^{n-1} k_j} B_n \tag{6}$$

$$\left\{ R(n) \quad B_n \xrightarrow{k_n} 2B_1 \right\} \xrightarrow{B_n = magi} (\text{II.2})\ B_n \xrightarrow{\kappa_2 = k_n} 2B_1$$

At the (quasi)steady state, assuming Temkin's "step steady state conditions" (Temkin, 1963) the overall growth rate equals the rates of each of its steps:

$$v = v_1 = v_2 = ... = v_{j-1} = v_j = ... = v_{n-1} = v_n \tag{7}$$

By means of eq.(7), the above particular cases result in formally identical growth rate equations, with only a switch between the "true" and "apparent" character of the pertaining two-step rate constants. This is an extension of previously demonstrated (Boudart and Djéga-Mariadassou, 1984) "simplified kinetics":

$$\begin{array}{lll}
(I) & \text{Binding "significant" ; } B_1 = magi \text{ ; } B_1, B_2 = gdi \text{ ; } qssa: & \\
(I.1) & B_1 + S \xrightarrow{\kappa_1} B_2 & v_I = \kappa_2 \mathbf{B} \dfrac{S}{K+S} \text{ ; } \mathbf{B} \cong B_1 + B_2 \\
(I.2) & B_2 \xrightarrow{\kappa_2} 2B_1 & K = \kappa_2 / \kappa_1 \, (\kappa_1 \, "true"; \kappa_2 \, "apparent") \\
 & & \\
(II) & \text{Fission "significant" ; } B_n = magi \text{ ; } B_1, B_n = gdi \text{ ; } qssa: & \\
(II.1) & B_1 + S \xrightarrow{\kappa_1} B_n & v_{II} = \kappa_2 \mathbf{B} \dfrac{S}{K+S} \text{ ; } \mathbf{B} \cong B_1 + B_n \\
(II.2) & B_n \xrightarrow{\kappa_2} 2B_1 & K = \kappa_2 / \kappa_1 \, (\kappa_2 \, "true"; \kappa_1 \, "apparent")
\end{array} \tag{8}$$

There is a notable mechanistic difference, evidenced in eq.(8): in the two-step sequence *(I)* growth is governed by a binding step, while in sequence *(II)* by a fission one.

We have to emphasize that, unlike the current practice in (heterogeneous) catalytic kinetics, where one often resorts to the **assumption** of an existing **rate determining step** (*rds*), **no such assumption was needed** for the derivation of eq.(8). (This is an important issue on which both Hinshelwood (Hinshelwood, 1946) and Monod (Monod, 1949) agreed from the very beginning: situations within which a **single step** is **rate determining** for the **whole process** are considerably less probable for bacterial growth.) On the contrary, the two particular cases of "binding significant" and "fission significant" emerged as **consequences** of the *qssa* and of the two different *magi* assumptions. Moreover, the obtained rate equation not only reproduces the form advanced in Monod's contribution, (Monod, 1942; 1949), but it also accounts for the common experimental observation that the two parameters of the growth equation are not independent of one another, reviewed in (Kovarova-Kovar and Egli, 1998) and (Lobry et al., 1992). Hyperbolic rate equations obtained are a consequence of the two-step steady state mechanisms and **biomass balance equations**:

$$\begin{array}{ll}
(I) & \text{Binding significant ; } B_1 = magi \text{ ; } qssa: \ \mathbf{B} \cong B_1 + B_2 \\
(II) & \text{Fission significant ; } B_n = magi \text{ ; } qssa: \ \mathbf{B} \cong B_1 + B_n
\end{array} \tag{9}$$

A natural generalization of the above approach is to take into consideration some steps (bacterial intermediates) preceding and succeeding to the binding of the growth limiting substrate and assume as *magi* either a pre-binding intermediate, as in eq.(10),

$$\left.\begin{cases} R(1) & B_1 \xrightarrow{k_1} B_2 \\ R(2) & B_2 \xrightarrow{k_2} B_3 \\ ------- & \\ R(s-1) & B_{s-1} \xrightarrow{k_{s-1}} B_s \end{cases}\right\} \xrightarrow{\sum_{j=1}^{s-1} R(j)} (1)\ B_1 \xrightarrow{\kappa_1 = \prod_{j=1}^{s-1} k_j} B_s$$
$$\left.\begin{cases} R(s) & B_s + S \xrightarrow{k_s} B_{s+1} \\ ------- & \\ R(n-1) & B_{n-1} \xrightarrow{k_{n-1}} B_n \\ R(n) & B_n \xrightarrow{k_n} 2B_1 \end{cases}\right\} \xrightarrow{\sum_{j=s}^{n} R(j)} (2)\ B_s + S \xrightarrow{\kappa_2 = \prod_{j=s}^{n} k_j} 2B_1 \qquad (10)$$

or a post-binding intermediate, as in eq.(11),

$$\left.\begin{cases} R(1) & B_1 \xrightarrow{k_1} B_2 \\ R(2) & B_2 \xrightarrow{k_2} B_3 \\ ------- & \\ R(s) & B_s + S \xrightarrow{k_s} B_{s+1} \end{cases}\right\} \xrightarrow{\sum_{j=1}^{s} R(j)} (1)\ B_1 + S \xrightarrow{\kappa_1 = \prod_{j=1}^{s} k_j} B_{s+1}$$
$$\left.\begin{cases} R(s+1) & B_{s+1} \xrightarrow{k_{s+1}} B_{s+2} \\ ------- & \\ R(n-1) & B_{n-1} \xrightarrow{k_{n-1}} B_n \\ R(n) & B_n \xrightarrow{k_n} 2B_1 \end{cases}\right\} \xrightarrow{\sum_{j=s+1}^{n} R(j)} (2)\ B_{s+1} \xrightarrow{\kappa_2 = \prod_{j=s+1}^{n} k_j} 2B_1 \qquad (11)$$

Application of the step steady state conditions and biomass balance result in:

(I) A pre-binding species, B_s, is *magi* ; $B_1, B_s = gdi$; *qssa*:

$$(I.1)\quad B_1 \xrightarrow{\kappa_1} B_s \qquad v_I = \kappa_1 \mathbf{B} \frac{S}{K+S}\ ; \mathbf{B} \cong B_1 + B_s$$

$$(I.2)\quad B_s + S \xrightarrow{\kappa_2} 2B_1 \qquad K = \kappa_1 / \kappa_2 \text{ (both } \kappa_1, \kappa_2 \text{ "apparent")}$$

(II) A post-binding species, B_{s+1}, is *magi* ; $B_1, B_{s+1} = gdi$; *qssa*: (12)

$$(II.1)\quad B_1 + S \xrightarrow{\kappa_1} B_{s+1} \qquad v_{II} = \kappa_2 \mathbf{B} \frac{S}{K+S}\ ; \mathbf{B} \cong B_1 + B_{s+1}$$

$$(II.2)\quad B_{s+1} \xrightarrow{\kappa_2} 2B_1 \qquad K = \kappa_2 / \kappa_1 \text{ (both } \kappa_1, \kappa_2 \text{ "apparent")}$$

Eq. (12) represents the two-step reduction of the detailed mechanisms (10) and (11) where all kinetic parameters (constants) are "apparent".

We may also examine the situation where more than one molecule of growth limiting or determining substrate is needed for the bacterial cell to reach its dream. Denote this number by *a*, meaning that *a* molecules of substrate *S* are needed for a complete growth turnover. We will consider the more general two-step mechanisms described in eqs. (10) and (11). If all binding processes of *S* take place after the formation of the *magi*, B_S, in eqs (10), or before the formation of the *magi*, B_{S+1}, in (11), equations (12) become:

$$(I)\ \text{A pre-binding species}, B_s, \text{ is } magi;\ B_1, B_s = gdi;\ qssa:$$

$$(I.1)\quad B_1 \xrightarrow{\kappa_1} B_s \qquad v_I = \kappa_1 \mathbf{B}\frac{S^a}{K+S^a};\ \mathbf{B} \cong B_1 + B_s$$

$$(I.2)\quad B_s + aS \xrightarrow{\kappa_2} 2B_1 \qquad \mathrm{K} = \kappa_1/\kappa_2\ (\text{both } \kappa_1, \kappa_2 \text{ "apparent"})$$

$$B_1 + aS \ = \ 2B_1$$

$$(II)\ \text{A post-binding species}, B_{s+1}, \text{ is } magi;\ B_1, B_{s+1} = gdi;\ qssa:$$

$$(II.1)\quad B_1 + aS \xrightarrow{\kappa_1} B_{s+1} \qquad v_{II} = \kappa_2 \mathbf{B}\frac{S^a}{K+S^a};\ \mathbf{B} \cong B_1 + B_{s+1}$$

$$(II.2)\quad B_{s+1} \xrightarrow{\kappa_2} 2B_1 \qquad \mathrm{K} = \kappa_2/\kappa_1\ (\text{both } \kappa_1, \kappa_2 \text{ "apparent"})$$

$$B_1 + aS \ = \ 2B_1 \tag{13}$$

A more complicated situation is the one where the substrate may bind (and be assimilated) **before and after the formation of *magi***. Suppose a_1 molecules of substrate bind in the sequence *R(1), ..., R(s)* and a_2 molecules in the remaining mechanistic steps, *R(s+1), ..., R(n)*, with $a = a_1 + a_2$. In this case there is no point in defining of a pre- or post-binding species: there is a *magi*, say B_S, and the two-step rate equation reads:

$$B_s, \text{ is } magi;\ B_1, B_s = gdi;\ \mathbf{B} \cong B_1 + B_s;\ a_1 + a_2 = a;\ qssa:$$

$$(X.1)\quad B_1 + a_1 S \xrightarrow{\kappa_1} B_s$$

$$(X.2)\quad B_s + a_2 S \xrightarrow{\kappa_2} 2B_1 \qquad (\text{both } \kappa_1, \kappa_2 \text{ "apparent"})$$

$$B_1 + aS \ = \ 2B_1$$

$$v_X = \mathbf{B}\frac{\kappa_1\kappa_2 S^{a_1+a_2}}{\kappa_1 S^{a_1} + \kappa_2 S^{a_2}} = \kappa_1 \mathbf{B}\frac{S^{a_2}}{\kappa_1/\kappa_2 + S^{a_2-a_1}} = \kappa_2 \mathbf{B}\frac{S^{a_1}}{\kappa_2/\kappa_1 + S^{a_1-a_2}} \tag{14}$$

Substrate dependence in eq. (14) is slightly more complicated than the classical Monod equation. If $a_1 = 0$ and $a_2 = a$, v_X from eq. (14) becomes identical with v_I from eq. (13); if $a_2 = 0$ and $a_1 = a$, v_X from eq. (14) becomes identical with v_{II} from eq. (13). This type of equation was advanced for the first time by Hill (Hill, 1910) for hemoglobin binding processes, before Michaelis-Menten (Michaelis and Menten, 1913) or Langmuir (Langmuir, 1918) ones! A comprehensive revisit (Weiss, 1997) of Hill's equation reviewed several instances of its "use and misuse". A rate equation similar to the ones given by eqs.(13) was derived by Liu (Liu, 2006), based on purely thermodynamic arguments.

Finally, we will examine in some detail the "reversibility issue". For the time being and within the simplified two-step description of the growth process, one may exclude the reversibility of the second step that involves the "bacterial dream" of fission. Thus, only the first step may be taken as a two-way one (Boudart and Tamaru, 1991) and we will keep the more general assumption that substrate binding may take place in both steps. The steady state rate equation of the two-step sequence takes a more complicated expression, given in eq. (15).

$$B_s\text{, is } magi\ ;\ B_1, B_s = gdi;\ \mathbf{B} \cong B_1 + B_s\ ;\ a_1 + a_2 = a\ ;\ qssa:$$

$$(Y.1)\quad B_1 + a_1 S \underset{\kappa_{-1}}{\overset{\kappa_{+1}}{\rightleftarrows}} B_s$$

$$(Y.2)\quad B_s + a_2 S \xrightarrow{\kappa_{+2}} 2B_1 \qquad \text{(both } \kappa_{\pm 1}, \kappa_{+2} \text{ "apparent")}$$

$$B_1 + aS = 2B_1$$

$$v_Y = \mathbf{B}\frac{\kappa_{+1}\kappa_{+2}S^{a_1+a_2}}{\kappa_{-1} + \kappa_{+1}S^{a_1} + \kappa_{+2}S^{a_2}} \tag{15}$$

Simpler alternatives of eq.(15) pertain to the same binding instances discussed above:

(i) Binding of substrate *S* takes place **after** the formation of the most abundant growth intermediate, *magi*:

$$B_s\text{, is } magi\ ;\ B_1, B_s = gdi;\ \mathbf{B} \cong B_1 + B_s\ ;\ a_1 = 0,\ a_2 = a\ ;\ qssa:$$

$$(Y.1)\quad B_1 \underset{\kappa_{-1}}{\overset{\kappa_{+1}}{\rightleftarrows}} B_s$$

$$(Y.2)\quad B_s + aS \xrightarrow{\kappa_{+2}} 2B_1 \quad \text{(both } \kappa_{\pm 1}, \kappa_{+2} \text{ "apparent")} \tag{16}$$

$$B_1 + aS = 2B_1$$

$$v_Y^{(i)} = \mathbf{B}\frac{\kappa_{+1}\kappa_{+2}S^{a}}{\kappa_{-1} + \kappa_{+1} + \kappa_{+2}S^{a_2}} = \mathbf{B}\kappa_{+1}\frac{S^a}{\dfrac{\kappa_{+1}+\kappa_{-1}}{\kappa_{+2}} + S^a}$$

(ii) Binding of substrate *S* takes place **before** the formation of the most abundant growth intermediate, *magi*:

$$B_s\text{, is } magi\text{ ; } B_1, B_s = gdi\text{; } \mathbf{B} \cong B_1 + B_s \text{ ; } a_1 = a\text{, } a_2 = 0 \text{ ; } qssa\text{:}$$

$$(Y.1) \quad B_1 + aS \underset{\kappa_{-1}}{\overset{\kappa_{+1}}{\rightleftarrows}} B_s$$

$$(Y.2) \quad B_s \xrightarrow{\kappa_{+2}} 2B_1 \qquad \text{(both } \kappa_{\pm 1}, \kappa_{+2} \text{ "apparent")} \tag{17}$$

$$B_1 + aS \;=\; 2B_1$$

$$v_Y^{(ii)} = \mathbf{B}\frac{\kappa_{+1}\kappa_{+2}S^a}{\kappa_{-1} + \kappa_{+1}S^a + \kappa_{+2}} = \mathbf{B}\kappa_{+2}\frac{S^a}{\dfrac{\kappa_{-1}+\kappa_{+2}}{\kappa_{+1}} + S^a}$$

Taking the substrate exponent as 1, we find the following correspondences between Monod's "maximum growth rate", μ_m, and "affinity constant" or "Monod constant", K_S, given in equation (2), and various apparent rate constants in the derived equations (8), (12), (16) and (17):

$$\text{eq.(8)}: \mu_m = \kappa_2 \text{ ; } K_S = \kappa_2 / \kappa_1\text{; either } B_1 \text{ or } B_n \; magi$$

$$\text{eq.(12)}: \begin{Bmatrix} \mu_m = \kappa_1 \text{ ; } K_S = \kappa_1 / \kappa_2\text{; } magi \text{ occurs before substrate binding} \\ \mu_m = \kappa_2 \text{ ; } K_S = \kappa_2 / \kappa_1\text{; } magi \text{ occurs after substrate binding} \end{Bmatrix}$$

$$\text{eq.(16)}: \mu_m = \kappa_{+1}\text{; } K_S = \frac{\kappa_{+1}+\kappa_{-1}}{\kappa_{+2}}\text{; } S \text{ binding after "reversible" } magi \text{ formation}$$

$$\text{eq.(17)}: \mu_m = \kappa_{+2}\text{; } K_S = \frac{\kappa_{-1}+\kappa_{+2}}{\kappa_{+1}}\text{; } S \text{ binding before "reversible" } magi \text{ formation} \tag{18}$$

Equations (18) offer some mechanistic insight, (not in "structural", elementary steps terms, but rather at a "mesoscopic" level) for various experimental facts:

1. The simultaneous variation of μ_m and K_S, reviewed in (Kovarova-Kovar and Egli, 1998). The observed fact that fitting various experimental data, on growth of various bacterial cultures, resulted in highly correlated parameters μ_m and K_S is clearly accounted for in eq.(18). Indeed, the detailed expressions of the two parameters in terms of apparent rate-constants pertaining to mechanistic steps given in (5), (6), (10) or (11) offer the needed information. μ_m is a term in the expression of K_S in eqs. (16) and (17). It is a factor of K_S in eqs. (8) and (9) that justifies the use of the ratio μ_m/K_S as a better fitting parameter to evaluate the competition for nutrient at low concentrations (Healey, 1980). This ratio was termed "specific affinity" by Button (Button, 1985; 1991; 1998) which extensively used this concept in modeling growth at extreme substrate dilutions (such as the ones present in the case arctic, sea-water bacterial strains that grow in "starving conditions"). In such cases substrate binding, or "uptake", is an essential first step in bacterial growth, structurally detailed and

extensively investigated in Button's works. It is the reason of the introduction of the quasi-chemical mechanism (5) or (17) in the present contribution. The complementary limiting case is given by mechanism (6) and pertains to abundant nutrient conditions that may be achieved in many laboratory studies. If only one-way (irreversible) steps are taken into consideration, both the above mechanisms result in formally identical expressions of the rate equation in terms of the apparent rate constants (with different "structure", i.e. meaning, of these constants).

2. Reported variations of several orders of magnitude for the Monod constant of the same bacteria on same substrates, but different laboratories, lead to a memorable formulation (Ferenci, 1999) of "an uncertainty principle instead of constants in bacterial growth kinetics"; various reasons such as inoculum history, genotype, substrate exposure, bacterial density, were invoked as possible causes of experimental discrepancies. Some apparent kinetic arguments may be added, as evidenced in eqs. (18). The interplay between the values of the apparent rate constants is the immediate, most convenient argument, but there is more. The actual expressions given in (18) are based on and result from some real mechanistic assumptions (even if quasi-chemical and mesoscopic). **If a most abundant growth intermediate exists, it not only leads to Monod-type growth kinetics. Substrate binding prior or subsequent to the formation of *magi* is expected to result in substantial variations of the overall Monod constant**. If the same culture, growing on the same substrate, is studied, even in the same laboratory, in experimental conditions that obey eq. (16) in one experiment, and eq. (17) in another one, differences in obtained K_S should be the norm; similarities may be regarded as a (doubtful) exception.
3. More complicated Monod-type rate equations, such as those obtained in eqs. (14) and (15), are common in heterogeneous catalysis and enzyme kinetics. Getting into the chemical details results in tremendous complexity, in which the method of King and Altman (King and Altman, 1956) developed by Hill (Hill, 1989) and Happel and Sellers (Happel and Sellers, 1992; 1995) offer (computed-aided) solutions. These are beyond the purpose of the present contribution, where the goal was to derive/explain the validity of "simple" growth rate equations, mainly that of Monod's. However, Monod equation alternatives such as those reviewed in (Kovarova-Kovar and Egli, 1998), or (Liu, 2007) are amenable to the quasi-chemical approach presented here.

Application to Microcalorimetric Growth Data

Among the wide range of microcalorimetry applications, an important and promising one is the direct measurement of heat generated by the biological processes within living cells. Microorganisms (including bacteria) are reported to produce heat to an average rate of 1-3 pW per cell (Braissant et al., 2010).The bacterial replication process can be monitored in real time due to the heat production associated with their metabolic activity recorded as heat flow versus time. Modern isothermal microcalorimeters (IMC) allow for the detection of less than one microwatt in power change.

As a result, as few as 10,000-100,000 active bacterial cells in a culture are sufficient to produce a measurable real-time signal, dynamically related to the number of cells present and their activity. For aerobic growth, a recent contribution (Maskow et al., 2012) used an extension of the above range to 1-4pW per cell based on earlier reported results (Beezer et al., 1974), thus pointing to a range of calorimetric detection of 6250 – 25000 cells per ml. Therefore, microcalorimetry may be considered as one of the most sensitive tools in the study of bacterial growth.

Recent microcalorimetric studies regarding the antibacterial effect or interaction of different compounds (chemical or biological) with certain bacterial strains further acknowledged the reliability and utility of this method (Li et al., 2011), (Kong et al., 2011; Wang et al., 2010). Applications of various growth models to microcalorimetric data were recently reviewed (Braissant et al., 2013). Although they acknowledge *the complex nature of microcalorimetric growth thermograms,* the approaches advanced thus far did not attempt to decompose the thermal signal into simpler growth stages and analyze the obtained components separately. An attempt to solve this issue was recently presented (Zaharia et al., 2013) and will be briefly resumed in the following.

Microcalorimetry

Two Setaram Differential Scanning Microcalorimeters (MicroDSC) were used in the referred study (Zaharia et al., 2013): the MicroDSC III and MicroDSC VII Evo. Both instruments were Joule effect factory calibrated and periodically checked with the factory naphthalene standard. Each calorimeter had an outer thermostatic loop provided by a Julabo F32-HE device operating in standard mode. 3D sensor protection was provided by a Nitrogen gas purge (99.99% SIAD - TP). The Calisto v1.077 software package was used for data acquisition and primary signal processing. This included baseline integration and export in Excel with equally spaced time increments. Heat values obtained were further analyzed in Excel and Origin 8.1. Exported baselines were further processed in Peakfit.

Peakfit Processing of the Thermograms

Complex growth (roughly two-peak) thermograms were obtained and their shape was found to vary with the sample volume, i.e. with the oxygen content of the calorimetric cell headspace (Zaharia et al., 2013). The natural extension of the thermograms analysis was to decompose the observed thermal signal into its components (by means of Peakfit® - Systat software) and examine their variation with (cell headspace) air volume. [The term "deconvolution" is often improperly used for various cases of complex signal analysis. The confusion is due to the fact that any measured signal is indeed a convolution between the intrinsic contribution of the investigated physical-chemical process and the instrumental contribution. The two may be separated, i.e. "deconvolved", by means of Fourier analysis, as in X-ray diffraction line broadening analysis. For most of the spectroscopy, chromatography and (micro)calorimetry data, the observed complex signal is a superposition (in fact, the sum) of various components (processes) that may be evidenced via Peakfit. Therefore, the term "decomposition" is the correct choice.]

Although experimental growth patterns were clearly more complex, the two-peak decomposition was chosen, as a first-order approximation; the two growth processes (peaks) were ascribed to dissolved oxygen and cell headspace diffused oxygen acting as a limiting factor for aerobic growth of both investigated strains: *Staphylococcus aureus* and *Escherichia coli*. Data exported from Calisto were processed in Peakfit by means of previously reported routines, (Popa and Segal, 2002; Popa, 2003). Whenever necessary, Savitsky-Golay smoothing was performed, generally with the "Al Expert" option. Calisto-generated baselines were imported and subtracted from the heat flow (HF, mW) signal. The time zero was changed for each thermogram by means of "Enter Calculation" option in Peakfit, allowing to a left shift of the whole data corresponding to the left intersection of the baseline and HF. This procedure brings all thermograms to a common X (time) scale, but definitely excludes any analysis of the growth lag time. The resulted data were subjected to "Area normalize" resulting in the "normalized heat flow" (NHF, h^{-1}). This brings all thermograms to a common Y (NHF) scale, with the advantage that areas of the component peaks represent their fraction to the overall thermal effect. All subsequent peak fitting may be performed on either HF – time or the NHF – time thermograms. Several built-in asymmetric peak functions were tried (EMG, GMG, LogN, Giddings, Pearson IV, HVL, etc.). The best one for the analyzed data proved to be the Haarhof – Van der Linde (HVL) chromatography function. This function resulted in both the best statistical criteria (r^2, F-statistic, standard errors, etc.) and most reliable variations of the fitting parameters among the member of each set. Peak parameters were allowed to vary independently through the "Vary Widths" and "Vary Shape" options. The "Medium (Lorentz Err.) Robust Minimization" procedure was applied instead of the classical least-squares one.

The culture medium utilized was a nutrient-rich one, containing a sufficient amount of glucose: a shift in the carbon source resulting in diauxic growth was therefore excluded within the experimental setup utilized. Supplementary physiological saline dilution and mineral oil addition experiments pointed to the reported interpretation (Zaharia et al., 2013) of oxygen acting as a "limiting substrate". The complex processes that take place inside the o-ring sealed batch cell is governed by oxygen as a limiting thermal growth factor (terminal electron acceptor): the first process (peak) may be ascribed to "dissolved oxygen growth" and the second one to "diffused oxygen growth". The following section illustrates the application of the Monod equation to the kinetic analysis of some of the thermal growth data reported in (Zaharia et al., 2013). To our best knowledge, the following approach is entirely novel.

Application of the Monod Growth Model

For the application of the Monod equation to microcalorimetric data the main model assumptions will be utilized in the form given by eq. (19).

Monod growth model

$$
\begin{gathered}
\text{1) Growth rate: } \frac{dB(t)}{dt} = B(t)\frac{\mu_m S(t)}{K_S + S(t)} \\
\text{2) Substrate consumption: } -\gamma\frac{dS(t)}{dt} = \frac{dN(t)}{dt}; \gamma = \text{"growth yield"} \\
\text{3) Limit conditions: } \begin{Bmatrix} t = t_0 : B(t_0) = B_0(\text{inoculum}) \; ; \; S(t_0) = S_0\,(\text{maximum}) \\ t = t_f : B(t_f) = B_f \; ; \; S(t_f) = 0\ (\text{substrate exhaustion}) \end{Bmatrix}
\end{gathered} \tag{19}
$$

There are several issues concerning thermal growth data, as previously reported (Zaharia et al., 2010; 2013). The main one concerns the thermal fingerprint of the inoculum, in situations where its size is not negligible (this happens whenever a faster growth, of say ~10 hours, is encountered, as the data analyzed in the present contribution). In such circumstances instrument equilibration interferes with the inoculum thermal signal, and baseline correction is a mandatory step in experimental data processing. The main issues, solutions, assumptions and available thermal quantities that can be evaluated from the raw and processed thermogram are presented in eq.(20-23):

$$
\begin{gathered}
\text{Main problems in } \mu\text{DSC: inoculum heat flow and heat} \\
HF(t_0) = ?\ ;\ \Delta H(t_0) = ? \\
\text{Solution: } HF(t)\,(\text{heat flow}) \xrightarrow{\text{baseline subtraction}} \underline{HF}(t) = \text{"net heat flow"} \\
\Rightarrow \underline{HF}(t_0) = \underline{HF}(t_f) = 0\ ;\ \underline{\Delta H}(t) = \int \underline{HF}(t)dt\ ;\ \underline{\Delta H}(t_0) = 0 \\
\underline{\Delta H}(t) = \int_{t_0}^{t} \underline{HF}(t)dt = \text{"net thermal growth"} \\
\int_{t_0}^{t_f} \underline{HF}(t)dt = \underline{\Delta H}(t_f) - 0 = \underline{\Delta H}_f = \text{"total net thermal growth"}
\end{gathered} \tag{20}
$$

Equation (20) states the lack of experimental information concerning thermal quantities associated with the inoculum. The thermal expression of the "growth lag", a fundamental parameter defined by Monod (Monod, 1949)] is, within the experimental conditions utilized in (Zaharia et al., 2013), unavailable. Meanwhile, another fundamental growth parameter termed by Monod "total growth" (*croissance totale)* (Monod, 1942) is clearly associated with the "total net thermal growth".

$$
\begin{gathered}
\text{Fundamental assumption: net heat flow scales with growth rate} \\
\underline{HF}(t) = \varphi\frac{dB(t)}{dt} = \text{"thermal growth rate"}\ ;\ \varphi = \text{constant} \\
\int_{t_0}^{t} \underline{HF}(t)dt = \varphi\int_{t_0}^{t} \frac{dB(t)}{dt}dt \Rightarrow \underline{\Delta H}(t) = \varphi[B(t) - B(t_0)] \\
B(t_0) = B_0\ ;\ B(t_f) = B_f\ ;\ \underline{\Delta H}(t) = \varphi[B(t) - B_0] \\
\varphi[B_f - B_0] = \underline{\Delta H}_f = \text{"total net thermal growth"}
\end{gathered} \tag{21}
$$

Equation (21) contains the standard assumption utilized in the kinetic analysis of calorimetric data (DSC, MicroDSC, Calvet, etc.): the proportionality between the investigated process rate and its calorimetric expression (heat flow or power). The thermal analogue of Monod's specific growth rate is also available:

$$\text{Specific quantities that can be evaluated from the thermogram:}$$
$$\underline{SHF}(t) = \frac{\underline{HF}(t)}{\underline{\Delta H}(t)} = \frac{dB(t)/dt}{B(t)-B_0} \neq \frac{dB(t)/dt}{B(t)}$$
$$\underline{SHF}(t) = \text{"specific net heat flow"}$$
$$\text{If } B_0 \text{ non-negligible:}$$
$$\text{"specific thermal growth rate"} \neq \text{"specific growth rate" !} \tag{22}$$

Equation (22) is the main reason for the application of the Monod growth equation in the form given by eq.(19), instead of the usual alternative eq.(2). With non-negligible inoculum, the available (i.e. computable from experimental data) specific thermal quantity is different from the actual specific growth rate. Normalized thermal quantities are also available, resulting in the adimensional "normalized net thermal growth", similar to a conversion variable:

$$\text{Normalized quantities that can be evaluated from the thermogram:}$$
$$\underline{NHF}(t) = \frac{\underline{HF}(t)}{\underline{\Delta H}_f} = \text{"normalized net heat flow" ; } \langle \underline{NHF}(t) \rangle = \text{time}^{-1}$$
$$\eta(t) = \frac{\underline{\Delta H}(t)}{\underline{\Delta H}_f} = \text{"normalized net thermal growth" ; } \langle \eta(t) \rangle = \text{adimensional} \tag{23}$$

Equation (23) is the most convenient form of the thermal quantities for the kinetic analysis of experimental data and its application will be further illustrated.

Combining equations (19), (20), and (21) we obtain:

$$-(\varphi\gamma)dS/dt = \underline{HF}(t) \Leftrightarrow \gamma_T dS/dt = \underline{HF}(t)$$
$$\gamma_T = \varphi\gamma = \text{"thermal growth yield"} \tag{24}$$
$$\underline{\Delta H}(t) = -\gamma_T[S(t)-S_0] \text{ ; } \underline{\Delta H}(t_f) = \underline{\Delta H}_f = \gamma_T S_0$$

Taking into account equations (23) and (24) and combining with (19) and (21) result in two alternative thermal expressions of Monod's growth model:

Monod "net thermal growth"

$$\underline{HF}(t) = \mu_m \frac{\left[\varphi N_0 + \underline{\Delta H}(t)\right]\left[\underline{\Delta H}_f - \underline{\Delta H}(t)\right]}{\underline{\Delta H}_f + \gamma_T K_S - \underline{\Delta H}(t)} \tag{25}$$

Monod "normalized net thermal growth"

$$\underline{NHF}(t) = \mu_m \frac{\left[(\varphi B_0 / \underline{\Delta H}_f) + \eta(t)\right]\left[1 - \eta(t)\right]}{1 + (\gamma_T K_S / \underline{\Delta H}_f) - \eta(t)} \tag{26}$$

Equation (26) will be applied as a TableCurve® user defined function (UDF) to test the Monod thermal growth model to microcalorimetric data. (The same equation may be used in Peakfit® with similar results, which will not be presented here.) The corresponding UDF reads:

Monod-NHF user defined function

$$y = \frac{a(b+x)(1-x)}{1+c-x} \; ; \; y = NHF(t) \; ; \; x = \eta(t) \tag{27}$$

Adjustable parameters:

$$a = \mu_m \; ; \; b = \frac{\varphi B_0}{\underline{\Delta H}_f} \; ; \; c = \frac{\gamma_T K_S}{\underline{\Delta H}_f}$$

The whole procedure outlined above will be illustrated for one of the runs performed in (Zaharia et al., 2013), where detailed experimental information is to be found: an *E coli* sample of 0.5ml. Figure 1 contains a typical result of the first stage of data processing. Baseline subtracted thermogram, obtained via the instrument's software, was imported in Peakfit and decomposed with the built-in (library) HVL peak function. The resulted fit components were exported and plotted in Excel, where Figure 1 was generated. As already mentioned, the two peaks (numbered in the order of their appearance on the time scale) were ascribed to dissolved oxygen aerobic growth (peak 1) and diffused oxygen (from the cell headspace) aerobic growth (peak 2). Multi-component thermograms were previously reported (Kabanova et al., 2012) and kinetically analyzed with respect to the ascending part(s) of the thermograms, taken as the exponential growth phases (Kabanova et al., 2013; Kabanova, 2013).

The present approach involves, in addition to the separation of the growth components, an attempt to kinetically analyze in "Monod thermal terms" *each component as a whole process*. In other words, the proposed method attempts to answer the following question: do the Peakfit-revealed thermal growth processes obey Monod-type kinetics over their whole time extension? The next step in data processing necessary for answering to the above question is illustrated in Figure 2 and 3. All thermal quantities involved in Eqs. (21) and (23) were calculated in Excel from the Peakfit-generated data. The procedure involved in obtaining Figs. 2 and 3 was then applied for each of the two peaks taken separately. This can be done either in Peakfit or in Excel and results in net heat flows normalized to each of the two peaks' corresponding areas. The obtained data, *NHF(t)* versus $\eta(t)$, are amenable to fit with the equation (27). The result, presented in Figs. 4 and 5, is a surprising answer "yes" to the above question: the Monod-NHF function offers a decent fit of the decomposed thermogram, for both growth components!

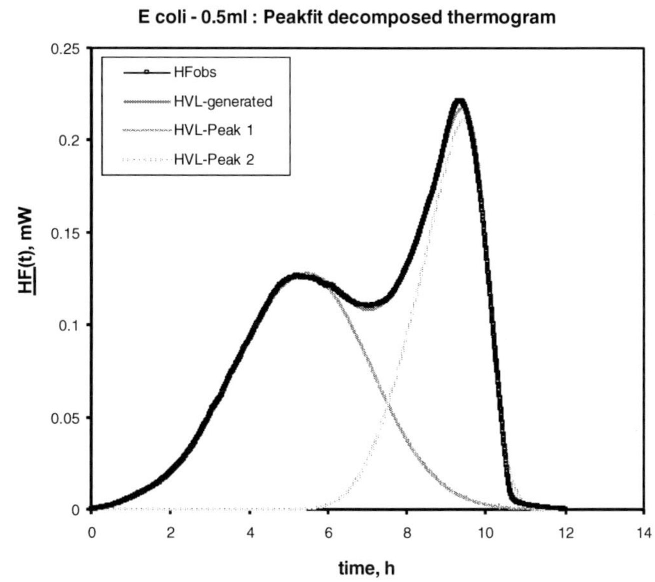

Figure 1. A Peakfit® decomposed thermogram using HVL (Haarhof – Van der Linde) functions.

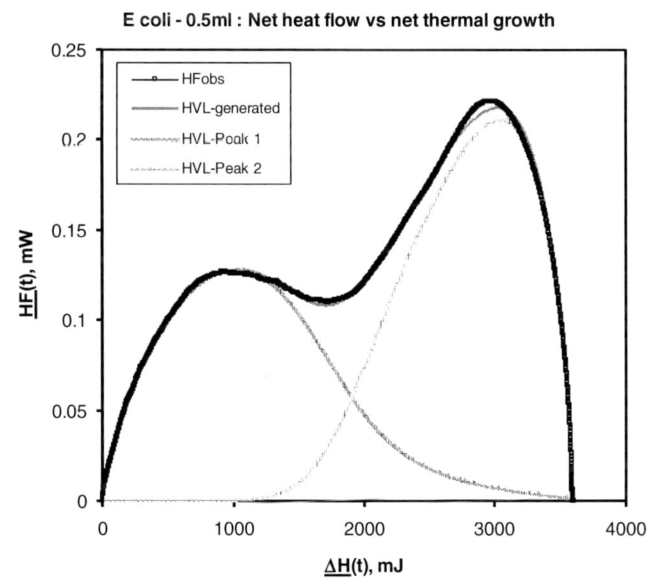

Figure 2. Net thermal growth representation of the data given in Figure 1.

This is a surprising result indeed, as the quasi-chemical kinetic derivation of the Monod equation presented in the preceding section involved, among other approximations and assumptions, the key one of a quasi-steady state growth process. This is clearly not the case with the bacterial growth in the hermetically sealed batch calorimetric cell. The actual situation is closer to the ones described in (Schnell and Maini, 2002). One possible explanation would be the following: with oxygen acting as a "limiting growth factor" (and not, literally speaking, a "limiting substrate") the actual form of the kinetic equation is preserved over the whole thermal growth range.

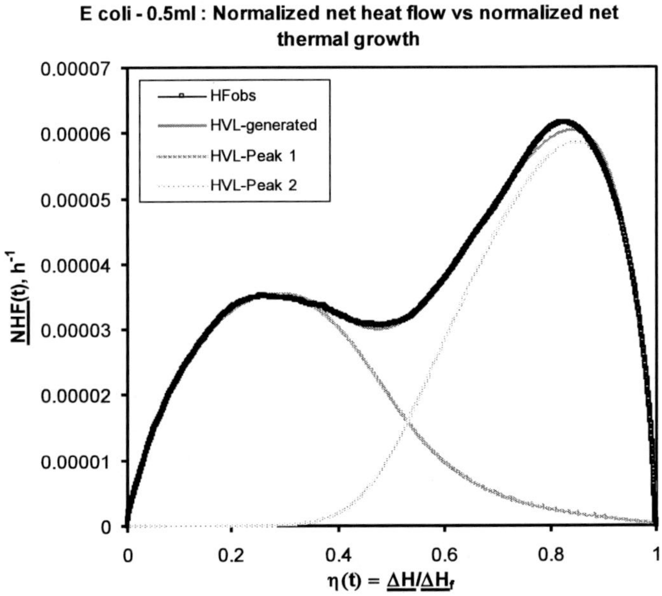

Figure 3. Normalized net thermal growth representation of the data given in Figure 1.

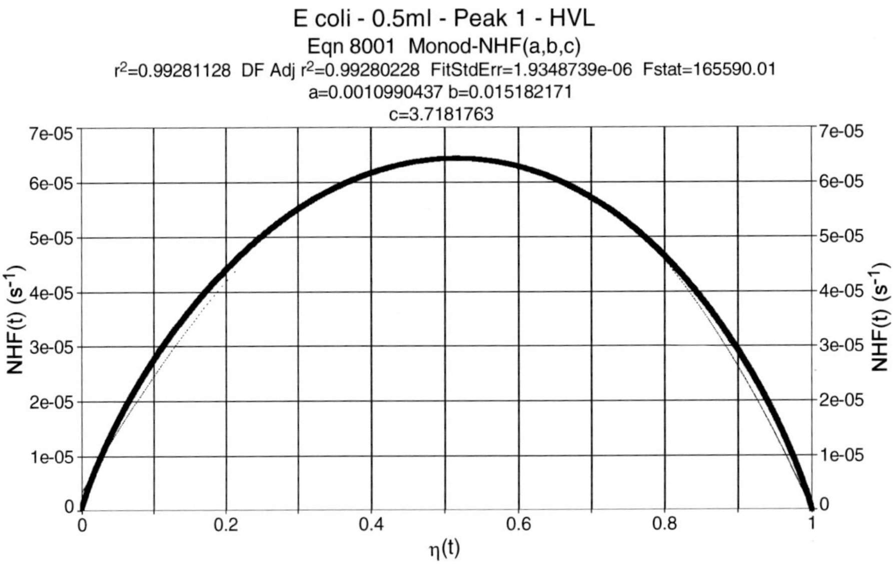

Figure 4. First peak fit with the Monod-NHF user defined function. ■ : processed data; ---: fit.

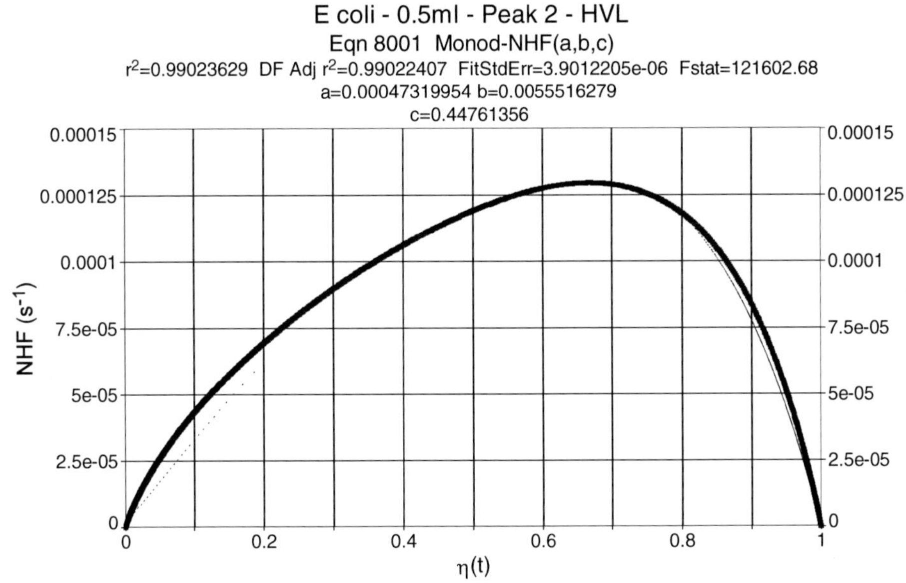

Figure 5. Second peak fit with the Monod-NHF user defined function. ■ : processed data; ---: fit.

As recently emphasized (Braissant et al., 2013), thermal growth may not always correspond to biomass growth. However, the thermal expression of bacterial activity still obeys Monod kinetics in certain experimental circumstances, such as those described in (Zaharia et al., 2013) and analyzed here. Whatever the real situation, the existence of a most abundant growth intermediate and its conservation over the whole range of the analyzed growth process seems to offer a plausible explanation of the Monod equation versatility.

Conclusion

The quasi-chemical approach to bacterial growth presented here was inspired by some important pioneers of heterogeneous catalysis (and later enzyme) kinetics: Juro Horiuti, Mikhail Temkin, John Happel and Michel Boudart. Experimental investigation of the whole bacterial growth process with detailed (bio)chemical means seems, even with modern high throughput techniques, hopeless. Various degrees of zooming have been applied in (micro)kinetic modeling of growth. In spite of considerable progress, the simple Monod model (or some of its modifications) still works unexpectedly well in describing overall growth data.

The present contribution is an attempt to explain this surprising fact. The proposed quasi-chemical approach involves a "zooming-out" of the detailed picture of the growing bacterial culture up to its main actors: the bacterial cell and the growth limiting (or controlling) substrate. It is restricted to a single growth limiting substrate.

The main idea is to follow the bacterial cell through its various significant transformation stages up to its dream defined by François Jacob: becoming two cells. These transformation stages were termed "bacterial species" in the quasi-chemical, not biological sense, and were treated as complex reaction intermediates. In one or several of these stages the bacterium has to encounter, uptake and process the substrate. With all other nutrients assumed in excess, (bio)mass balance involves only bacterial species and limiting substrate. The other assumptions involved are that the growth process is taking place at (quasi)steady state and the existence of a most abundant growth intermediate. Under these circumstances, with standard methods of catalytic kinetics, several versions of "detailed" mechanisms were proven to reduce to two-step sequences resulting in Monod-type rate equations.

As the resolution of the utilized "zooming lens" is restricted to the bacterial cell, all detailed biochemistry involved is embodied in apparent kinetic parameters of the steps that link various bacterial species. In analogy with Happel's reaction path concept, following the bacterial cell through its dream is similar with following a radioactive tracer from reactant to product species. We may thus adequately term as "bacterial paths" all versions of mechanisms presented here.

This approach carries all the advantages and shortcomings of what Boudart termed "simplified kinetics":

- it is simple, as its derived result, the Monod equation (or similar ones);
- it embodies all reported "ambiguities" of this equation on data analysis;
- but it may offer some reasonable explanations of these ambiguities.

In addition to that, the proposed quasi-chemical approach is easily amenable to an increase in resolution. Indeed, with a stronger "zooming lens" one may introduce further mechanistic details and obtain rate equations that differ from Monod's one. Extension to nonuniform bacterial reactivity is also possible.

Application of the Monod model to microcalorimetric data requires the decomposition of the raw thermal signal as a mandatory step for further processing. In spite of the complexity of the initial signal, the thermal version of Monod's growth model describes the components of the experimental thermogram surprisingly well. For the analyzed experimental data, which involves *E coli* aerobic growth with oxygen as a limiting growth factor, the versatility of the Monod thermal growth model may be explained by the existence and conservation of the same most abundant growth intermediate, over the whole time-extension of the process.

References

Beezer, A. E., Bettelheim, K. A., Newell, R. D. & Stevens, J. (1974). Diagnosis of bacteriuria by flow microcalorimetry: preliminary report. *Sci. Tools.*, *21*, 13-16.

Boudart, M. & Tamaru, K. (1991). The step that determines the rate of a single catalytic cycle. *Catal. Lett.*, *9* (1-2), 15-22.

Boudart, M. & Djéga-Mariadassou, G. (1984). *Kinetics of Heterogeneous Catalytic Reactions*. Princeton University Press, Princeton, NJ.

Boudart, M. (1968). *Kinetics of Chemical Processes*. Prentice Hall Inc., Englewood Cliffs, N. J.

Boudart, M. (1975). Heterogeneous Catalysis, in Eyring H., Editor, *Reactions in Condensed Phases: Physical Chemistry. An Advanced Treatise*. Academic Press, New York, NY.

Braissant O., Wirz D., Göpfert B. & Daniels A. U. (2010). Use of isothermal microcalorimetry to monitor microbial activities. *FEMS Microbiol. Lett.*, *303*, 1-8.

Braissant, O., Bonkat, G., Wirz, D. & Bachmann, A. (2013). Microbial growth and isothermal microcalorimetry: Growth models and their application to microcalorimetric data. *Thermochim. Acta.*, *555*, 64-71.

Button, D. K. (1985). Kinetics of Nutrient-Limited Transport and Microbial Growth. *Microbiol. Rev.*, *49* (3), 270-297.

Button, D. K. (1991). Biochemical Basis for Whole-Cell Uptake Kinetics: Specific Affinity, Oligotrophic Capacity, and the Meaning of the Michaelis Constant. *Appl. Environ. Microb.*, *57* (7), 2033-2038.

Button, D. K. (1998). Nutrient Uptake by Microorganisms according to Kinetic Parameters from Theory Related to Cytoarchitecture. *Microbiol Mol. Biol. R.*, *62* (3), 636-645.

Button, D. K. (2000). Abandon Michaelis-Menten? *ASM News*, *66*, 510.

Button, D. K. (2010). Kinetics and Physiology at Vanishingly Small Substrate Concentrations. In *Handbook of Hydrocarbon and Lipid Microbiology*, Kenneth N. Timmis, Editor, Springer-Verlag Berlin Heidelberg, 1635-1647.

Button, D. K. & Garver, J. C. (1966). Continuous Culture of *Torulopsis utilis*: A kinetic Study of Oxygen Limited Growth. *J. gen. Microbiol.* 45, 195-204.

Button, D. K., Robertson, B., Gustafson, E. & Zhao, X. (2004). Experimental and Theoretical Bases of Specific Affinity, a Cytoarchitecture-Based Formulation of Nutrient Collection Proposed To Supercede the Michaelis-Menten Paradigm of Microbial Kinetics. *Appl. Environ. Microb.*, *70* (9), 5511-5521.

Chebotareva, N. A., Kurganov, B. I. & Livanova, N. B. (2004). Biochemical Effects of Molecular Crowding. *Biochemistry (Moscov)*, *69* (11), 1522-1536.

Csuha, R. S. & Happel, J. (1971). Structure of rate equations for heterogeneous catalysis. *AIChE J.*, *17* (4), 927–932.

de Crécy, E., Metzgar, D., Allen, C., Pénicaud, M., Lyons, B., Hamsen, C. J. & de Crécy-Lagard, V. (2007). Development of a novel continuous culture device for experimental evolution of bacterial populations. *Appl. Microbiol. Biotechnol.*, *77*, 489-496.

Dean, A. C. R. & Hinshelwood, C. (1966). *Growth, Function and Regulation in Bacterial Cells*. Oxford University Press, London.

Derris, J. B., Kim, M., Zhang, Z., Okano, H., Hermsen, R., Groisman, A. & Hwa, T. (2013). The Innate Growth Bistability and Fitness Landscapes of Antibiotic-Resistant Bacteria. *Science*, 342, 1237435. DOI: 10. 1126/science. 12367435.

Dubnau, D. & Losick, R. (2006). Bistability in bacteria. *Mol. Microbiol.*, *61* (3), 564-572.

Ferenci, T. (1999). Growth of bacterial cultures' 50 years on: towards an uncertainty principle instead of constants in bacterial growth kinetics. *Res. Microbiol.*, *150*, 431-438.

Happel, J. (1968). Heterogeneous Catalysis with a Single Overall Reaction. *J. Res. Inst. Catal. Hokkaido Univ.*, *16* (1), 305-322.

Happel, J. (1981). Comments on Horiuti's Stoichiometric Number Concept. *J. Res. Inst. Catal. Hokkaido Univ.*, *28* (3), 185-188.

Happel, J. (1986). *Isotopic Assessment of Heterogeneous Catalysis*. Academic Press Inc., Orlando, FL.

Happel, J. & Sellers, P. H. (1992). Systemization for the King-Altman-Hill diagram method in chemical kinetics. *J. Phys. Chem.*, *96* (6), 2593–2597.

Happel, J. & Sellers, P. H. (1995). New Perspective on the Kinetics of Enzyme Catalysis. *J. Phys. Chem.*, *99*, 6595-6600.

Healey, F. P. (1980). Slope of the Monod equation as an indicator of advantage in nutrient competition. *Microb. Ecol.*, *5*, 281-286.

Hill, A. V. (1910). The possible effects of the aggregation of the molecules of hæmoglobin on its dissociation curves. *Proc. Physiol. Soc.*, Jan. 22, iv-vii.

Hill, T. L. (1989). *Free Energy Transduction and Biochemical Cycle Kinetics*. Springer-Verlag, New York, INC.

Hinshelwood, C. N. (1946). *The Chemical Kinetics of the Bacterial Cell*. Oxford University Press, London.

Horiuti, J. (1948). A Method of Statistical Mechanical Treatment of Equilibrium and Chemical Reactions. *J. Res. Inst. Catal. Hokkaido Univ.*, *1* (1), 8-79.

Horiuti, J. (1957). Stoichiometrische Zahlen und die Kinetik der Chemischen Reaktionen. *J. Res. Inst. Catal. Hokkaido Univ.*, *5* (1), 1-26.

Horiuti, J. (1973). Theory of Reaction Rates as Based on the Stoichiometric Number Concept. *Ann. NY Acad. Sci.*, *213*, 5-30.

Horiuti, J. & Nakamura, T. (1967). On the Theory of Heterogeneous Catalysis. *Adv. Catal.*, *17*, 1-74.

Jin, Q. & Bethke, C. M. (2002). Kinetics of Electron Transfer through the Respiratory Chain. *Biophys. J.*, *83*, 1797-1808.

Jin, Q. & Bethke, C. M. (2003). A New Rate Law Describing Microbial Respiration. *AEM*, 69 (4), 2340-2348.

Jin, Q. & Bethke, C. M. (2007). The Thermodynamics and Kinetics of Microbial Metabolism. *AJS*, *307*, 643-677.

Kabanova, N. (2013) *Development of a Microcalorimetric Method for the Study of Fermentation Processes*. PhD Thesis, Tallin University of Technology, Faculty of Science, Department of Chemistry.

Kabanova, N., Stulova, I. & Vilu, R. (2012). Microcalorimetric study of the growth of bacterial colonies of *Lactococcus lactis* IL1403 in agar gels. *Food Microbiol.*, *29*, 67-79.

Kabanova, N., Stulova, I. & Vilu, R. (2013). Microcalorimetric study of growth of *Lactococcus lactis* IL1403 at low glucose concentration in liquids and solid agar gels. *Thermochim. Acta*, *559*, 69-75.

King, E. L. & Altman, C. (1956). A Schematic Method of Deriving the Rate Laws for Enzyme-Catalyzed Reactions. *J. Phys. Chem.*, *60*, 1375-1378.

Kong, W., Wang, J., Xing, X., Jin, C., Xiao, X., Zhao, Y., Zhang, P., Zang, Q. & Li, Z. (2011). Screening for novel antibacterial agents based on the activities of compounds on metabolism of *Escherichia coli*: A microcalorimetric study. *J. Hazard. Mater.*, *185*, 346–352.

Kovarova-Kovar, K. & Egli, T. (1998). Growth Kinetics of Suspended Microbial Cells: From Single-Substrate-Controlled Growth to Mixed-Substrate Kinetics. *Microbiol. Mol. Biol. R.*, *62* (3), 646-666.

Langmuir, I. (1918). The Adsorption of Gases on Plane Surfaces of Glass, Mica and Platinum. *J. Am. Chem. Soc.*, *40* (9), 1361-1403.

Li, X., Zhang, Z., Wang, C., Zhang, T., He, K. & Deng, F. (2011). Synthesis, crystal structure and action on *Escherichia coli* by microcalorimetry of copper complexes with 1,10-phenanthroline and amino acid. *J. Inorg. Biochem.*, *105*, 23–30.

Liu, Y. (2006). A simple thermodynamic approach for derivation of a general Monod equation for microbial growth. *Biochem. Eng. J.*, *31*, 102-105.

Liu, Y. (2007). Overview of some theoretical approaches for derivation of the Monod equation. *Appl. Microbiol. Biot.*, *73*, 1241-1250.

Liu, Y., Lin. Y.-M. & Yang, S.-F. (2003). A Thermodynamic Interpretation of the Monod Equation. *Curr. Microbiol.*, *46*, 233-234.

Lobry, J. R., Flandrois, J. P., Carret, G. & Pave, A. (1992). Monod's Bacterial Growth Model Revisited. *Bull. Math. Biol.*, *54*(1), 117-122.

Maskow, T., Wolf, K., Kunze, W., Enders, S. & Harms, H. (2012). Rapid analysis of bacterial contamination of tap water using isothermal calorimetry. *Thermochim. Acta.*, *543*, 273-280.

Michaelis, L. & Menten, M. L. (1913). Die Kinetic der Invertinwirkung. *Biochem. Z.*, *49*, 334-369.

Monod, J. (1942). *Recherches sur la Croissance des Cultures Bactériennes*. Hermann et Cie., Paris. This is the published version of Monod's PhD thesis with the same title, defended in 1941.

Monod, J. (1946). Kinetics of bacterial growth. *Nature*, *160*, 105-106.

Monod, J. (1949). The Growth of Bacterial Cultures. *Annu. Rev. Microbiol.*, *3*, 371-394.

Monod, J. (1950). La téchnique de culture continue: théorie et applications. *Ann. Inst. Pasteur (Paris)*, *79*, 390-410.

Nielsen, J. & Villadsen, J. (1992). Modelling Microbial Kinetics. *Chem. Eng. Sci.*, *47* (17-18), 4225-4270.

Popa, V. T. (2003). Apparent kinetics of ice melting revealed by DSC endotherms. *Rev. Roum. Chim.*, *48*(12), 987-995.

Popa, V. T. & Segal E,. (2002). Shape analysis of DSC ice melting endotherms. Towards the estimation of the instrumental profile. *J. Therm. Anal. Cal.*, *69*, 149-161.

Roth, D., Finkelshtein, A., Ingham, C., Helman, Y., Sirota-Madi, A., Brodski, L. & Ben-Jacob, E. (2013). Identification and characterization of a highly motile and antibiotic refractory subpopulation involved in the expansion of swarming colonies of Paenibacillus vortex. *Environ. Microbiol.*, *15* (9), 2532-2544.

Ryall, B., Eydallin G. & Ferenci, T. (2012). Culture History and Population Heterogeneity as Determinants of Bacterial Adaptation: the Adaptomics of a Single Environmental Transition. *Microbiol. Mol. Biol. R.*, *76* (3), 597-625.

Schnell, S. & Maini, P. K. (2002). Enzyme Kinetics Far From the Standard Quasi-Steady-State and Equilibrium Approximations. *Mat. Comput. Model.*, *35*, 137-144.

Scott, M., Gunderson, C. W., Mateescu, E. M., Zhang, Z. & Hwa, T. (2010). Interdependence of Cell Growth and Gene Expression: Origins and Consequences. *Science*, *330*, 1099-1102.

Temkin, M. I. (1963). Kinetics of stationary reactions (in Russian). *Dokl. Acad. Nauk SSSR*, *152*, 156-159.

Temkin, M. I. (1968). Kinetics of stationary reactions (in Russian). *J. Res. Inst. Catal. Hokkaido Univ.*, *16* (1), 355-365.

Temkin, M. I. (1971). The kinetics of steady-state complex reactions. *Int. Chem. Eng.*, *11*, 709-717.

Temkin, M. I. (1979). The kinetics of some heterogeneous catalytic reactions. *Adv. Catal.*, *28*, 173-281.

Verhulst, P. F. (1845). Recherches mathématiques sur la loi d'accroissement de la population. *Mém. Acad. R. Sci. Lett. Belg.*, *18*, 1-38.

Vlad, M. O., Popa, V. T., Segal, E. & Ross, J. (2005). Multiple Rate-Determining Steps for Nonideal and Fractal Kinetics. *J. Phys. Chem. B.*, *109*, 2455-2460.

Wang, J., Zhao, H., Kong, W., Jin, C., Zhao, Y., Qu, Y. & Xiao, X. (2010). Microcalorimetric assay on the antimicrobial property of hydroxyanthraquinone derivatives in rhubarb (*Rheum palmatum* L.) to *Bifidobacterium adolescentis*. *Phytomedicine.*, *17*, 684–689.

Weiss, J. N. (1997). The Hill equation revisited: uses and misuses. *FASEB J.*, *11*, 835-841.

Xu, F. & Ding, H. (2007). A new kinetic model for heterogeneous (or spatially confined) enzymatic catalysis: contributions from the fractal and jamming (overcrowding) effects. *Appl. Catal. A.*, *317*, 70-81.

Zaharia, D. C., Iancu, C., Steriade, A. T., Muntean, A. A., Balint, O., Popa, V. T., Popa, M. I. & Bogdan, M. A. (2010). MicroDSC study of Staphylococcus epidermidis growth. *BMC Microbiology*, *10*, 322.

Zaharia, D. C., Muntean, A. A., Popa, M. G., Steriade, A. T., Balint, O., Micut, R., Iftene, C., Tofolean, I., Popa, V. T., Baicus, C., Bogdan, M. A. & Popa, M. I. (2013). Comparative analysis of Staphylococcus aureus and Escherichia coli microcalorimetric growth. *BMC Microbiology*, *13*, 171.

Zhang, Z., Boccazzi, P., Choi, H. -G., Perozziello, G., Sinskey, A. J. & Jensen, K. F. (2006). Microchemostat - microbial continuous culture in a polymer-based, instrumented microbioreactor. *Lab Chip*, *6*, 906-913.

Editor's Contact Information

Dr. Marina Nisnevitch
Department of Chemical Engineering,
Biotechnology and Materials
Ariel University, Israel
Tel: +972-39066606/7
Fax: +972-39066323
Email: marinan@ariel.ac.il

Index

#

10-methyloctanoic acid, 79
3-methylbutan-1-ol, 81
4-chlorophenol, 79, 81
7-oxocholic acid, 64
9-cis-hexadecenoic acid, 79

α

α-helical region, 10
α-helix, 10, 92, 96
α-type peptide, 10

β

β-1,6-N-acetyl-D-glucosamine, 42
β-barrel domain, 85
β-hemolysin, 8
β-lactam antibiotic, 66, 183
β-proteobacteria, 95
β-strand, 85, 96
β-type peptide, 10

δ

δ-toxin, 10

A

A. actinomycetemcomitans, 114, 117, 120, 121, 124, 128, 129, 131, 132, 133
A. aegypti, 151
*Aa*CDT, 111, 114, 115, 117, 120, 121, 124, 134
ABC, 16, 56, 58, 63, 66, 67, 72, 83
abiotic surface, vii, 1, 2, 4, 67
abolition, 133
absorption spectra, 206
abundant nutrient conditions, 233
accessory gene regulator, 6, 26
accessory regulator, 6
accounting, 92, 187
acetylcholine, 166
acetylcholine neurotransmitter, 166
acetyltransferase, 66, 72
achromobactin transporter, 63
acid, 8, 13, 24, 28, 41, 47, 49, 50, 51, 59, 60, 61, 63, 64, 68, 71, 72, 78, 79, 82, 86, 87, 88, 89, 92, 95, 96, 97, 105, 120, 123, 132, 159, 195, 199, 205, 206, 216
acidic, 61, 67, 93, 205
Acinetobacter baumannii, 59, 64, 74, 214
acne, 215, 218
acne vulgaris, 215, 218
acoustic cavitation, 210
acquisitions, 165
acridine orange, 198
acrylodan, 154
actin, 125, 131, 132, 135
activation state, 126
active compound, 24
active site, 38, 132, 133, 134
active transport, 57
activity level, 168
acute infection, 9, 14
acute intestinal disease, 110
acute septic arthritis, vii, 2
adaptation(s), vii, viii, 16, 51, 52, 73, 77, 78, 79, 80, 82, 164, 187, 222, 224
adaptation mechanism, viii, 77, 78, 79
adaptation to antibiotics, 224
adaptive response, 81
adenine, 67

adenine transport, 67
adenosine, 97, 103
adenovirus, 147
adhesion, vii, 1, 2, 4, 15, 21, 22, 25, 33, 36, 40, 41, 42, 44, 47, 50, 112, 113
adhesive factor, 8, 39
adhesive matrix molecule, 4, 11
adhesive properties, 14
adjustment, 46
ADP, 116
ADR, 29
adsorption, x, 36, 221, 223
adsorption of cells, 36
adults, 165
Aedes aegypti, 151, 157, 158, 159, 161
aerobic growth, 234, 235, 238, 242
aesthetic, 165
aetiology, 22
affinity constant, 224, 232
agar, 207, 209, 244
aggregate, 36
Aggregatibacter actinomycetemcomitans, 111, 114, 136, 137, 140, 142, 144, 145, 147
aggregation, 11, 14, 31, 40, 41, 63, 80, 153, 157, 244
agricultural pest, 150, 151
agriculture, 151
AIDS, 12
Akt signaling pathway, 134
ALA, 199, 206
alcagin transporter, 63
alginate, 94
alkaline pH homeostasis, 60, 61, 70
alkaline pH tolerance, 61
alkaline resistance, 61, 62
alkane, 79
alkane solvent, 79
alters, 86, 114, 147
amikacin, 95
amino, 4, 7, 8, 10, 38, 56, 60, 61, 65, 75, 84, 92, 95, 96, 97, 98, 120, 123, 132, 151, 152, 157, 160, 166, 168, 173, 204, 244
amino acid(s), 7, 8, 10, 38, 56, 60, 61, 65, 75, 84, 92, 95, 96, 97, 98, 120, 151, 152, 157, 166, 168, 173, 204, 244
amino acid transporter, 173
aminoglycosides, 42
amphipathicity, 12
amphiphilic molecule, 83
amylase, 157
amyloid-like fiber, 11
anabolism, 66
anaerobic bacteria, vii, 195
anaerobic conditions, viii, 67, 77, 78
anchorage, 133
anchoring, 4
animal botulism, 166
anion-selective porin, 84
annealing, 100
anti-apoptotic protein, 128
anti-bacterial activity, 154
anti-bacterial agent, 154
antibiotic, vii, viii, x, 2, 6, 11, 16, 18, 20, 29, 38, 41, 42, 46, 47, 49, 55, 65, 66, 72, 164, 183, 184, 186, 197, 205, 209, 213, 214, 217, 218, 224, 245
antibiotic resistance, viii, 11, 29, 41, 42, 46, 47, 49, 55, 164, 184, 186, 214, 217
antibiotic resistance mechanism, viii, 55
antibiotic resistant strain, vii, 2
antibiotic therapy, 6
antibody, 18, 20, 143, 195
anti-cancer, 155, 213
anti-cancer drug, 155, 213
antigen, 19, 28, 92, 112, 123, 143
antigenicity, 166
antimicrobial, viii, x, 11, 12, 14, 16, 17, 18, 25, 27, 28, 31, 33, 36, 41, 45, 46, 49, 55, 56, 61, 67, 74, 75, 105, 127, 197, 198, 209, 210, 214, 215, 216, 217, 218, 219, 246
antimicrobial chemotherapy, 210, 214, 216, 217
antimicrobial efflux activity, 56
antimicrobial peptide, 12, 28, 33
antimicrobial transport, 61
antimicrobial treatment, x, 11, 197
antiparallel β-sheets, 92
antisense, 104, 107
antisense RNA, 104, 107
antitumor, 142, 155, 161, 219
antitumor activity, 155, 161
apoptosis, 26, 110, 125, 127, 128, 138, 141, 145, 146, 147, 210
apoptotic pathways, 139
apoptotic response, 125, 128, 131, 134
apparent rate constant, 232, 233
aqueous solutions, 210
archaea, 58, 75, 92, 161, 188, 192
archaeal species, 96
Archaeoglobus fulgidus, 96
aromatic compounds, 77, 81, 89
aromatic hydrocarbon(s), 78, 81, 82, 83, 84
arrest(s), ix, 109, 124, 125, 126, 128, 129, 131, 133, 134, 136, 137, 138, 147, 152, 154
arsenic, 173
arthroplasty, 13, 20
aspartate, 4, 28
assessment, 191
assimilation, 65

astrocytes, 31
ataxia, 129
atmosphere, 31
atomic force, 71
ATP, viii, 27, 55, 56, 57, 71, 82, 83, 85
ATP-binding cassette, 56, 83
attachment, vii, 1, 2, 3, 4, 6, 8, 14, 25, 36, 45, 46, 51, 93, 119
autocrine, 117, 118
autoimmunity, 106
auto-inducing peptide, 11
autolysin, 4, 25, 30, 80, 87
autolysis, 20
autolytic lysozyme, 173
autophosphorylation, 130, 135
avian, 187
avoidance, 16

B

B. cenocepacia, 95, 96, 97, 98, 99, 101, 102
B. multivorans, 97
B. sphaericus, 151
B. vietnamiensis, 97
bacilibactin transporter, 63
Bacillus anthracis, 96
Bacillus sphaericus, 151, 161
Bacillus subtilis, 58, 59, 60, 68, 69, 70, 71, 72, 73, 75, 76, 81
Bacillus thuringiensis israelensis, 158
Bacilus licheniformis, 81
bacterial adherence, 4
bacterial cell(s), vii, viii, 2, 32, 36, 38, 39, 44, 46, 52, 62, 63, 66, 77, 78, 80, 81, 82, 83, 84, 85, 86, 205, 211, 212, 222, 225, 226, 230, 233, 241, 242
bacterial cell membrane, viii, 63, 77, 78, 80
bacterial cell size, 78
bacterial cellulose synthesis operons, 44
bacterial colonies, 244
bacterial density, 233
bacterial genome, 57, 60, 97, 113, 163, 164, 183
bacterial growth, vii, 22, 30, 80, 222, 223, 224, 225, 226, 228, 232, 233, 234, 240, 241, 243, 245
bacterial infection, 12, 14, 18, 20, 63, 112, 198, 205, 211
bacterial intermediate, 229
bacterial pathogenicity, 63, 135
bacterial pathogens, 26, 93, 100, 103, 127, 134, 139
bacterial proteome, viii, 77, 78
bacterial resistance, x, 70, 197, 198, 212, 217
bacterial response, 49, 63, 213, 224
bacterial solvent tolerance, 78
bacterial strains, x, 64, 83, 84, 101, 110, 111, 112, 115, 118, 216, 221, 232, 234
bactericidal activity, 13, 205
bactericide, 64
bacteriocin(s), 164, 190
bacteriophage, viii, 91, 104, 173, 174, 175
bacteriostatic, 154
bacterium, ix, 45, 67, 81, 84, 89, 92, 93, 94, 98, 137, 149, 150, 154, 226, 241
bai operon, 64
BaiG transporter, 64
base, 28, 100
base pair, 100
basic amino acid-specific porin, 84
batch culture, 226
batch growth, 94
Bcc, 91, 94, 95, 96, 97, 102
bcsABZC, 44
bcsEFG, 44
benefits, 36, 164, 165
benzene, 77, 81, 86, 90
BIA, 103
Bifidobacterium breve, 59, 64, 74
Bifidobacterium longum, 59, 60, 64, 70, 74
bile, 58, 59, 60, 63, 64, 68, 69, 70, 71, 72, 73, 74, 75
bile acid resistance, 63
bile acids, 64, 69
bile salt, 58, 59, 60, 64, 69, 72, 73, 75
bile salt efflux, 58, 59, 60, 64
binding step, 228
bioavailability, 63
biochemistry, ix, 51, 68, 74, 87, 110, 242
biofilm accumulation, vii, 2, 5, 28, 30, 31
biofilm associated disease, vii, 3
biofilm culture, 205, 206, 216
biofilm dispersal, 3, 8, 10, 11
biofilm extracellular matrix, 67
biofilm formation, vii, viii, 1, 3, 4, 5, 6, 7, 8, 9, 10, 11, 12, 13, 14, 17, 18, 19, 20, 21, 22, 23, 26, 30, 32, 33, 36, 37, 40, 41, 42, 44, 46, 47, 48, 49, 50, 51, 52, 55, 67, 70, 72, 94, 95, 101, 103, 105, 107
biofilm inhibition, 10
biofilm matrix, 4, 13, 22, 36, 37, 41, 45, 52
biofilm maturation, 4, 5, 6, 10, 27, 28, 32
biofilm mode of growth, 16
biofilm regulator, 39, 41, 43
biofilm-related disease, 11
biofilm-resistant material, 14, 19
biogenesis, 46, 87, 97, 107, 116, 117
bioinformatics, 98, 184, 186
biological activities, 158
biological control, 150
biological niche, 119

biological processes, 233
biological roles, 101
biomass, 93, 152, 154, 223, 224, 226, 227, 228, 229, 241
biomass balance, 227, 228, 229
biomass balance equation, 228
biomass growth, 152, 154, 241
biomass yield, 93
biomaterials, 15, 22
biomolecules, 62
biopolymers, 88
biosynthesis, 12, 14, 23, 41, 43, 44, 46, 47, 65, 66, 67, 70, 74, 79, 82
biotechnology, 72
biothreat agent, 165
biotic, vii, 1, 2, 67
bismuth, 18
black fly, 150
black fly larvae, 150
bladder, 93
bleomycin, 155, 161
blood, 11, 12, 33
bloodstream, 33
blue-light sensory domain, 38
B-lymphocyte, 112
body weight, 165
bonds, 150, 157
bone, 14, 23, 114
bone marrow, 23
BoNT, 164, 165, 166, 167, 168, 169, 170, 171, 173, 174, 175, 176, 177, 178, 179, 180, 181, 182, 184, 185, 186, 187, 192, 193, 194
bont gene, 170, 171, 172, 173, 175, 179, 180, 182, 187
Bordetella pertussis, 63
botulinum, x, 163, 164, 165, 169, 171, 172, 173, 174, 175, 176, 177, 178, 179, 181, 182, 184, 185, 186, 187, 188, 189, 190, 191, 192, 193, 194, 195
botulinum neurotoxin, x, 163, 164, 188, 189, 190, 191, 192, 193, 194, 195
botulinum neurotoxin genes, x, 163, 189
botulism, x, 163, 165, 166, 177, 178, 182, 183, 186, 187, 188, 189, 190, 193, 195
bowel syndrome, 113
breast cancer, 155, 215
Brucella abortus, viii, 91, 93, 106
brucellosis, 93
Bt kyushuenesis, 152
Bti, ix, 149, 150, 151, 160
Burkholderia cepacia, viii, 91, 94, 95, 102, 103, 104, 105, 106, 205, 214
Burkholderia cepacia complex, 91, 94, 95, 102, 103, 105, 205, 214
burn, 25
butan-1-ol, 80
by-products, 65, 66

C

C. argentinense, 164, 169, 170, 174, 183
C. baratii, 169, 170, 172, 173
C. botulinum, x, 163, 164, 166, 169, 170, 171, 172, 173, 174, 175, 176, 177, 178, 179, 180, 181, 182, 183, 184, 185, 186, 187, 194
C. butyricum, x, 163, 164, 169, 170, 172, 174, 179, 182, 183, 184, 185, 186, 187, 191
C. elegans, 99, 101
C. jejuni, 113, 117, 120, 121, 124, 128, 129, 132
C. perfringens, 175, 182, 183
C. tetani, 175
Ca^{2+}, 59, 137
Caenorhabditis elegans, 21, 95, 106
caffeine, 131
CagA, 111
calcein, 153
calcium, 18, 63, 75, 132
calorimetric data, 236
calorimetry, 234
cAMP, 38, 40, 104
Campylobacter jejuni, 110, 111, 113, 140, 141, 142, 147
cancer, ix, 85, 109, 135, 138, 139, 143, 150, 155, 210, 211, 213, 215, 216, 218, 219
cancer cells, ix, 150, 155, 210, 211, 213, 216
cancer therapy, 138, 210, 219
cancerogenesis, 134
capsule, 81
carbohydrate(s), 81, 84, 120, 152
carbohydrate extracellular capsule, 81
carbon, viii, 42, 77, 78, 81, 93, 235
carbon source, viii, 77, 78, 81, 93, 235
carbon storage regulator, 42
carboxyl, 10, 160
carcinogen, ix, 109, 133
carcinogenesis, 135
caregivers, 95
cargoes, 118
cascades, 39, 40, 43, 46
case study, 71
caspase(s), 127, 128, 138
catabolism, 66, 70, 82, 103, 104
catalysis, 144, 225, 246
catalyst, 208
catalytic activity, 127, 132
catalytic domain, 166, 187
catalytic properties, 166, 195

cathelicidin, 21
catheter, 30
cation, 48, 61, 62, 67, 71, 75, 84, 132, 133, 152, 153, 158
cation symporter, 67
cation/H^+ antiporter, 61
cation-selective pore, 152
Caulobacter crescentus, 45
caveolae, 122
caveolin, 122
CDC, 36, 181
c-di-GMP, viii, 35, 36, 37, 38, 39, 40, 41, 42, 43, 44, 45, 46, 48, 49, 50, 51, 52, 53, 94
CDK inhibitor, 127
CDT, ix, 109, 110, 111, 112, 113, 114, 115, 116, 117, 118, 119, 120, 121, 122, 123, 124, 125, 126, 127, 128, 129, 131, 132, 133, 134, 135, 136, 141, 142, 143, 144, 145, 147
CDT defective strain, 117
ceftazidime, 205
cell adherence, 8
cell aggregation, 31
cell culture, 195
cell cycle, ix, 39, 109, 110, 124, 125, 126, 127, 128, 131, 133, 134, 135, 136, 137, 138, 139, 140, 145, 147
cell cycle arrest, ix, 109, 124, 125, 126, 128, 129, 131, 134, 136, 137, 138, 147
cell cycle checkpoints, 110, 138
cell cycle modulation, ix, 110
cell death, ix, 6, 23, 77, 109, 110, 124, 125, 127, 128, 129, 133, 204
cell density, 3, 8, 9, 10
cell dissemination, 8, 10
cell division, 81, 82, 83, 101, 135, 144, 226
cell envelope, x, 40, 43, 83, 85, 94, 103, 105, 197, 204
cell fate, 23
cell filamentation, 81
cell growth, 8, 66, 71, 79, 87
cell killing, 215
cell line(s), 26, 52, 121, 124, 125, 126, 128, 131, 134, 141, 147, 155, 215
cell membrane(s), 62, 64, 77, 78, 81, 116, 118, 119, 124, 151, 152, 153, 154, 155, 157, 204
cell metabolism, 83, 110
cell motility, 82, 95
cell signaling, 50, 66
cell size, 78, 81, 82, 93
cell surface, vii, ix, 2, 5, 49, 52, 81, 82, 85, 118, 119, 120, 121, 122, 123, 181, 205
cell surface roughness, 205
cell swelling, 151, 153
cellular aggregation, 80
cellular energy, 67
cellular homeostasis, viii, 55
cellular membrane permeation, 223
cellular response, 38, 45, 85, 100, 126, 129, 133
cellular tropism, 121
cellulose, 39, 41, 42, 43, 44, 45, 47, 49, 50, 51, 52, 53, 70, 71
cellulose biosynthesis, 41, 43, 44, 47, 70
cellulose fibre, 51
centrosome, 126
ceramic, 206
ceramide, 120
CF, 94, 95
cGMP, 38
challenges, 15, 28, 62, 155
chancroid, 114, 137, 147
chaperones, 99
chelates, 49
chemical, viii, x, 4, 14, 18, 66, 77, 131, 150, 198, 208, 221, 222, 223, 225, 226, 233, 234, 240, 241, 242, 243
chemical kinetics, x, 221, 243
chemiluminescence, 209
chemiluminescent illumination, x, 197, 206, 213
Chemiluminescent Photodynamic Antimicrobial Therapy, 208
chemokine(s), 13, 112
chemotaxis, 49
chemotherapeutic drug, 57
chemotherapy, vii, x, 14, 155, 197, 210, 214, 215, 216, 217
chenodeoxycholic acid, 60, 64
chitosan, 17, 18, 23, 25
chitosan-containing nanoparticle, 17
chloramphenicol, 73, 79, 95, 205
chloramphenicol resistance, 73
chlorin derivative, 198
chlorin e_6, 206, 217
chlorination, 51
chlorophenol, 79
CHO cells, 120, 122, 124
cholesterol, 117, 120, 122
cholic acid, 60, 64, 74
cholinergic nerve ending, 166
cholylglycine, 64
chromatin, 125, 126, 130
chromatography, 234, 235
chromosome, 70, 97, 129, 173, 174, 175, 176, 177, 178, 179, 183, 184, 186, 191, 192
chronic hepatitis, 115
chronic inflammation, 135, 139
chronic myeloid leukemia, 121

ciprofloxacin, 46, 95, 205
circular DNA, 176
circular plasmid, 177, 178
circulation, 110, 166, 169
cis-trans isomerase, 79, 87
cis-trans isomerization, 79, 89
cis-vaccenic acid, 79
citrate, 63, 70
classes, 84, 115, 210
classification, 57, 215
clathrin, 122
cleavage, 100, 105, 129, 152, 166, 167, 168, 192
clinical application, 208, 219
clinical isolate, 5, 19, 22, 93, 95, 111, 213
clinical symptoms, 98
clinical trials, 15
clone, 136
cloning, 71, 102, 103, 144, 193
closed channel, 85
clostridia, 161, 163, 164, 165, 169, 171, 173, 175, 179, 182, 183, 186, 187, 191
clostridia Groups, 163, 169, 173, 179, 182, 187
Clostridium, x, 65, 163, 164, 168, 169, 170, 178, 183, 188, 189, 190, 191, 192, 193, 194, 195, 196
Clostridium acetobutylicum, 65
Clostridium argentinense, x, 163, 195
Clostridium botulinum, 163, 164, 169, 178, 188, 189, 190, 191, 192, 193, 194, 195, 196
clumping factor, 4, 18, 23, 28
clustering, 58
clusters, vii, 2, 9, 171, 172, 173, 175, 178, 179, 187, 189, 190, 191, 194, 196
coagulase-negative staphylococci, 11, 22
coatings, 27
cobalt, 69
coding, viii, x, 91, 99, 104, 106, 156, 163, 164, 170, 171, 173, 174, 177, 179, 181, 182, 187
co-enzyme, 97
co-factor, 79
colanic acid, 41, 50, 51
Coleoptera, 150
colibactin, 134
collagen, 4, 11, 20, 29
collagen-binding protein, 4
collateral, 12
colloid-osmotic lysis, 153
colon, 113, 114
colon mucosa, 113
colonization, 3, 10, 11, 21, 24, 26, 48, 93, 98, 113, 115, 127, 186
common sense, 223
community(s), vii, 2, 7, 9, 14, 15, 32, 35, 36, 40, 48, 51, 67, 68
compaction, 125, 126, 154
comparative analysis, 194
competition, 111, 232, 244
complement, 12, 23, 27, 67
complementarity, 100
complex reaction theory, 222
complexity, viii, 6, 9, 31, 40, 91, 226, 233, 242
composition, 7, 62, 70, 71, 78, 79, 86, 88, 89, 101, 104, 163, 193
compounds, vii, viii, 14, 41, 56, 57, 58, 63, 64, 65, 72, 77, 78, 80, 81, 86, 210, 211, 217, 234, 244
computing, 181
concentration ratios, 211
concomitant downregulation, 11
condensation, 126, 127
configuration, 79
conformational change, ix, 133, 149, 152, 153, 154, 157
conjugation, 155, 163, 164, 182, 186
conjunctivitis, 28
consensus, 27, 119, 120, 141, 183, 185
conservation, xi, 56, 75, 134, 173, 222, 241, 242
constituents, 80, 205
constitutive enzyme, 79
consumption, 224
contaminated food, 168
contamination, 245
contig, 178
continuous culture, 86, 223, 226, 243, 246
control cascade, 40, 43, 44
convergence, 75
coordination, 36, 50, 85
copper, 40, 52, 53, 244
copper stress, 40
correlation, 30, 78, 79
cosmetic, 166
cost, 56, 75, 164
CPAT, 208, 209
Cpx, 43, 50, 52
CpxAR, 40, 42
CpxR, 45, 52
CRISPR, 163, 184, 185, 186, 187, 188, 191, 193, 194, 195
CRISPR loci, 163, 184, 186, 187, 191
crop, 159
CRP, 104
crust, 65
Cry, ix, 149, 150, 151, 152, 155, 157, 158, 159, 160
Cry 4Aa, ix, 150
Cry10Aa, ix, 149, 150
Cry11Aa, ix, 149, 150, 151, 155, 157, 158, 159
Cry4Ba, ix, 149, 150, 151, 157
crystal structure, 56, 115, 132, 157, 166, 244

crystalline, ix, 78, 149, 150
crystallization, 150
crystals, 160, 161
CsgD, 40, 41, 43, 44, 47, 49, 104
CsrA, 39, 42, 48, 52, 105
C-terminal periplasmic α-helix domain, 85
C-terminus, 4, 93, 96, 97, 152
Cti, 79
cues, vii, 6, 10, 35, 37, 40, 46
Culex quinquefasciatus, 151, 161
culture, 7, 24, 33, 80, 114, 117, 225, 233, 235, 241, 243, 245, 246
culture medium, 7, 80, 235
curli fimbriae, 40, 41, 43, 44, 51, 69
curli structural operon, 41
curli subunit gene D, 40
curli-mediated adhesion, 41, 50
cyclic dimeric guanosine, viii, 35, 36
cyclic dimeric guanosine monophosphate, viii, 35, 36
cyclin(s), 126, 127, 131
cycling, 100
cyclohexane, 81
cyclomodulin, ix, 109, 132
Cyprus, 160
Cys, 65, 154
cys regulon, 65
cystathionine, 69
cysteine, 8, 65, 70, 156
cysteine biosynthesis, 65, 70
cysteine protease, 8
cystic fibrosis, vii, 2, 6, 23, 51, 70, 94, 103, 104, 105, 214
Cyt, v, ix, 149, 150, 151, 152, 153, 154, 155, 157, 160
Cyt1Aa, ix, 149, 150, 151, 152, 153, 154, 155, 156, 157, 158, 159, 160
Cyt1Ca, ix, 149, 150, 152, 158
Cyt2Ba, ix, 149, 150, 151, 152, 155, 157, 159
cytokine(s), 13, 20, 28, 33, 112, 136
cytolethal distending toxin, ix, 109, 111, 135, 136, 137, 138, 139, 140, 141, 142, 143, 144, 145, 147
cytolysin, 152, 153
cytolytic activity, 154, 159
cytometry, 119, 126
cytoplasm, 83, 101, 118, 123, 125, 126, 127, 154, 205
cytoplasmic membrane, 56, 79, 81, 154
cytoplasmic tail, 4
cytoskeleton, 139
cytosol, 38, 63, 118, 121, 123, 166
cytotoxic activity, 117, 133, 155
cytotoxic agents, 213
cytotoxicity, ix, 109, 115, 123, 124, 131, 134, 155, 161, 210
cytotoxin, 137, 155

D

damages, 112, 114, 137
data analysis, 242
data processing, 236, 238
database, 92, 96
DCA, 60
DDR, 124, 128, 129, 131, 133, 134
deacetylase, 5
deacetylation, 5
decomposition, xi, 222, 234, 242
deconvolution, 234
defects, 45, 94, 124, 128, 131, 132
defence, 24
defense mechanisms, 218
deficiencies, 10
degenerate, 38, 42
degradation, 7, 9, 28, 40, 42, 66, 84, 85, 86, 92, 98, 100, 105, 123, 126, 127, 132, 171
degranulation, 12
delivery system, 80
denaturation, 18
dendritic cell, 12, 13, 25
denitrifying, 89
Denmark, 30
dental caries, 216
deoxycholic acid, 64
deoxyribonuclease, 132
dephosphorylation, 126, 127, 137
deposition, 12, 27
depth, 208
derivatives, 62, 198, 205, 217, 218, 246
dermatology, 208, 216
dermis, 18
desiccation, 44
destruction, ix, 103, 149, 153, 211, 216
desulfonation, 65
detachment, 4, 8, 10, 26, 208
detection, 111, 193, 233, 234
detergent(s), ix, 10, 64, 84, 149, 153, 159
deuteroporphyrin, 205
developmental process, 36
DGC, 38, 39, 41, 42, 43, 44, 45
diarrhea, 110, 112, 113, 114
differentiation factor, 13
diffusion, 41, 84
digestive epithelium, 110
diguanylate cyclase, 37, 38, 46, 51, 52, 53, 94
diguanylate cyclase HmsT, 94

diode laser, 208
diphtheria toxin, 121
Diptera, 150, 157, 158, 160, 161
disclosure, 68
disease progression, 114
diseases, vii, 2, 3, 11, 25, 27, 48, 110, 111, 112, 127, 143, 150, 151, 159, 216
disinfection, 215
disorder, 204
dispersion, 8, 10
dissociation, 129, 135, 189, 244
dissolved oxygen, 204, 210, 235, 238
distribution, 87, 187
disulfide bond, 116, 150, 157
divalent cation, 83, 132, 133
divergence, 168, 171
diversification, 168
diversity, 36, 57, 164, 165, 168, 171, 173, 191, 193, 194
DNA binding groove, 133
DNA damage, ix, 18, 63, 109, 110, 124, 127, 128, 131, 133, 134, 135, 136, 137, 138, 139, 140, 141, 142
DNA damage response, 63, 124
DNA lesions, 129, 133
DNA polymerase, 181
DNA repair, 63, 131
DNA strand breaks, 133, 145
DNA-binding domain, 97
DnaK heat shock protein, 213
DNAs, 177
DNase, 6, 132, 133, 138, 144
DOI, 243
dominance, 161
donors, 181
doping, 14
DOT, 82, 83, 88, 89
double bonds, 198
double-strand breaks, ix, 109, 110, 130, 138, 147
down-regulated protein, 82
down-regulation, 104
Drosophila, 154, 159
drug delivery, 211
drug efflux, 60, 69
drug resistance, 16, 58, 72, 83, 155, 212, 217, 218
drug target, 16
drug/H^+ antiporter, 56, 57
drugs, 14, 16, 18, 57, 60, 66, 83, 142, 155, 210, 211, 213, 218
DSC, 236, 245
DsrA, 100, 105
duodenal ulcer, 111
duodenum, 64
DupA, 111
dyes, 198, 215, 216, 219
dynamin, 122
dysentery, 98, 113
dysfunction, 57, 67
dysplasia, 139

E

E. coli, viii, 38, 39, 40, 41, 42, 43, 44, 45, 53, 57, 60, 61, 62, 64, 65, 66, 67, 83, 84, 91, 92, 93, 96, 97, 98, 100, 101, 112, 113, 117, 120, 121, 124, 130, 131, 132, 134, 152, 154, 205, 206, 208, 210, 211, 212
ecology, 72, 191
EcoRI, 96
eDNA, 3, 4, 5, 6, 7, 10, 13, 20, 28, 41
effector, 8, 10, 39, 40, 43, 44, 126, 131, 135
efflux activity, 56
efflux of drug, 16
efflux of toxin, 83
efflux pump, 57, 63, 68, 69, 72, 73, 74, 75, 77, 78, 83, 85, 86, 87, 88, 89, 212
efflux transporter(s), 66, 70, 71, 73, 83
elastase, 94
elastin, 11, 27
electron(s), 18, 83, 158, 204, 225, 235
electron transfer, 18, 225
electrophoresis, 82, 86, 98, 133, 169, 191, 193
ELISA, 143
elucidation, 187
EMG, 235
emission, 206, 208, 209, 210
emulsan, 81, 88
encapsulation, 155
endocarditis, vii, 2, 11, 22, 30
endocytosis,, 122
endopeptidase, 168
endoplasmic reticulum, 122
endoribonuclease, 100, 103
endosome, 118, 122
endothelial cell(s), 128
endotherms, 245
endotoxin, 156, 158, 159, 160, 161
endotoxins, ix, 110, 149, 150, 156, 158, 160
endotracheal tube, 10, 19
energy, viii, x, 57, 64, 77, 78, 79, 81, 82, 83, 89, 197, 204, 208
energy expenditure, 57
energy metabolism, 78
energy source, 81
EngA, 97, 107
engineering, 192, 223

enlargement, 124
enteric bacteria, 64, 72
enteritis, 113
Enterobacter, 81, 88
enterobacteria, 50, 67
Enterobacteriaceae, 44, 50
enterotoxin, 141
environment(s), viii, ix, x, 10, 14, 36, 41, 45, 48, 61, 62, 65, 69, 77, 78, 81, 84, 86, 110, 117, 149, 150, 154, 164, 168, 178, 183, 186, 221, 222
environmental change, 224
environmental conditions, 6, 79, 94
environmental factors, 87
environmental stimulus, 39
environmental stress(s), 7, 40, 41, 44, 60, 82, 224
enzymatic activity, 38, 132
enzymatic catalysis, 225, 246
enzymatic reaction, 66
enzymatic toxin, 168
enzyme(s), x, 8, 10, 39, 79, 86, 87, 97, 150, 174, 181, 182, 183, 221, 223, 233, 241
epidemic, 95
epidemiology, 159
epidermis, 42
epithelia, 120
epithelial barrier, 112, 114, 190
epithelial cell(s), ix, 26, 93, 112, 124, 128, 129, 149, 150, 171
epithelium, 110, 111, 114, 137
epitopes, 168
EPS, 16, 39, 41, 44, 96
EPS-deficient mutant, 96
equilibrium, 127, 153
ER, 118, 121, 122, 123
Erwinia carotovora, 154, 159
Erwinia chrystanthemi, 63
Erwinia virulence factor, ix, 149, 152
erythema, 114
erythrocytes, 93
ester, 65
ethanol, 7, 27, 81, 86
ethanol-induced biofilm, 7
ethlylbenzene, 83
ethyl acetate, 209
etiological agent, 2, 11
etoposide, 129
Eubacterium, 59, 64, 72
eukaryotic, ix, 48, 66, 92, 110, 111, 112, 117, 118, 122, 126, 139, 141, 168
eukaryotic cell, ix, 48, 66, 110, 111, 112, 118, 122, 126, 139, 141
European Commission, 182, 190
Evf, ix, 149, 150, 152, 154, 159, 160
evidence, viii, 6, 10, 55, 57, 63, 91, 93, 94, 103, 122, 124, 173, 174, 183, 187, 189, 190, 213, 216, 222, 224
evolution, 23, 58, 73, 161, 164, 173, 186, 188, 190, 193, 214, 222, 225, 226, 243
excision, 174
excitation, x, 197, 206, 208, 210, 211, 217
excretion, 65, 189
exocytosis, 166
exopolysaccharide(s), 3, 4, 27, 32, 35, 37, 40, 41, 42, 44, 45, 46, 51, 81, 94, 96, 103
exopolysaccharide production, 37, 40
exotoxin(s), 94, 110, 111, 115, 116, 146
experimental condition, 233, 236
exploitation, 102, 151, 155
exported protein, 97
exporter(s), 5, 65, 74
exposure, ix, 60, 62, 64, 81, 82, 83, 84, 85, 87, 88, 109, 125, 126, 128, 129, 133, 134, 139, 153, 183, 186, 205, 213, 217, 233
external environment, 61
extracellular DNA, 4, 13, 26, 42
extracellular input signal, 39
extracellular matrix, vii, 1, 2, 7, 9, 20, 36, 41, 53, 67
extracellular polymeric substance, 16
extraction, 96
extracts, 132
extrusion, 56, 64, 83

F

families, viii, 55, 56, 57, 58, 61, 67, 150
fatty acid(s), 71, 78, 79, 80, 82, 86, 87, 88, 89, 90, 204
fauna, viii, 77
febrile gastroenteritis, 98
feces, 193
feedback inhibition, 38
feedstocks, 88
fever, 110, 113, 115
fiber(s), 11, 48, 50, 125, 131
fibrin, 11
fibrinogen, 4, 11, 15, 23, 25, 28, 29
fibrinogen-binding protein, 4
fibroblasts, 117, 128, 129, 131, 134, 136
fibronectin, 4, 11, 15, 25, 27, 29, 31, 48
fibrosis, vii, 2, 6, 14, 23, 51, 70, 94, 103, 104, 105, 110, 214
financial, 18, 68
financial support, 18, 68
first generation, 208
fish, 178
fission, xi, 221, 228, 231

fitness, 47, 94, 105, 164
flagellar function, 40
flagellin, 169
flank, 174
flexibility, 154, 179
FliG, 40
FliM, 40
flora, viii, 15, 24, 31, 77, 114
flow-cytometry, 126
fluctuations, 7
fluid, 78
fluorescence, 44, 86, 154
fluorophor, 154
food, 21, 64, 98, 110, 113, 176, 177, 183, 188, 193
food intake, 21
food-born pathogen, 98
food-borne contamination, 110
force, 62, 64, 83, 164, 218
formation of vesicles, 78
formylmethionyl peptide, 11
Fourier analysis, 234
fragments, 152, 153, 160, 161
Francisellla tularensis,, 94
free organic radical, 204
free radicals, 18
fructose, 79
fullerene, 218
functional analysis, 70, 102, 184
functional changes, 210
functional duplication, 57
fungi, 12
fusion, 83, 117, 118, 122
Fusobacterium nucleatum, 206

G

G2/M transition, ix, 109, 126, 137
Galleria melonella, 94
gallium, 203, 219
ganglioside, 120, 166, 167
gastric enzyme, 150
gastroenteritis, 98, 110, 112, 113
gastrointestinal infection, 110
gastrointestinal mucosa, 115
gastrointestinal stress, 64
gastrointestinal tract, 63, 168
gdi, 227
GDP, 107
gel, 78, 82, 86, 98, 133, 169, 191, 193
gel electrophoresis, 82, 86, 98, 133
gel phase, 78
gene amplification, 75
gene cluster, 33, 51, 171, 172, 173, 174, 175, 177, 178, 179, 183, 187, 189, 190, 191, 194, 196
gene duplication, 58
gene expression, viii, 6, 7, 8, 16, 19, 20, 21, 33, 36, 37, 68, 72, 81, 91, 103, 106, 107, 224
gene pool, 193
gene promoter, 101
gene regulation, vii, viii, 91, 184
gene transfer, 173, 182, 194
general porin, 84
genetic background, 110
genetic disease, 94
genetic diversity, 173, 179
genetic engineering, 166
genetic information, 222
genetic instability, ix, 109
genetic traits, 187
genetics, 22, 32, 146
genital mucosa, 114, 120
genome, viii, 10, 23, 39, 57, 69, 73, 74, 91, 95, 96, 97, 113, 139, 164, 169, 171, 173, 175, 177, 178, 180, 182, 183, 188, 194, 195
genome sequencing, 57
genomic instability, 129, 134, 137, 139, 140
genomics, 74, 108, 218
genotoxic process, 135
genotoxic stress, 133
genotoxin, ix, 109, 117, 133, 134, 138, 139, 140, 143
genotype, 127, 233
genotyping, 186, 190, 195
genus, 96, 113, 117, 168
GFP, 123
gingival, 114, 117, 136, 137, 147
gingival bone, 114
global signalling network, 37
global stress response regulator, 7
Gluconoacetobacter xylinum, 44
glucosaminidase, 16
glucose, 26, 81, 84, 101, 104, 107, 235, 244
glucose-specific porin, 84
glycans, 120
glycine, 97
glycoprotein(s), 120, 121
glycosylation, 123
Golgi compartment, 123
Gram-negative, vii, viii, x, 42, 44, 50, 64, 65, 66, 73, 78, 80, 81, 84, 86, 87, 88, 90, 91, 109, 110, 111, 112, 117, 140, 161, 197, 198, 204, 205, 206, 208, 209, 212, 213, 214, 217, 218
Gram-negative bacteria, x, 44, 50, 64, 73, 78, 80, 81, 84, 87, 88, 90, 109, 110, 112, 117, 140, 161, 197, 198, 204, 205, 208, 209, 213, 217, 218

Gram-positive, vii, viii, ix, x, 2, 62, 64, 65, 66, 70, 78, 80, 81, 91, 111, 149, 150, 154, 168, 197, 198, 204, 205, 206, 208, 209, 212, 213, 214, 217, 218
Gram-positive bacteria, 64, 78, 80, 111, 198, 204, 212, 214
green fluorescent protein, 123
ground state, 204
grouping, 58
growth determining intermediate, 227
growth equation, x, 221, 223, 228, 237
growth factor, 235, 240, 242
growth intermediate, 225, 226
growth lag, 235, 236
growth limiting substrate, 227, 229, 241
growth model(s), 223, 224, 226, 234, 237, 238, 242
growth rate, 93, 94, 223, 224, 228, 232, 233, 236, 237
growth specific rate, 94
growth temperature, 45, 80
growth thermogram, 234
growth turnover, 230
GTP, 38, 47, 107
GTPase, 97, 123

H

H. ducreyi, 114, 120, 121, 124, 128, 129, 131, 132, 133
H. hepaticus, 115, 124, 128, 129, 132, 133, 143
habitats, 95
habituation, 213
haemagglutinin, 171
haematopoietic cell, 125, 128
Haemophilus ducreyi, ix, 109, 111, 114, 137, 138, 139, 142, 146, 147
Haemophilus influenza, 57
halogen, 206, 217
halogenated xanthene, 198
*Hd*CDT, 111, 114, 115, 120, 121, 122, 123, 129, 133
HE, 234
health, 2, 10, 56, 95
health care, 2
health status, 95
heat flow, 233, 235, 236, 238
heat shock, 98, 107, 214, 219
heat shock protein, 213, 219
heavy metal(s), 75, 164
HeLa cell, 117, 119, 120, 123, 129, 131, 137, 144
helicase, 181
Helicobacter hepaticus, ix, 109, 111, 115, 137, 142, 143, 147
Helicobacter pylori, 111, 147, 214
hemagglutinin(s), 30, 171
heme, 47
Hemiptera, 150
hemoglobin, 230
hemoglobin binding, 230
hemolysin, 160
hemolytic activity, 153, 159
hepatic dysplastic nodules, 115
hepatitis, 115
hepato-biliary channel, 115
hepatocarcinogenesis, 139
heterogeneity, 169, 223
heterogeneous catalysis, x, 221, 222, 226, 233, 241, 243
Hfq, v, viii, 91, 92, 93, 94, 95, 96, 97, 98, 99, 100, 102, 103, 104, 105, 106, 107, 108
hfq deletion mutant, 93, 94, 95
hfq mutant, viii, 91, 93, 94, 95, 106
Hfq protein, viii, 91, 92, 96, 97, 98, 102
Hfq-like protein, viii, 91, 92, 95, 97
HGF, 117
HhCDT, 111, 115
histidine, 93
histone, 92, 129
history, 194, 198, 216, 233
holotoxin, 115, 116, 117, 119, 120, 122, 123, 124, 132, 142
homeostasis, viii, 55, 58, 59, 60, 61, 62, 63, 65, 67, 70, 71, 72, 73, 75, 94, 127
homogeneity, 132
homolog, 22, 72, 73, 105
homology, ix, 62, 64, 120, 149, 160, 169, 182, 193
homophilic interaction, 6
Homoptera, 150
host, vii, viii, ix, 1, 2, 3, 11, 12, 13, 14, 15, 23, 24, 29, 30, 36, 41, 46, 71, 91, 94, 101, 102, 104, 106, 110, 111, 112, 116, 117, 118, 119, 120, 121, 122, 123, 124, 130, 132, 133, 134, 135, 140, 146, 168, 174, 182, 186, 187, 218
host defence, 168
host factor, vii, viii, 1, 2, 91, 104, 106, 120, 121
host immune response, vii, 2, 3, 12
hot spots, 173
hub, 122
human, vii, viii, ix, x, 1, 2, 3, 11, 13, 19, 22, 24, 28, 31, 55, 64, 68, 92, 95, 98, 106, 109, 113, 114, 115, 117, 124, 132, 136, 138, 144, 147, 150, 163, 164, 165, 166, 168, 182, 187, 215, 218
human body, 115
human botulism, x, 163, 165, 166, 182, 187
human gingival fibroblasts, 117, 136
human gut, 64
human neutrophils, 28
human skin, 11

hybridization, 174, 175, 177, 178, 196
hybridoma, 155
hybridoma cell, 155
hydrocarbons, viii, 77, 80, 81, 87, 89, 223
hydrogels, 215
hydrogen, 17, 63, 93, 204, 208, 209
hydrogen peroxide, 63, 93, 204, 208, 209
hydrolysis, 57
hydrophobic molecule, 78
hydrophobic pocket, 154
hydrophobicity, 4, 77, 79, 80, 211
hydroxyl, 204
hyperbolic equation, 223, 224
hypothesis, xi, 26, 134, 153, 154, 175, 184, 187, 210, 221
hypothetical lipoprotein, 16

I

iatrogenic, 165
iatrogenic botulism, 165
ica operon, vii, 2, 5, 7, 21
IceT, 59, 63
ID, 190, 217
ideal, 81, 208
identification, 21, 22, 38, 57, 92, 100, 101, 102, 106, 121, 133, 141, 158, 189, 191, 192, 193
identity, 7, 61, 96, 115, 152, 175, 194, 223
IL-8, 13, 31
illumination, x, 197, 198, 206, 208, 209, 210, 213
images, 2
immune cell, 13, 14, 115
immune defense, vii, 2
immune effector, 112
immune response, vii, 2, 3, 12, 14, 20, 27, 33, 63, 112, 127, 138
immune system, ix, 12, 16, 22, 94, 109, 111, 112, 143
immunity, 13, 20, 24, 28, 30, 31, 63, 71, 112, 188
immunoavoidance factor, 11
immunodepletion, 112
in vitro, 5, 6, 7, 25, 42, 48, 98, 100, 115, 126, 132, 134, 142, 147, 151, 155, 160, 164, 168, 195, 215, 218, 219
in vivo, ix, 7, 9, 11, 15, 21, 25, 26, 28, 32, 46, 48, 109, 114, 115, 133, 134, 137, 155, 160, 211, 216
individuals, x, 12, 110, 113, 114, 115, 221, 225
inducer, 68
induction, 7, 13, 14, 42, 46, 85, 86, 112, 113, 115, 127, 129, 140, 145, 210
inefficiency, 211
infant botulism, 177, 186, 188, 189, 195
infants, 165
infection, vii, viii, 1, 2, 3, 4, 5, 8, 9, 11, 12, 13, 14, 15, 16, 18, 19, 20, 22, 24, 27, 29, 30, 31, 32, 55, 70, 93, 94, 95, 98, 101, 110, 112, 113, 114, 127, 140, 143, 150, 174, 186, 205, 211, 214
inflammation, ix, 31, 32, 109, 110, 114, 115, 134, 139, 140
inflammatory bowel disease, 195
inflammatory cells, 218
inflammatory mediators, 11
influenza, 57
influenza a, 57
influx of solvent, 79
ingest, 14
ingestion, 150
inhalation botulism, 165
inheritance, 174
inhibition, 10, 18, 30, 63, 68, 79, 82, 100, 130, 131, 132, 133, 210
inhibition of growth, 82
inhibition of multidrug resistance, 63
inhibitor, 16, 27, 103, 127, 130, 131, 138
initiation, 50, 75, 98, 100, 114, 126
innate immunity, 13, 30, 31, 112
inner lipid bilayer, 78
inner membrane, 42, 84, 116, 117, 205
inoculation, 11, 12
inoculum, 233, 236, 237
inorganic sulphur, 65
inositol, 134
insect, ix, 149, 153, 159, 160
insects, 151
insertion, 4, 7, 84, 90, 107, 154, 157, 158, 173, 179, 181, 191
insertion sequence element, 7
insulin, 155, 156
integral membrane protein, 55, 121
integral membrane transporter protein, viii, 55
integration, 36, 38, 42, 44, 45, 113, 174, 222, 234
integrity, 41, 42, 48, 51, 72, 80, 85, 151
intensive care unit, 11
intercellular interaction, 5, 6, 9
interface, 51, 71, 119, 144, 210
interference, 26, 81, 110, 218
intermediate, xi, 63, 83, 129, 221, 226
internalization, 48, 119, 120, 124, 135, 139, 140
intestinal cell, 117
intestinal flora, 112
intestinal tract, ix, 64, 109, 114
intestine, 113, 168
intoxication, ix, 110, 119, 120, 121, 122, 123, 124, 125, 126, 127, 129, 130, 131, 134, 136
intracellular messenger, 37
intracellular pathogen, 93, 94

intracellular signalling cascade, 39
intravenous drug abusers, 12
intravenously, 168
intrinsic bacterial infection, 205
intron, 181
invertase, 186
invertebrates, 154
ion channels, 65
ion transport, 63, 72
ionizing radiation, 129
ions, 61, 62, 63, 64, 208
iron, 7, 15, 33, 63, 70, 72, 100
iron transport, 70
iron-chelating compound, 63
irradiation, 18, 129, 206
irritable bowel syndrome, 113
islands, 74
isolation, 96, 177, 215
isomerization, 79, 80, 89, 90
isothermal calorimetry, 245
isothermal microcalorimeter, 233
issues, 184, 236

K

K. pneumonia, 94
K^+, 59, 60, 61, 62, 69, 70
K^+ gradient, 62
K^+/H^+ antiport, 62
keratinocyte(s), 13, 128
kidney(s), 93, 219
kill, 17, 28, 31, 110, 217
kinase, 127, 129, 131, 135
kinase activity, 131
kinetic analysis, vii, 235, 236, 237
kinetic model, 225, 241, 246
kinetic parameter(s), 223, 230, 242
kinetics, x, 84, 221, 222, 223, 227, 228, 233, 238, 241, 242, 243, 245
Klebsiella pneumonia, 46, 94, 103
Krebs cycle, 82

L

L. acidophilus, 64
L. lactis, 63, 64, 65
lactate dehydrogenase, 30
lactic acid, 69, 72, 191
Lactobacillus, 59, 71, 74
Lactococcus lactis, 59, 63, 69, 244
lactoferrin, 12
lambdoid prophage, 113, 135
Langmuir adsorption, x, 221
larvae, 150, 151, 156, 157, 158
larval midgut, ix, 149
larvicidal activity, 152, 161
lasers, 206
lead, ix, x, 22, 42, 45, 77, 79, 82, 94, 109, 119, 122, 124, 131, 133, 134, 149, 154, 164, 187, 197, 213, 233
leakage, 153, 205
lectin, 120, 152
LED, 206
LEE, 93
Legionella, 69, 94, 105
legume, 98
lens, 69, 242
Lepidoptera, 150
lesions, 81, 93, 129, 147
lethality, 165, 187
leukemia, 121
leukotoxin D, 8
ligament, 136
ligand, 4, 23, 24, 27, 29, 155, 216
ligation, 96, 129
light, x, 38, 52, 93, 121, 165, 166, 197, 198, 204, 205, 206, 207, 208, 209, 210, 213, 214, 215, 216, 218
light-emitting diodes, 206
limiting substrate, 225, 235, 240, 242
lincomycin, 65
lipase, 4, 20
lipid, ix, 78, 79, 81, 86, 116, 117, 121, 124, 149, 153, 154, 156, 158, 159, 160, 196, 204
lipid bilayer, ix, 149, 153, 158, 204
lipid metabolism, 196
lipids, 65, 77, 79, 86, 117, 153, 154, 204
lipophilic cationic compound, 63, 72
lipophilic cationic drug, 83
lipopolysaccharide, 112, 138, 205
lipoprotein(s), x, 13, 24, 27, 117, 197, 205
liposome(s), 120, 153, 217
liquid phase, 86, 210
liquid-crystalline phase, 78
liquids, 64, 210, 244
Listeria monocytogenes, 59, 64, 68, 74, 98, 103
lithium, 62
liver, ix, 98, 109, 115, 143
liver cancer(s), 115
LmrP, 59, 63, 72, 74, 75
lobulation, 126
localization, 40, 49, 104, 121, 122, 123, 126, 127, 143, 191, 214
loci, 21, 22, 163, 175, 184, 185, 186, 187, 188, 191

locus, 3, 5, 6, 7, 8, 21, 26, 27, 32, 42, 47, 52, 68, 93, 182, 184, 185, 186, 190, 192
longitudinal study, 23
low temperatures, 72
LPS, 51, 78, 80, 112
LTA, 13
LTD, 149
lumen, 93
luminol, 208, 209, 215
lung disease, 70
Luo, 18, 28, 142
lupus, 92
LuxR, 40
lymph, 114
lymphadenopathy, 114
lymphoblastoid, 131
lymphocytes, 112, 126, 134
lymphoid, 134
lysis, ix, 6, 12, 149, 153, 154
lysogeny, 174, 194
lysosome, 122
lysozyme, 12, 173

M

machinery, 44, 131, 134
macrophage(s), 12, 13, 14, 23, 26, 31, 32, 94, 112, 122
macrophage infection, 94
macrophage inflammatory protein, 13
magi, xi, 221, 227, 228, 229, 230, 231, 233
magnesium, 132
Magnetospirillum magnetotacticum, 96
magnitude, 207, 233
major facilitator superfamily, viii, 55, 56, 69, 70, 72, 73, 74, 75, 89
majority, 40, 102, 113, 117, 124, 225
malachite green, 206, 219
malaria, 150, 157, 160
mammalian cell(s), ix, 17, 137, 141, 145, 146, 149, 151, 155, 160, 205
mammals, 11
mari, 227
marrow, 20
mass, x, 26, 82, 84, 171, 221, 226, 242
mass spectrometry, 26, 82, 171
mast cell(s), 112
mastitis, 5
MATE, 56, 58, 67, 71, 73, 83
materials, 14, 19, 26, 49, 71
matrix, 3, 4, 6, 7, 9, 10, 11, 13, 22, 36, 37, 41, 42, 44, 45, 52, 67, 68
matter, 128, 223, 227
maturation of biofilm, 8, 41
maturation phase, 3
maximum growth rate, 232
MB, 198, 204, 205, 206, 207, 208, 209, 210, 211, 213
MBPp, 155
MCP, 136
measurement, 75
media, 64, 65, 81, 86, 94
medical, vii, 2, 4, 11, 12, 22, 32, 188, 211, 217
medical device, 2, 4, 11, 12, 32
megaplasmid, 184, 186
melting, 78, 245
membrane composition, 78
membrane fluidity, 79, 80, 82
membrane fusion protein, 83
membrane lipid composition, 79
membrane lysis, ix, 149
membrane permeability, 79, 85, 88, 127, 152
membrane protein, 18, 68, 87, 97, 121
membrane receptor, 153
membrane transporter, viii, 55
membranes, 12, 78, 86, 89, 156
meningitis, 98, 112
meso-tetra (N-methyl-4-pyridyl) porphine tetra tosylate, 205, 215, 216
messenger(s), viii, 35, 36, 37, 38, 41, 45, 46, 49, 50, 53, 94
metabolic activity, 233
metabolic intermediate(s), 82
metabolism, viii, 37, 38, 48, 49, 58, 59, 62, 63, 65, 66, 67, 69, 70, 78, 82, 88, 91, 100, 225, 244
metabolites, viii, 55, 83, 90
metabolizing, viii, 35, 38
metal acquisition, 63
metal homeostasis, viii, 55
metal ion(s), 58, 62, 63, 133, 144
metal ion homeostasis, 58, 62, 63
metalloprotease aureolysin, 8
metals, 62, 69
Methanobacterium thermoautotrophicum, 96
Methanococcus janneschii, 58
methicillin, 5, 29
methicillin susceptibility, 5, 29
methicillin-resistant *S. aureus*, 206
methicillin-resistant *Staphylococcus aureus*, 17, 19, 22, 23, 24, 216
methionine, 65, 70
methyl group, 74
methylation, 65
methylene blue, 204, 214, 215
MFS, 56, 57, 58, 59, 60, 61, 62, 63, 64, 65, 66, 67, 68, 72, 74

Mg^{2+}, 38, 137
mice, 23, 32, 94, 98, 106, 114, 115, 161, 168, 195
Michaelis constant, 223
microarray analysis, 72
microbial cells, x, 197
microbial community(s), 35, 45, 48
microbial efflux pump, 212
microbial membrane, 12
microbial metabolism, 225
microbiota, 111, 196
microcalorimeters, 233
microcalorimetry, 233, 234, 242, 243, 244
microinjection, 123
micronucleation, 125, 126
microorganism(s), viii, x, 14, 29, 36, 44, 55, 75, 78, 79, 80, 87, 89, 94, 111, 160, 164, 168, 186, 187, 197, 198, 213, 216
microscopy, 71, 86, 117, 121, 123
MIP, 13
mitochondrial membrane, 127, 128
mitosis, 125, 126, 128, 129
model system, 134
models, 3, 16, 93, 94, 114, 153, 154, 168, 223, 243
modifications, 80, 85, 166, 223, 241
modules, 4, 39, 40, 41, 44
molecular biology, 68, 87, 217
molecular dynamics, 133
molecular mass, 78
molecular oxygen, x, 38, 197, 204
molecular weight, 84, 85, 96, 171
molecules, viii, 4, 5, 8, 10, 11, 12, 13, 17, 32, 33, 38, 41, 55, 64, 77, 78, 80, 82, 83, 84, 92, 99, 107, 122, 176, 184, 198, 204, 205, 210, 211, 230, 244
monoclonal antibody, 143
monocyte, 134
Monod constant, 225, 232, 233
Monod equation, x, 221, 224, 225, 230, 233, 235, 240, 241, 242, 244, 245
Monod kinetics, 223, 241
monomers, 84, 129, 130, 152, 153
monovalent cation/H^+ transport, 61
morphological abnormalities, 126
morphological alteration, 125, 129, 205, 218
morphological change, ix, 78, 81, 82, 110, 125, 129
morphology, 49, 71, 126
mosaic, 193
mosquito, 150, 151, 156, 157, 159, 161
mosquito larvae, 150, 151, 156, 157
mosquitocidal strain, ix, 149
mosquitoes, 151, 158
most abundant growth intermediate, xi, 221, 222, 227, 231, 233, 241, 242
most abundant reaction intermediate, x, 221, 227
motif, 4, 19, 38, 70, 92, 97, 132, 152, 166
movement disorders, 191
MR, 83, 217
mRNA(s), viii, 42, 63, 66, 91, 92, 94, 98, 99, 100, 101, 103, 104, 105, 106
MRSA, 5, 17, 18, 20, 32, 206, 207, 208, 209
mucoid, 104
mucosa, 112, 113, 114, 115, 120
mucous membrane(s), 12
mucous surface, 11
mucus, 15, 119, 214
multicellular organisms, 127
multidrug and toxic compound extrusion, 56, 83
multidrug efflux protein, viii, 55, 56, 57, 58, 61, 67, 72
multidrug resistance, viii, 55, 56, 57, 60, 61, 62, 63, 64, 65, 66, 68, 69, 72, 73, 155, 219
multidrug transporter, vii, 56, 57, 58, 60, 61, 62, 63, 64, 65, 66, 67, 68, 70, 72, 73, 74, 75, 76, 88, 104
multidrug transporter protein, vii
multilayered cell cluster, vii, 2
multiplication, 82
multivalent vaccine, 16
muscle contraction, 165, 166
mushroom, 3, 5, 152
mutagenesis, 6, 65, 70, 71, 96, 158, 181
mutant, 7, 8, 45, 61, 74, 93, 94, 95, 96, 106, 115, 133, 134, 136, 146, 151, 181
mutant protein, 61
mutation(s), 7, 9, 25, 84, 93, 94, 98, 107, 119, 120, 133, 153, 222
mutational analysis, 119
m-xylene, 84
Mycobacterium, 80, 81, 86, 195
myelin, 155
myelin basic protein, 155
myelin basic protein peptide, 155
myeloid, 13, 31

N

Na^+, 59, 60, 61, 62, 64, 69, 70, 73
Na^+/H^+ antiport, 61, 70, 73
N-acetylglucosamine, 5, 26, 47, 52
N-acetylglucosamine transferase, 5
NADPH oxidase, 13
nanomaterials, 25
nanoparticle(s), 16, 17, 18, 23, 24, 25, 27, 28, 30, 32
nanotechnology, 2, 3, 16, 20, 32
naphthalene, 234
National Academy of Sciences, 137, 140
native valve endocarditis, vii, 2
n-butanol, 81

NCTC, 19, 194
necrosis, 110, 112, 113, 125, 128, 147
negative supercoiling, 93
nematode, 95, 99, 101, 106
nerve, 165, 166
nerve cell, 165
nerve pulse transmission, 166
Netherlands, 25, 141
N-ethylmaleimide-sensitive factor, 166
neuroexocytosis, 168, 194
neurological disease, 114
neuronal cell, 168, 195
neurons, 166
neurotoxigenic clostridia, x, 163, 164, 165, 168, 169, 172, 173, 185, 186, 187
neurotoxin, 165, 187, 188, 189, 190, 191, 192, 193, 194, 195, 196
neurotransmitter, 166
neutral, 137, 206
neutral red, 206
neutrophil(s), 12, 13, 20, 23, 24, 27, 31, 33, 112
NhaR, 42, 48
NIR, 208
nitric oxide, 12, 17, 25, 30, 31, 63
nitric oxide-releasing nanoparticle, 17
nitrogen, 17, 208
NLS, 123
n-octanol, 78
nodules, ix, 109, 115
nonequilibrium, 225
non-lethal concentration, 81
non-pathogenic bacteria, 60
non-specific interaction, 155
Northern blot, 99
nosocomial pathogen, 11
novobiocin, 66
Novosphingobium aromativcivorans, 96
NPS, 16
NR, 199, 206
N-terminal β-barrel domain, 85
N-terminus, 38, 45, 92, 96, 133, 152, 153
NTNH, 171
nuclease, 7, 132, 133, 134, 137
nucleating, 3
nucleic acid, 5, 12, 41, 65, 66
nucleoid, 154, 159
nucleoid DNA, 154
nucleoside, 88
nucleotide, 26, 37, 38, 74, 82, 133, 171, 174, 178, 180, 181
nucleotide biosynthesis, 74, 82
nucleotide sequence, 26, 174, 178, 180, 181
nucleotides, 38, 66, 99, 100
nucleus, ix, 118, 121, 122, 123, 124, 125, 126, 134
null, 128
nutrient(s), x, xi, 6, 10, 15, 63, 84, 107, 221, 223, 232, 235, 242, 244
nutrient restriction, 63
nutrient-deficient conditions, 84
nutrition, viii, 55, 80

O

octane, 79, 86
ocular infection, 11
oil, 217, 235
OipA, 111
oligomerization, 151, 152, 153, 160
oligopeptide, 173
oligopeptide permease, 173
oligosaccharide, 120
OmpA, 45, 49, 89, 107
OMV, 80, 117, 124
open channel, 85
operon, vii, 2, 5, 7, 8, 10, 21, 27, 29, 40, 41, 42, 43, 50, 52, 64, 65, 66, 69, 83, 103, 115, 116, 150, 171, 173, 179, 183, 185
opportunistic pathogen, 12, 95, 102
opportunities, 28
opsonization, 12, 22
oral cavity, ix, 15, 109, 120
oral flora, 114
organ(s), ix, 109, 110, 111
organic compounds, 66, 79, 88
organic solvent(s), viii, 77, 78, 79, 80, 81, 83, 85, 86, 87, 89, 90
organic substrate, 204
organism, 38, 44, 100, 112, 187, 193
organosulphate, 65
osmolarity, 6, 7, 88
osmotic balance, ix, 149, 151
osmotic stress, 93
osteomyelitis, 2, 16, 33
osteomyelitis infection, 16
osteopontin, 11
otitis media, vii, 2
outer membrane, 45, 77, 78, 80, 81, 83, 84, 85, 86, 87, 88, 89, 101, 104, 110, 116, 117, 136, 138, 154, 205
outer membrane protein, 45, 80, 83, 88, 104
outer membrane vesicle, 77, 87, 88, 116, 117, 136, 138
overall growth rate, 228
overproduction, 40, 65, 66
oxalate, 209
oxidation, 65, 93, 208

oxidation products, 65
oxidative damage, 82, 204
oxidative stress, 49, 63, 82, 90, 95, 100
oxidative stress response, 49
oxide nanoparticles, 31
oxygen, x, 6, 7, 10, 13, 38, 39, 47, 52, 68, 197, 204, 210, 218, 234, 235, 238, 240, 242

P

P. aeruginosa, 38, 39, 42, 46, 66, 80, 84, 85, 206, 207, 208, 213
P. putida, 79, 83, 84, 85
p53, 125, 126, 127, 128, 129, 130, 131, 145
PACT, 198, 204, 205, 206, 207, 208, 209, 211, 212, 213, 214
Paenibacillus vortex, 224, 245
pairing, 100, 103
palmitate, 154
palmitoleic acid, 79
Paracoccus pantotrophus, 65, 74
paracrine, 117, 118
paradigm shift, 26, 35
parallel, 39, 56, 107, 130
paralysis, 165
Paramecium caudatum, 198
parasite(s), 12, 135
parasporal crystal, 150, 161
partition, 78, 81
pathogen, vii, 2, 11, 13, 15, 18, 32, 71, 72, 93, 94, 95, 103, 106, 111, 112, 114, 115, 145, 154, 159, 188
pathogenesis, 10, 22, 30, 41, 50, 104, 107, 127, 184
pathogenicity, viii, 71, 91, 93, 110, 111, 112, 113, 114, 115, 116, 138, 174, 188
pathogens, vii, viii, 1, 2, 5, 11, 12, 15, 23, 31, 86, 91, 94, 95, 102, 105, 111, 112, 120, 164, 165, 187, 188, 206, 214, 217, 218
pathophysiological, 33
pathophysiology, 33
pathways, 7, 13, 38, 41, 45, 50, 63, 65, 117, 122, 123, 124, 127, 140, 173, 177
pBtoxis, 150
pBtoxis plasmid, 150
PCR, 19, 113, 141, 169, 176, 178, 184, 190, 191, 193
p-cresol, 85
PDE, 38, 39, 41, 43, 44
PDEs, 38, 45
penicillin, 198, 219
peptide(s), 8, 9, 10, 11, 12, 20, 21, 26, 28, 32, 33, 101, 116, 153, 155, 166, 167, 205
peptidoglycan, 11, 13, 22, 78, 80, 86, 88, 154, 205
peptidoglycan hydrolase, 80
peptidoglycan layer, 80, 154, 205
perforation, 152
peri-implantitis, 11, 25
periodontal, 72, 114, 136, 145, 147, 216
periodontal disease, 72, 114, 145, 147, 216
periodontitis, 11, 25, 114, 140, 218
periplasm, 78, 83, 116, 117
periplasmic compartment, 117
periplasmic extract, 132
periplasmic orientation, 39
periplasmic protein, 43
periplasmic space, 78, 79, 83, 84
permeability, 18, 78, 85
permeation, 205, 223
permission, 17
peroxyl radical, 210
pertussis, 59, 63, 116
pertussis toxin, 116
pest control, 150
pesticide, 150
pests, 150, 151
PFGE, 169, 174, 175, 176, 177, 178, 182, 183, 185, 186, 194
PGA, 42
PgaD, 42, 43
PGN, 13
pH, viii, 18, 55, 59, 60, 61, 62, 63, 67, 68, 69, 70, 71, 72, 73, 74, 75, 101, 104, 154, 168
phage, 104, 113, 173, 174, 182, 194
phagocyte, 12, 31
phagocytosis, 12, 24, 31, 32, 122
pharmaceuticals, 168
pharmacology, 190
phase transition temperature, 78
phenol, 8, 24, 31, 32, 33, 77, 79, 81, 82, 84, 85, 86, 87, 89
phenol-soluble modulin, 8, 24, 32, 33
phenothiazine, 198, 205, 219
phenothiazine dye, 198, 219
phenothiazinium, 212, 216
phenotype(s), viii, 10, 24, 25, 29, 32, 35, 36, 40, 45, 61, 66, 67, 91, 93, 94, 98, 99, 107, 125, 134, 136
Philadelphia, 189
phosphate, 84, 106, 205
phosphatidylethanolamine, 78
phosphatidylinositol-3,4,5-triphosphate, 134
phosphatidylinositol-3,4-diphosphate, 134
phosphodiesterase, 47, 49, 50, 51, 52, 94
phosphodiesterase HmsP, 94
phosphoenolpyruvate, 107
phospholipase C, 80
phospholipid(s), x, 78, 80, 82, 154, 157, 197, 205
phosphorylation, 38, 126, 127, 131

phosphorylation of proteins, 38
photodynamic antimicrobial chemotheraphy, 198
photodynamic chemotherapy, vii, x, 197
photodynamic therapy (PDT), 215
Photogem®, 202, 209
photoporphyrin IX, 206
photosensitization, 204, 205, 206, 208, 216
photosensitizer(s), x, 197, 198, 214, 215, 216, 217, 218, 219
phthalocyanine, 202, 203, 218
phylogenetic analysis, 58, 60
phylogenetic tree, 179
physicochemical properties, 84
physiological adaptation, viii, 77, 78
physiological process, viii, 8, 46, 55, 56, 58, 59, 60
physiology, viii, 46, 48, 69, 72, 84, 91, 93, 94
physiopathology, ix, 109, 110, 111
PIA, 5, 6, 7, 8, 13, 42
pigmentation, 94
PIKK, 129, 131
pilot study, 214
PilZ, 39, 40, 44, 46, 50
Pimelobacter, 79, 88
pinocytosis, 122
PIP3, 134
plague, 21, 94, 103
planktonic cells, 12, 16, 30, 35, 37, 51, 67
planktonic culture, 7
planktonic growth, 6
planktonic mode of growth, 16
plants, 71
plaque, 72
plasma membrane, 117, 119, 121, 122, 123
plasma protein(s), vii, 1, 2
plasmid, x, 11, 61, 113, 132, 150, 156, 163, 164, 170, 173, 174, 175, 176, 177, 178, 179, 181, 182, 183, 185, 186, 187, 190, 191, 196
plasmid conjugation, 163, 182, 186
plasmid DNA, 132
plasmid-specific marker, 178
plasposon, 96
platform, 119
Platinum, 244
playing, 67, 93, 102, 134
pleiotropic phenotype, viii, 91, 93, 94, 98, 99, 107
PNAG, 5, 6, 7, 41, 42, 43, 45, 52
pneumonia, 10, 19, 66, 94, 219
point mutation, 119, 141, 155, 171
poison, 129
polarity, 84
polarization, 14, 20
polyacrylamide, 86, 98
polyadenylated RNA, 93
polyamine, 65, 66, 71, 75, 76
polyamine catabolism, 66
polyamines, 66, 74
polymer(s), 5, 13, 18, 42, 48, 81, 115, 219, 246
polymer molecule, 18
polymerase, 171
polymicrobial biofilm, 11
polymorphism, 133, 169, 192
polymorphonuclear neutrophil, 12, 20, 24
polypeptide(s), 38, 160, 165, 168
polypeptide chain, 165, 168
polyphosphate-specific porin, 84
polysaccharide(s), x, 5, 13, 21, 22, 23, 26, 30, 31, 32, 36, 40, 41, 42, 44, 45, 50, 51, 52, 94, 197, 205
polystyrene, 25
population, 36, 80, 85, 136, 146, 175, 223, 224, 246
population diversity, 36
pore-forming toxin, ix, 149, 150, 156
porin, 84, 85, 88, 89
porphycene, 216
porphyrin(s), 198, 201, 205, 210, 214, 217, 218
Porphyromonas gingivalis, 206
Portugal, 91, 103, 159
post-binding intermediate, 229
post-transcriptional regulation, viii, 91, 104, 107
post-translational control, 42
potassium, 61, 74
potential benefits, x, 163
ppGpp, 38, 42
pre-binding intermediate, 229
pre-mRNA, 92
preparation, 213
pre-toxin, ix
prevention, 14, 15
primary cells, 128
primase, 178, 181
principles, vii
pro-apoptotic protein, 127
probability, 213
probe, 178, 183
probiotic(s), 182, 187, 195
prochlorphyllide, 205
proelastase, 80
pro-inflammatory, 13, 32, 112, 115
pro-inflammatory cytokine, 112
project, 18
prokaryotes, viii, 55, 56, 57, 58, 59, 96, 188, 190
prokaryotic cell, 37, 66, 225
prokaryotic gene regulation, viii, 91
prokaryotic multidrug transporter, 57, 61, 68
proliferation, 4, 6, 11, 126, 133, 145, 164, 187
proliferation stage, 11
promoter, 8, 39, 46, 48, 96, 99, 101

prophage, 113, 173
prophage infection, 113
prophylactic, 14, 16
proportionality, 236
propylbenzene, 83
prostheses, 2
protease, 7, 10, 101, 103, 107, 159, 187
protease inhibitor(s), 7
protection, 14, 36, 41, 74, 79, 82, 86, 159, 234
protective mechanism(s), 57, 82, 85
protective role, 36
protein components, 138
protein family, 28, 47, 70
protein kinase(s), 129, 135, 137
protein misfolding, 63
protein receptor, 166
protein synthesis, 66, 104
proteinaceous adhesin, vii, 1, 2
proteinaceous fibre, 41
proteinase, 153
proteinase K, 153
protein-protein interaction(s), 40
proteolysis, 103
proteolytic activation, 152
proteolytic activity, 27, 153
proteolytic cleavage, 168
proteolytic enzyme, 84
proteolytic extracellular enzyme, 8
proteome, viii, 10, 30, 77, 78, 82, 89
proteomics, 89, 192
Proteus mirabilis, 80
protons, 61, 62
pseudolysogeny, 174
Pseudomonas aeruginosa, 19, 33, 47, 49, 50, 51, 52, 81, 87, 88, 89, 94, 104, 106, 138, 206, 214, 216
Pseudomonas fluorescence, 44
Pseudomonas oleovorans, 79, 86
Pseudomonas putida, 80, 81, 82, 86, 87, 88, 89, 90
PTEN, 134
public health, 36, 164, 188
pulmonary pathogen, 206, 214
pulsed-field gel electrophoresis, 169, 191, 193
pumps, 16, 57, 64, 66, 68, 72, 73, 74, 75, 77, 78, 83, 85, 86, 88, 89, 212, 219
purification, 28, 102, 132
purine biosynthesis, 82
purine metabolism, 67
purine transporter, 67
PurR transcription factor, 67
putrescine, 66, 70
pyocyanin, 94
pyrimidine, 66
pyrimidine nucleotide, 66
pyrolysis, 210

Q

quality of life, 165
quantification, 113
quasi-chemical approach, 222, 233, 241, 242
quasi-chemical mechanism, 233
quorum sensing, 11, 40, 69, 100

R

Radachlorin®, 206, 215
radiation, 28, 129, 138, 147
radical formation, 215
radicals, 18, 204, 210, 216
radioactive tracer, 242
radiotherapy, 215
rate determining step, 228
rate equation, xi, 222, 226, 228, 231, 233, 242, 243
RB, 199, 206, 210, 211, 212
rdar morphotype, 49, 50
RE, 97
reactant(s), 225, 226, 242
reactions, 62, 66, 82, 204, 213, 222, 225, 226, 245
reactive oxygen, x, 12, 17, 18, 82, 128, 197, 204
reactive oxygen species, x, 12, 17, 18, 82, 128, 197, 204
reactivity, x, 221, 224, 242
reading, 158, 171
real time, 233
receptor(s), ix, 6, 13, 24, 25, 26, 27, 28, 50, 110, 118, 119, 120, 121, 122, 123, 127, 149, 151, 152, 153, 155, 157, 158, 159, 166, 167
receptor-mediated endocytosis, 122, 166
recognition, 13, 19, 32, 44, 66, 72, 120, 144, 160, 176, 192
recombination, 90, 131, 173, 179
recovery, 82, 219
recovery mechanism, 82
redox reaction, 62
redundancy, 39, 57, 62
regulation, vii, viii, 6, 7, 9, 19, 21, 22, 28, 29, 32, 35, 36, 37, 39, 40, 41, 42, 43, 45, 47, 48, 49, 50, 51, 52, 63, 65, 67, 68, 69, 70, 73, 82, 91, 93, 98, 99, 103, 104, 105, 106, 107, 127, 138, 145, 184, 188
regulatory cascade, 39
regulatory changes, 9
regulatory control, 38
regulatory function, 38
regulatory gene, 5, 175
regulatory RNA, viii, 8, 9, 91, 99, 104, 105, 106

regulatory system(s), 6, 8, 39, 93, 184
regulon gene, 83
relatives, 169
relevance, 57, 63
reliability, 234
remission, 114
remodelling, 49
repair, ix, 14, 82, 107, 109, 129, 131, 142
replication, viii, 91, 92, 129, 181, 184, 233
repression, 8, 40, 45, 52
repressor, 7, 21, 52
requirements, 20, 192
researchers, vii, 10, 16, 39
residues, 56, 72, 92, 95, 96, 97, 132, 133, 134, 138, 152, 156, 205
resilience, 36
resistance nodulation and cell division, 83
resistance to antibiotics, 11, 164
resistance-nodulation-division, 56
resolution, 152, 157, 242
resolvase, 173, 179, 183
respiration, 62, 225
respiratory chain, 85, 225
respiratory tract, 94
response effector, 11
responsiveness, 19
restoration, 158
restriction enzyme, 10
restriction fragment length polymorphis, 176
restrictions, 208
restrictive barrier, 41
restructuring, 100, 106
reticulum, 122
retrograde traffic, 110, 118, 124
rhinosinusitis, 11, 22, 31
Rhodococcus opacus, 79, 89
ribosomal RNA, 99, 175, 183
ribosome, 18, 97, 100, 107
ribosome denaturation, 18
ricin toxin, 120
risk, 134, 165, 190
risk factors, 190
RNA, v, viii, 8, 9, 10, 39, 42, 48, 70, 91, 92, 93, 96, 99, 100, 102, 103, 104, 105, 106, 107, 108, 171, 183, 184, 218
RNA chaperone Hfq, 91, 105, 106
RNA genome, viii, 91
RNA polymerase, 171
RNA processing, 92
RNA-binding protein, 39, 103
RNAs, v, viii, 91, 93, 99, 100, 102, 103, 104, 105, 106, 107, 108, 188
RNase, 98, 100, 103, 105
RNase E, 98, 100, 103
RND, 56, 58, 61, 66, 67, 73, 83
ROI, 13
ROS, 17, 18, 82, 204, 213
Rose Bengal, 199, 206, 214, 217, 219
roughness, 81, 205
routes, 118, 122
routines, 235
rRNA, 99, 169, 183
ruthenium, 206, 217

S

S. aureus, 2, 3, 4, 5, 6, 7, 8, 10, 11, 12, 13, 14, 15, 16, 17, 18, 24, 65, 66, 154, 205, 206, 208, 209, 210, 211, 212, 213
S. cerevisiae, 124
S. dysenteriae, 113
S. enterica, 65, 66
S. flexneri, 207
S. mutans, 206
S. pneumonia, 66
S. typhi, 63
S. typhimurium, 63
Salmonella, 39, 40, 44, 47, 49, 50, 51, 52, 53, 59, 61, 67, 68, 69, 70, 71, 73, 88, 89, 110, 111, 114, 118, 140, 146, 194, 207, 217
Salmonella enterica, 39, 49, 50, 51, 52, 59, 61, 68, 69, 73, 88, 110, 111, 114, 194
salt concentration, 42, 64
salts, 64, 72, 75
SAM, 65
Sarcina lutea, 206
saturation, 79
science, 14, 48, 136, 138, 141, 143, 147, 243
scope, 217
SDT, 210
second generation, 208, 214
secondary efflux pump, 83
secondary messenger, 38
secondary structure, 63, 92, 96
secrete, 13, 63
secretion, 8, 42, 60, 63, 72, 80, 87, 93, 116, 117, 147, 173, 182
secretion system, 80, 93, 173, 182
seed, 100, 103
segregation, 129, 181
selectivity, 52, 84, 216
senescence, 128, 136
senses, 39, 128
sensing, vii, 2, 3, 8, 9, 11, 29, 32, 33, 36, 38, 40, 45, 52, 68, 69, 71, 99, 100, 105, 138

sensitivity, vii, x, 68, 93, 121, 125, 127, 197, 204, 211, 213
sensors, 33, 142
sensory domain, 38, 40
sepsis, 10
septic arthritis, vii, 2
septicaemia, 98
sequence conservation, 56, 134
sequence homology, 123, 132, 134, 152, 177, 178, 179
sequencing, 20, 57, 68, 85, 100, 102, 144, 145, 169, 184
serine, 4, 8, 28, 129
serine protease, 8
serotype, 113, 166, 167, 169, 171, 190, 193
Serratia marcescens, 80
sessile mode of growth, 36, 41
shape, 81, 84, 88, 205, 234
shear, 4, 9
shiga toxin, 93, 113
Shigella dysenteriae, 113
Shigella flexneri, 61, 71, 98, 106, 207
shigellosis, 113
shock, 84, 98, 100, 101, 107, 214, 217
showing, 12, 96, 99
sialic acid, 120
side effects, 211
siderophore, 68, 70, 75
sigma factor, 26, 39, 52, 94, 171
sigma factor RpoS, 39
signal peptide, 116
signal transduction, 38, 46, 47
signal transduction system, 38, 47
signaling, 8, 13, 23, 47, 48, 50, 51, 66, 112, 128, 129, 131, 134
signaling cascade, 128
signaling pathway, 112, 131, 134
signalling, viii, 37, 38, 39, 41, 42, 44, 45, 46, 48, 51, 52, 55, 62, 63, 136
signalling molecule, viii, 38, 42, 55
signals, 40, 44, 111, 130, 133, 177
signs, 16, 24
silica, 25
silver, 17, 18, 25, 27
simulations, 133
single-stranded DNA, 129
singlet oxygen, 210
Sinorhizobium meliloti, 98, 106
skin, 10, 12, 198, 208
skin diseases, 198
S-layer, 80
slime, 4, 10, 19, 36, 205
slime layer, 36
Sm/LSm, 92
Sm/LSm protein superfamily, 92
small intestine, 64
small multidrug resistance, 56
Sm-like protein, 92, 104, 106, 107
smoothing, 235
SMR, 56, 58, 67, 83
SNAP, 167
SNARE, 166
sodium, 61, 86
software, 58, 92, 234, 238
soil bacterial strain, 84
solid surfaces, 44, 48
solid tumors, 161
solubility, 63
solution, 152, 209
solvent adaptation mechanism, 79
solvent molecules, 79, 82
solvents, viii, 77, 78, 79, 81, 82, 83, 89
sonodynamic activation, x, 197, 198, 206, 210, 211, 213
sonodynamic therapy, 216
Southern blot, 176, 177, 178, 182, 183
Southern hybridization, 174, 175, 178
spatial structure, ix, 149, 152
specific affinity, 223, 232
specific ligand, 155
specific porin, 84, 89
specific receptor, ix, 149, 152, 153, 155
specific targeting, ix, 150
spectroscopic techniques, 215
spectroscopy, 234
spending, 82
spermidine, 58, 66, 69, 70, 71, 72, 76
spermidine efflux, 58, 66
spermidine efflux pump, 66
spermidine metabolism, 66
spermine, 66
sphingomyelin synthase, 121
spindle, 126
spleen, 98
spore, ix, 71, 149, 150, 168, 224
spore-forming bacterium, ix, 149
sporulation, ix, 149, 150, 161
sporulation phase, ix, 149, 150
squamous cell, 147
squamous cell carcinoma, 147
sRNA, 100, 101, 102, 103, 106, 107
sRNA-Hfq complexe, 102
ssDNA, 129, 147
stability, 7, 40, 78, 119, 147, 161, 193
stabilization, 6, 43, 85, 130, 131, 150
standard error, 235

staphopain, 8
staphylococci, vii, 2, 4, 6, 7, 9, 11, 22, 24, 31
Staphylococcus aureus, vii, 2, 18, 19, 20, 21, 22, 23, 24, 25, 26, 27, 28, 29, 30, 31, 32, 42, 47, 60, 62, 70, 81, 205, 218, 219, 235, 246
Staphylococcus epidermidis, vii, 2, 19, 20, 21, 22, 23, 24, 25, 26, 27, 28, 29, 30, 31, 32, 33, 206, 246
starvation, 93, 98
starving conditions, x, 221, 232
state(s), x, 14, 32, 36, 37, 44, 45, 78, 107, 154, 157, 204, 221, 222, 224, 225, 227, 228, 229, 231, 236, 240, 242, 245
stationary phase, 8, 45, 93, 98, 105, 106
statistical criteria, 235
*St*CDT, 111
steady state, x, 221, 222, 224, 225, 227, 228, 229, 231, 240, 242
steady state conditions, 228, 229
steady state equations, 225
steady state rate equation, 231
Stenotrophomonas maltophilia, 94, 95, 106
stimulus, 28, 39
stomach, 113
storage, 42, 63
strategy use, 102
stratified epithelia, 120
Streptococcus mutans, 206, 214, 215, 218
stress, viii, 7, 27, 36, 37, 39, 40, 41, 43, 44, 45, 46, 49, 50, 52, 62, 63, 68, 70, 73, 78, 81, 82, 85, 87, 88, 89, 90, 91, 93, 98, 99, 100, 101, 103, 105, 106, 125, 131, 133, 213, 222, 223, 224
stress factors, 224
stress protein, 78
stress resistance, viii, 41, 44, 68, 70, 91, 106
stress response, 7, 37, 39, 40, 41, 45, 50, 52, 68, 82, 88, 89, 101, 105
stress response regulator, 40, 45
stressors, vii, 7, 35
structural gene, 181
structure, ix, 14, 29, 41, 49, 50, 56, 84, 85, 87, 92, 93, 96, 98, 115, 116, 119, 120, 126, 147, 149, 150, 152, 153, 154, 155, 159, 168, 175, 178, 179, 183, 192, 204, 211, 233
structuring, 10
style, 10, 36, 50
styrene, 77, 83
subdomains, 4
sub-lethal concentration, 151
sublethal dose, 133
substitutes, 205
substitution(s), 38, 96, 159
substrate(s), x, 4, 6, 27, 55, 57, 58, 59, 60, 62, 64, 65, 66, 67, 71, 80, 81, 84, 87, 89, 129, 134, 166, 168, 187, 188, 192, 204, 212, 219, 221, 222, 223, 224, 225, 227, 229, 230, 231, 232, 233, 235, 240, 241, 242
substrate binding, x, 55, 84, 166, 221, 231, 232
substrate binding site, 84
substrate uptake, 223
subunit, 42, 45, 97, 115, 116, 118, 120, 121, 123, 124, 131, 143
succession, 129
succinate, 85
sugar, 82, 100, 106, 120
sugar transport, 82
sulbactam, 65
sulfadiazine, 65
sulfamethoxazole, 25, 65
sulfate, 74, 86
sulfur, 69, 75
sulphate, 65, 72
sulphonate, 65, 75
sulphur, 58, 59, 65, 75
sulphur ester, 65
sulphur metabolism, 58, 59, 65
sulphur transporter, 65
sulphur-containing lincosamide, 65
supercoiled plasmid, 132
superoperon, 98, 107
suppression, 211, 216
suppurative lymphadenopathy, 114
surface area, 81
surface component, 3, 4
surface hydrophobicity, 4, 80
surface protein, 5, 8, 21, 23, 31
surface-anchored protein, 4
surface-attached cell, 36, 37
surfactant(s), 3, 8, 10, 32
surfactant peptide, 10, 32
surgical intervention, 16
surgical removal, 12, 14
survival, vii, 16, 35, 36, 46, 47, 52, 75, 80, 98, 106, 111, 139, 164, 205
susceptibility, 5, 24, 29, 46, 66, 72, 93, 94, 95, 105, 125, 128, 134, 158, 207, 211, 213, 214, 219
swelling, 151, 153
symptoms, 113, 114, 115, 168
synaptic vesicles, 166
synaptobrevin-2, 168, 192
synaptogyrin, 121
Synechocystis, 58
synergistic activity, 151
synergistic effect, 218
synteny, 175, 177, 178, 179

synthesis, 5, 6, 8, 11, 22, 32, 36, 38, 39, 40, 41, 42, 43, 44, 45, 46, 48, 49, 50, 52, 71, 79, 80, 86, 103, 104, 107, 127, 188
systemic lupus erythematous, 92

T

T cell(s), 128, 145
T-antigen, 123
target, viii, 12, 16, 18, 39, 78, 81, 91, 99, 100, 101, 103, 104, 106, 110, 120, 121, 124, 134, 141, 155, 166, 167, 168
target peptide, 166
taxonomy, 169, 188
TBO, 199, 205, 206, 208
techniques, x, 197, 217, 241
technological advances, 217
technology, 68, 208
teichoic acid, 3, 4, 5, 24
teichuronic acid, 205
teicoplanin, 26
telangiectasia, 129
telomere, 92
telomere replication, 92
temperature, 45, 78, 87, 114, 207, 209
testing, 210
Tet protein, 62, 70
Tetanus, 189, 193
tetracycline(s), 62, 69, 70, 71, 74, 75, 95, 182
tetracycline antibiotics, 182
tetracycline efflux, 62, 69, 71
TGN, 122, 123
theoretical approaches, 245
therapeutic approaches, 155
therapeutics, 23
therapy, 6, 12, 14, 16, 23, 166, 210, 214, 215, 216, 217, 218, 219
thermograms, 234, 235, 238
thermonuclease, 10
thermostable nuclease, 7
time increment, 234
tissue, 2, 11, 98, 111, 112, 127, 208, 211
tissue homeostasis, 127
tissue infection, 2
TLR, 19, 24
TLR2, 13, 27, 31, 32
TLR4, 32
TLR9, 13
TMP, 200, 205, 206, 213
TMS, 56, 57, 58, 59, 60, 62, 64
TNF, 13
TNF-α, 13
tobramycin, 38, 95
TolC protein, 83
toluene, 77, 79, 80, 81, 82, 83, 85, 86, 87, 88, 89, 90
toluene stress, 85
toluidine blue O, 205
topoisomerase, 129, 142
Torulopsis utilis, 223, 243
toxic effect, 78, 81, 89, 211
toxic metals, 62
toxic products, 10
toxic substances, 164
toxicity, ix, 17, 66, 72, 77, 87, 89, 120, 132, 133, 149, 150, 151, 152, 153, 155, 156, 157, 158, 159, 166, 192, 198
toxicity of aromatic compound, 77
trafficking, ix, 110, 118, 119, 121, 122, 123, 124, 133, 146
traits, 93, 145, 164, 186, 187
trajectory, 226
trans fatty acid, 79
trans/cis ratio, 79, 87
transactions, 107
trans-configuration, 79
transcription, 6, 7, 22, 39, 40, 42, 44, 48, 50, 65, 67, 70, 75, 98, 127, 171
transcriptional activator, 83, 107
transcripts, 93, 98, 99
transduction, 29, 39, 46, 182
transfer RNA, 99
transformation(s), 96, 134, 158, 182, 204, 241
transformed cell, 133
trans-Golgi network, 122
trans-hexadecenoic acid, 79
transition metal, 63, 71
transition temperature, 78, 86
translation, 42, 98, 100, 105
translation-inhibitory antibiotic, 42
translocation, 56, 71, 123, 124, 139, 165, 166, 192
translocation pore, 56
transmembranal pore-forming protein, 84
transmembrane protein, 65
transmembrane segment, 4, 39
transmembrane subunit, 153
transmission, 95, 104, 106, 110, 166
transport, 55, 57, 59, 61, 62, 63, 65, 66, 67, 68, 69, 70, 71, 72, 75, 82, 83, 88, 101, 118, 121, 123, 124, 187, 223
transport protein, 57, 71, 75
transporter, 16, 27, 56, 57, 61, 65, 66, 67, 68, 69, 70, 72, 73, 74, 75, 83, 104, 173
transporter lipoprotein, 16
transposase, 174, 186
transposon, 6, 25, 174, 179, 182
transposon-associated enzyme, 174

trauma, 11, 12
treatment, x, 6, 11, 14, 16, 17, 18, 20, 31, 105, 114, 123, 124, 125, 126, 128, 131, 135, 191, 195, 197, 198, 205, 207, 208, 209, 211, 212, 213, 214, 215, 216, 218, 219
Trematoda, 150
tricarboxylic acid, 63
tricarboxylic acid cycle, 63
triggers, 10, 117, 128, 137, 139, 145
triplet state, 204
tRNA, 97, 99, 103, 107
tropism, 120, 121
tuberculosis, 195
tularemia, 94
tumor, 126, 134, 139, 155, 211, 219
tumor cell(s), 155
tumor development, 134
tumor growth, 211
tumor progression, 139
tumor suppressor, 126
turnover, 38, 39, 40, 45, 50, 226, 230
turnover protein, 38, 39, 40, 45
two-phase fermentation, 79
two-step mechanism, 227, 230
two-step rate equation, 230
two-step steady state mechanism, 228
Type I reaction, 204
Type II reaction, 204
typhoid, 110, 114, 115, 116, 117, 118, 132, 146
typhoid fever, 110, 114, 115
typhoid toxin, 115, 116, 117, 118, 132, 146
tyrosine, 25, 52

U

ulcer, 111, 114
ultrasound, 210, 211, 218, 219
ultraviolet light, 93
unique features, 120
unsaturated esterified fatty acid, 79
unsaturated fatty acid, 80, 82, 86, 87, 89
unsaturated lipid, 153
unsaturated membrane fatty acid, 79
urinary tract, vii, 2, 93, 112
urinary tract infection, vii, 2, 93, 112
UTR, 98, 101

V

VacA, 111
vaccine, vii, 2, 3, 15, 16, 19, 28, 29
vaccine design, 19
valve, vii, 2
vancomycin, 16, 17, 18, 26, 205
variations, 82, 104, 166, 233, 235
vector, 147, 151, 157, 159
velocity, 4, 46
versatility, 241, 242
vesicle, 122, 142, 166, 167
Vibrio cholera, 39, 40, 52, 195
Vibrio parahaemolyticus, 58, 60, 63, 72, 75
virioferrin transporter, 63
virulence, vii, viii, ix, 2, 3, 5, 6, 7, 8, 9, 10, 11, 15, 19, 23, 24, 32, 33, 37, 39, 41, 52, 80, 91, 93, 94, 99, 103, 105, 106, 107, 109, 110, 111, 112, 113, 115, 117, 140, 145, 147, 159, 164, 186, 193, 194
virulence factor, vii, ix, 2, 3, 5, 10, 11, 15, 19, 39, 80, 94, 105, 109, 110, 111, 112, 115, 117, 147, 159, 164, 186
virulence of bacteria, viii, 91, 110
virulence phenotype, viii, 91, 94
virus(s), 12, 150
virus infection, 150
visible light, x, 197, 198, 204, 205, 208
visualization, 133
Volvariella volvacea, 152
volvatoxin, ix, 149, 152, 158, 161

W

water, 78, 84, 88, 153, 158, 210, 217, 232, 245
wavelengths, 206
weight loss, 115
western blot, 143
wild type, 117, 134
worldwide, 98, 114, 150, 166
worms, 95
wound botulism, 165
wound healing, 31
wound infection, 11

X

xenon, 206
X-ray, 18, 28, 56, 234
X-ray diffraction, 234
xylene, 77, 83, 86, 88

Y

Y. pestis, 94
YaiC, 44
YddV, 38, 43, 51
YdeH, 42, 43, 53

yeast, 71, 74, 124, 131, 134
Yersinia pestis, 42, 94, 103, 106
yield, 94, 177, 182
YjiO, 61
YoaD, 44

Z

zinc, 18, 21, 30, 132, 166, 168, 193, 218
zinc endopeptidase, 166
zinc oxide, 18, 30
zinc-protease, 168
ZnO, 18
zoonotic pathogen, viii, 91, 93
zwitterionic phospholipid, 154